Advances in 21st Century Human Settlements

Indexed by SCOPUS

This Series focuses on the entire spectrum of human settlements – from rural to urban, in different regions of the world, with questions such as: What factors cause and guide the process of change in human settlements from rural to urban in character, from hamlets and villages to towns, cities and megacities? Is this process different across time and space, how and why? Is there a future for rural life? Is it possible or not to have industrial development in rural settlements, and how? Why does 'urban shrinkage' occur? Are the rural areas urbanizing or is that urban areas are undergoing 'ruralisation' (in form of underserviced slums)? What are the challenges faced by 'mega urban regions', and how they can be/are being addressed? What drives economic dynamism in human settlements? Is the urban-based economic growth paradigm the only answer to the quest for sustainable development, or is there an urgent need to balance between economic growth on one hand and ecosystem restoration and conservation on the other – for the future sustainability of human habitats? How and what new technology is helping to achieve sustainable development in human settlements? What sort of changes in the current planning, management and governance of human settlements are needed to face the changing environment including the climate and increasing disaster risks? What is the uniqueness of the new 'socio-cultural spaces' that emerge in human settlements, and how they change over time? As rural settlements become urban, are the new 'urban spaces' resulting in the loss of rural life and 'socio-cultural spaces'? What is leading the preservation of rural 'socio-cultural spaces' within the urbanizing world, and how? What is the emerging nature of the rural-urban interface, and what factors influence it? What are the emerging perspectives that help understand the human-environment-culture complex through the study of human settlements and the related ecosystems, and how do they transform our understanding of cultural landscapes and 'waterscapes' in the 21st Century? What else is and/or likely to be new vis-à-vis human settlements – now and in the future? The Series, therefore, welcomes contributions with fresh cognitive perspectives to understand the new and emerging realities of the 21st Century human settlements. Such perspectives will include a multidisciplinary analysis, constituting of the demographic, spatio-economic, environmental, technological, and planning, management and governance lenses.

If you are interested in submitting a proposal for this series, please contact the Series Editor, or the Publishing Editor:

Bharat Dahiya (bharatdahiya@gmail.com) or
Loyola D'Silva (loyola.dsilva@springer.com)

Anjali Krishan Sharma · Ashutosh Joshi
Editors

Proceedings of the 3rd International Conference on Opportunities and Challenges for a Resilient Future

ICOCRF-2024; 22–24 April; Dehradun, India

Springer

Editors
Anjali Krishan Sharma
School of Architecture, Planning
and Design (SoAPD)
DIT University
Dehradun, Uttarakhand, India

Ashutosh Joshi
School of Architecture, Planning
and Design (SoAPD)
DIT University
Dehradun, Uttarakhand, India

ISSN 2198-2546 ISSN 2198-2554 (electronic)
Advances in 21st Century Human Settlements
ISBN 978-981-96-0930-7 ISBN 978-981-96-0931-4 (eBook)
https://doi.org/10.1007/978-981-96-0931-4

This Springer imprint is published by the registered company Springer Nature Singapore Pte Ltd.
The registered company address is: 152 Beach Road, #21-01/04 Gateway East, Singapore 189721, Singapore

Foreword by Priyadarsan Patra

It is both an honor and a privilege to contribute this Foreword to the proceedings of the SoAPD Conference 2024—Challenges and Opportunities for a Resilient Future. Earlier this April, I witnessed firsthand the remarkable synergy that unfolded during our conference. Our collective commitment to sustainable development, innovative design and analysis, and resilient solutions resonated throughout the conference room and the virtual platform. The gathering of architects, designers, and visionaries explored crucial themes that transcended borders: ecological and inclusive development, energy and sustainability, and the vital role of governance and policy in shaping a more resilient world.

The electric energy that permeated SoAPD 2024, both in-person and virtually, fueled a collective spirit of innovation. Participants exchanged ideas, forged collaborations, and ignited a spark of change that continues to illuminate our paths forward. These proceedings capture the essence of those discussions, offering a valuable resource for anyone dedicated to building a resilient future. Beyond these pages, may our ideas find expression in the structures we build, the policies we help draft and the futures we envision for our communities.

The Himalayan region, where our conference unfolded, served as a poignant backdrop. Its rugged beauty and ecological significance reminded us of our responsibility as stewards of the Earth. The urgency of our mission cannot be overstated. Natural disasters, climate shifts, and societal challenges demand immediate action. Let the architects and designers, entrusted with safeguarding our planet, weave effective resilience into the fabric of our/their creations.

With gratitude to the organizers, speakers, and tireless contributors, I invite you to immerse yourselves in the knowledge shared within these proceedings. May our collective wisdom inspire transformative change.

Priyadarsan Patra
Distinguished Professor and Pro
Vice-Chancellor
DIT University
Dehradun, Uttarakhand, India

Acknowledgements

The proceeding of the said Conference is the outcome of our previous conference 'CHANGING DIMENSIONS OF THE BUILT ENVIRONMENT'.

We acknowledge the support and constant encouragement of our Hon. Chairman Mr. Anuj Aggarwal, Chancellor Mr. N. Ravi Shanker, Vice-Chancellor Mr. G. Raghurama and Pro-Vice Chancellor Mr. Priyadarsan Patra of the university from the inception till the execution of the International Conference.

This book would not have been complete without the effective participation of the presenters, who produced quality work, and the young faculty members of SoAPD. We acknowledge the efforts of Ar. Rajarshi Roy in efficiently compiling the work, as well as the invaluable support of the office staffs.

Dehradun, India

Dr. Anjali Krishan Sharma
Dr. Ashutosh Joshi

Keynote Address by Prof. Anne Feenestra

In his speech, Prof. Feenestra emphasized the importance of experiential knowledge in architecture, underscoring that true understanding arises from engaging directly with the environment rather than relying solely on theoretical study. He advocated for spending time on-site, interacting with local communities, and observing local practices to gain meaningful insights. He shared his observations about different cultures, contrasting the rich cultural heritage of Indian cities like Ahmedabad with places like Dubai and certain parts of America, which he described as lacking cultural depth and history.

Professor Feenestra also highlighted the distinct architectural styles and cultural heritage of various regions, emphasizing the need for urban environments to preserve their historical and cultural identities. He discussed the challenges within academic environments, such as internal politics, and stressed the importance of focusing on significant work rather than getting caught up in unproductive activities. He urged for serious efforts in heritage identification and stabilization, suggesting that universities should conduct comprehensive heritage mapping and proactive conservation efforts. He proposed organizing international conferences and involving experts to promote heritage conservation, emphasizing that cultural heritage plays a crucial role in shaping resilient futures. He concluded by calling on academics and professionals to lead these efforts, highlighting the real need to value and preserve cultural heritage for future generations.

His comments:

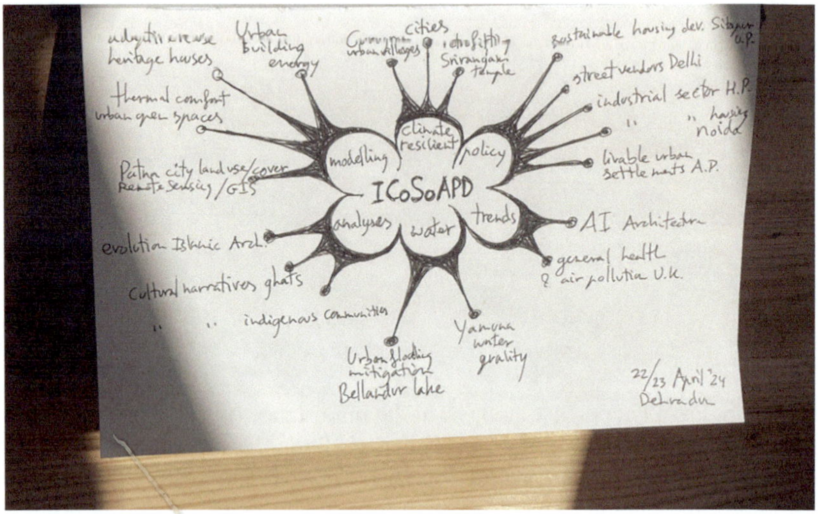

Thank you for having me over at DIT in Dehradun at the foothills of my beloved Himalayas.

Two full days of dialogues, presentations, feed-back sessions followed. Double plus! New encounters (for me) with Prof. Habeeb Khan and Dr. Neena Zutshi were deeply inspiring and added to my personal learning process within the ICoSoAPD conference.

Many interactions took place with community members and the team led by Prof. Anne Feenstra (seated with notebook) before, during and after the Bagh-e Jehan Nama garden and palace restoration project in Kholm, North Afghanistan

Architects, designers, planners often forget to take time to listen.

Listen to the community, to the users.

I believe this is one of the pillars of 'Tikau' (Hindi for durable/long-lasting).

Apart from the fact that one learns a lot and that one gets buckets of information. Not everything, but a lot is relevant at some stage in the design process. It informs and most of the time demands an adjustment to the project approach. Rather than starting with the restoration of the palace, the original plan, we first restored the garden. We understood clearly from the many community sessions, that for local people, a safe, publicly accessible and beautiful garden would be of greater value than the restored building. We got the irrigation system going, rebuilt/repaired the garden walls and we planted 72 varieties of fruit and nut trees. People started using and enjoying the Bagh especially on Fridays and Saturdays.

Even the first exhibition (co-curated by Dr. Kala Gurung, Prof. Anne Feenstra and architect Khalid Dawari) in the palace building was about the biodiversity of the garden. UNESCO Netherlands has called this project (Bagh-e Jehan Nama, Kholm) during a conference in The Hague (2010) a Best Practice Community-based restoration project and the project approach.

Earthen architecture and earthen construction is very common in South and Central Asia. In Afghanistan the use of Pakhsa, the carefully prepared and densified mud mixture, normally goes up to 3 to 3.5 m. For the garden walls of the former palace garden (1892) of Emir Abdur Rahman the walls are 5.2 m high.

It took two building seasons to repair and re-construct nearly 800 m of garden walls. It is not possible to do this work in the winter as temperatures dip to -10, sometimes $-15°$.

The heavy Afghan bulls make rounds through the prepared mud fields, stamping out the air bubbles from the mud mixture. After that, the building team does another round of kneading by feet. The clumps are handed over till they arrive at the Ustad who is in charge of making the wall. Pakhsa walls are made in layers of approx. 50–60 cm height.

Applying of the last layer of densified mud, during the rebuilding of the garden walls of Bagh-e Jehan Nama, Kholm

I would like to share a most relevant quote from Prof. Jana Revedin:

Nachhaltigkeit, Nach-halten, halten, in the sense of to hold,

but also in the sense of to hold out, to continue or,

to put it simply: to last long and age well !

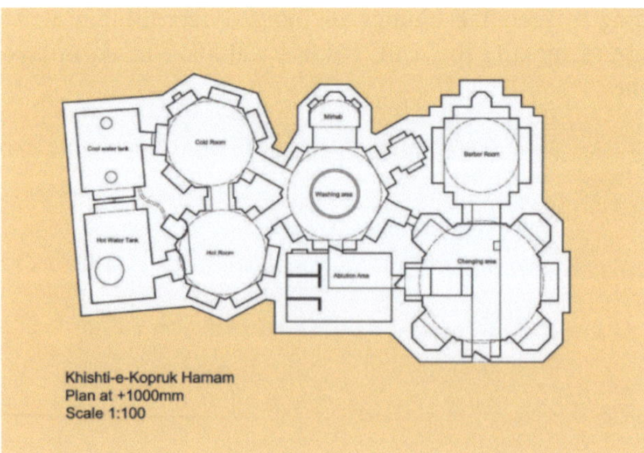

The plan of Hammam Khisht-i-Kupruk, Kholm, North Afghanistan. Only the domes of the building are above the ground level

In 1972 Prof. Pierre Centlivres (Neuchatel, Switzerland) had described in his book about Central Asia, that this underground bathhouse as a unique structure. He observed that the mehrab, the niche which indicates the direction of Mecca, was inside the central washing area, while normally the ablution and praying area are separate spaces. He was elated by the spatial quality of the interconnecting heptagon, septagons, octagon, oblong and square. Most of the spaces only have one aperture in the domed roof which offers a unique natural light quality in the subterranean structure. In my conversations with Prof. B. V. Doshi, he had called this building the great-grandmother of the Gufa!

During the restoration of the 1845 hammam, we had to convince the local kiln owners to make the 26 cm x 26 cm x 5 cm (thickness) bricks which where used in the original building. Grinded ash and lime was prepared for the mortar and the plaster. As the grey color of this mixture is close to the standard cement mortar, the most knowledgeable masons had to pacify so-called experts, who thought we had used this modern material.

The social-cultural function of this small building is most important. This is where local people bathe, talk and gossip during the cold winter months. Some days in the week the hammam is used by women, some days by the men. The caretaker obviously has the same gender as the users of the day!

A carpenter's crafted Mulberry timber door, with star bolts from the local black smith. Hammam Khisht-i-Kupruk

There is no pool, or larger tub, but the users take buckets, mixing the steamy hot water with the cold water. With the extremely high humidity the timber from the Mulberry tree is best. Usage of other wood could lead to cracks or worse a 'banana door'. It was a humbling experience to see the incredible amount of knowledge and skills the craftsmen of the war-torn country had. With great pride they created, for their own community a much improved bathhouse. Many of my colleagues complain that the crafts of South Asia and Central Asia are rapidly disappearing. For Afghanistan, while going through yet another terrible period, it is not too late to re-connect with it's rich cultural roots.

Prof. Anne Feenstra
Laureate of the Global Award for Sustainable Architecture 2012 (Paris), finalist TERRA Award 2016 (Lyon finalist MATERIA Award 2024 (Paris) and former Dean of Architecture, CEPT University Ahmedabad (2015–2017) Prof. Anne Feenstra has been teaching sustainable architecture in Delhi, Kabul, Kathmandu and Mysore.

He practices a pro-ecology, pro-people, material up-cycle, research-based form of architecture. His work of fundamentally re-thinking the idea of 'Making City' through urban processes like DELHi2050, Mussoorie2040 and his ecological sensitive projects have been embraced by communities all over South Asia especially in the Hindu Kush-Himalayan region.

Keynote Address by Ar. Habib Khan

The keynote speaker for the International Conference was Ar. Habib Khan, former President of the Council of Architecture and current Director at PIADS Nagpur and LTIADS Mumbai. The SoAPD department felt privileged to have him as the keynote speaker.

Ar. Khan addressed the conference by emphasizing the broad spectrum of discussions and deliberations covered. He highlighted the importance of integrating architectural practice with academia, as both enrich and reinforce each other, leading to improved architectural outcomes and a better quality of life for citizens.

He stressed the need to break the compartmentalized approach to sustainability and environmental friendliness, advocating for a holistic perspective that transcends traditional rating systems like LEED, IGBC, and GEM. Ar. Khan pointed out that many buildings with high sustainability ratings often fail to consider vital factors such as the empowerment of the local economy and the preservation of local culture, art, and crafts.

Discussing the concept of net-zero buildings, he called for a positive approach to understanding and mitigating global environmental damage. He briefly touched on the concept of degrowth, urging a critical examination of our actual needs and sustainable living practices.

Ar. Khan also spoke animatedly about the rapid urban growth in India, driven by migration from rural areas. He criticized the smart city concept as a means of offloading obsolete technology and hardware from the Western world and China onto developing countries like India and Brazil.

In conclusion, Ar. Khan expressed his hope that professionals and intellectuals attending such conferences will focus on empowering rural economies and agricultural sectors. By addressing the root causes of urban migration, he believes we can alleviate the growing strain on urban infrastructure.

Prof. Habib Khan
Chairperson, BoG, SPA, Delhi, Former President, COA, Director PIADS, Ngp, LTIADS Mumbai, Principal Arch. SHK ATP Architects, Prof. Emeritus, MSAJAA Chennai, Member Advisory Board, Ar edu, National Board of Accreditation (NBA).

Contents

Ecological and Inclusive Development

Advancements in Urban Building Energy Modelling: Innovations, Smart Technologies, and Sustainability

Ami Mangaldas

Abstract Purpose: Ecological and Inclusive development is a multifaceted approach that emphasizes sustainable practices, social equity, and responsible resource management. At its core, it aims to promote economic growth while safeguarding the environment and ensuring benefits of growth that are shared equitably among stakeholders. Methodology: The approach will acknowledge the interdependence of environmental, social, and economic factors and integrate sustainability principles into economic activities. Methods are classified under:

1. Sustainable Building Practices: Encouraging the adoption practices that minimize the environmental impact. This will impact "Earth Offshoot Day"; which marks the day in a given year when humanity has consumed all the renewable resources the Earth can regenerate in that year. Beyond this date, we are depleting the planet's resources, contributing to ecological degradation..
2. Biodiversity Conservation: Consisting of forests, oceans, soil, and biodiversity, which provide a wide range of ecosystem services essential for human well-being. These services include clean water, air purification, climate regulation, and raw materials for industries like agriculture, forestry, and construction.
3. Resource Efficiency: Natural capital refers to the Earth's stock of natural assets. The building and construction industry is a significant consumer of natural resources. By promoting the R's; Reuse, Renew, Recycle, Repurpose and Recondition; it can reduce its impact on "Natural Capital" and help extend the availability of these resources for future generations.
4. Social Inclusivity: This includes providing equitable access to housing, jobs, and opportunities for all.

Keywords "Sustainability" · "Social Equity" · "Construction Industry" · "Earth Offshoot Day" · "Natural Capital"

A. Mangaldas (✉)
Balwant Sheth School of Architecture, Vaikunthlal Mehta Rd, Vile Parle West, Mumbai, Maharashtra 400056, India
e-mail: ami.mangaldas@nmims.edu

A. K. Sharma and A. Joshi (eds.), *Proceedings of the 3rd International Conference on Opportunities and Challenges for a Resilient Future*, Advances in 21st Century Human Settlements, https://doi.org/10.1007/978-981-96-0931-4_1

1 Introduction

Urban Building Energy Modelling (UBEM) is a computational methodology that employs advanced technologies to simulate and assess the energy performance of buildings within urban environments [2]. It integrates data analytics, simulation tools, and smart technologies to analyse factors such as building design, materials, energy systems, and occupant behavior. The study aims to investigate recent advancements in UBEM, specifically focusing on innovations, smart technologies, and sustainability. It seeks to scrutinise the need for inclusive and sustainable development in urban areas, emphasising the role of UBEM in achieving these goals. Furthermore, the research aims to explore the adoption of UBEM and sustainable building practices concerning Sustainable Development Goals (SDGs) 9, 11, and 12, while also addressing challenges that may impact their successful implementation.

2 Aim and Objectives of the Study

The aim of the study is to analyse recent advancements in Urban Building Energy Modelling with a focus on Innovations, Smart Technologies, and Sustainability to achieve different SDGs.

- To analyse the recent advancement in urban building energy modelling systems with reference to innovation, smart technologies, and Sustainability
- To scrutinise the need for inclusive and sustainable development in urban regions and the role of UBEM in the achievement of such objectives
- To investigate the adoption of UBEM and sustainable building practices with respect to SDG 9, 11, and 12
- To explore the different challenges which can influence the adoption of UBEM and the achievement of SDG in a current urban environment
- To suggest policies which can lead to efficient UBEM and achievement of SDG

3 Literature Review

3.1 SDG (Sustainable Development Goals)

Sustainable Development Goals (SDG) 9 focuses on both engaging in innovation as well as development of sustainable infrastructure. According to [12], cities and real estate developers can engage in innovation and also develop sustainable infrastructure which is developed based on sustainable materials and at the same time, also utilise energy in an efficient manner (in which energy wastage is reduced as far as practicable). Thus, UBEM can be considered to aid in the development of both inclusive as well as sustainable infrastructure projects which aim to adopt innovative ideas which will further aid in improving energy optimisation by reducing energy consumption needs as well as energy wastage levels (Fig. 1).

The development of inclusive as well as sustainable cities is SDG 11, in which UBEM can aid in goal achievement as covered in a prior study conducted by [16]. UBEM aims to make building projects energy efficient which can reduce costs for residents (which is expected both during construction as well as operation). Thus, the adoption of UBEM can lead to both the achievement of environment-friendly projects and also economic as well as socially inclusive projects which hence aids in the achievement of SDG 11. Further, resident well-being is also achieved through UBEM and sustainable building which further aid in goal achievement.

SDG 12 has been developed to take into consideration both responsible consumption as well as production practices when developing a project, and such can be covered with the aid of UBEM strategy deployment and utilisation of sustainable building strategies as discussed in a prior study conducted by [20]. UBEM aids in

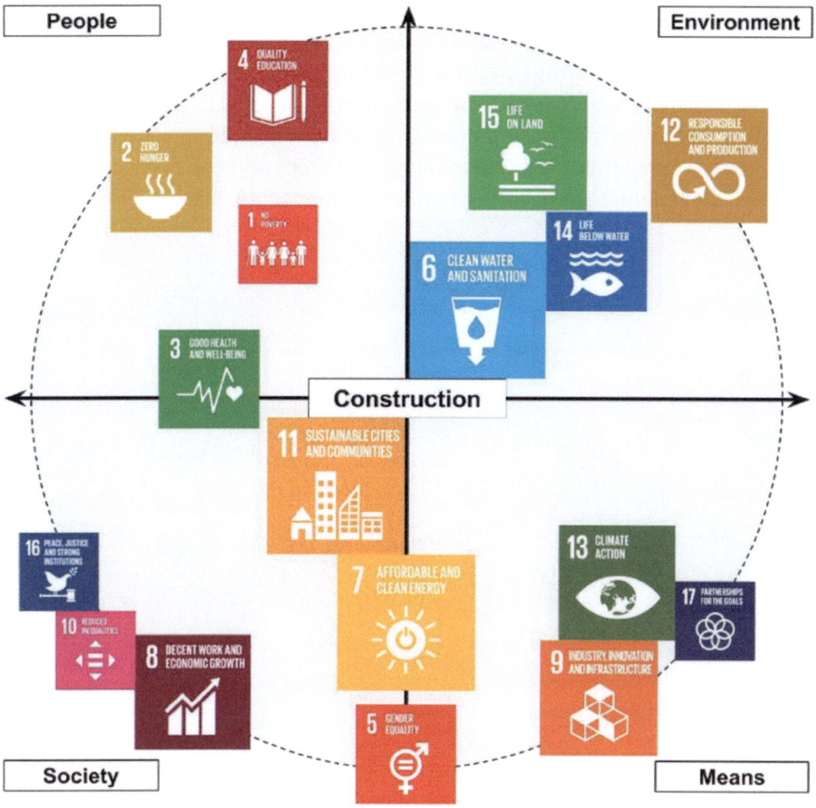

Fig. 1 SDG (Sustainable Development Goals) and UBEM (Adapted from [7])

ensuring energy efficiency by taking into consideration each stage of the process ranging from the material being used in construction and land design to waste generation which occurs during utilisation of property and then determines how to ensure energy optimisation in each stage, leading to achievement of SDG 12.

3.2 *Sustainable Building Practices*

UBEM stands as a pivotal aspect of sustainable building, with scholars like [18] arguing that it promotes sustainable practices. Sustainable building entails utilizing eco-friendly materials and minimizing resource wastage, objectives achievable through UBEM modeling. UBEM analyzes energy needs and material usage during construction and transportation, enabling firms to procure environmentally friendly materials efficiently [18]. Scholars, such as [17], advocate for efficient integration of renewable energy resources like solar, wind, geothermal, and biomass energy, facilitated by UBEM modeling. UBEM addresses challenges like availability uncertainties and storage issues, allowing firms to plan and transition to renewable energy sources effectively [17]. Amidst rising concerns about climate change, there is a growing demand for properties that mitigate its impacts [3]. UBEM strategies, incorporating ventilation, thermal mass concepts, and energy-efficient designs, cater to such demands. These strategies reduce energy demands and ensure resilience against extreme weather events, contributing to climate change adaptation [3] (Fig. 2).

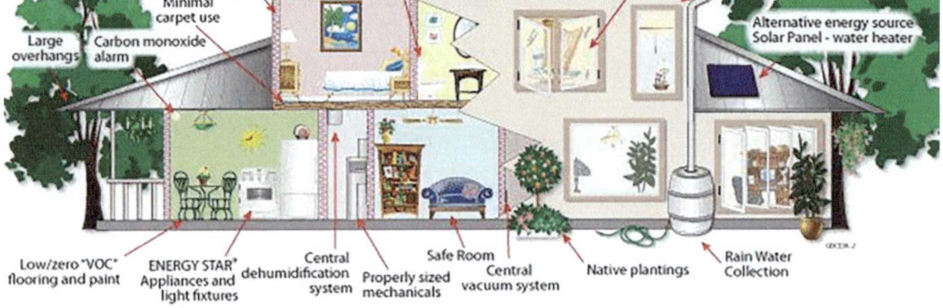

Fig. 2 Sustainable building practices (Adapted from Pacific Northwest National [22])

3.3 Biodiversity Conservation

UBEM practices, highlighted by scholars like [2], play a crucial role in energy management, enabling the deployment of simulation experiments incorporating green building elements such as green energy solutions, energy-efficient windows, and renewable energy systems [2]. Green building design not only enhances urban green spaces but also fosters habitats for flora and fauna, with features like green roofs enriching biodiversity [2, 24]. Addressing challenges like the heat island effect, which exacerbates urban temperatures, UBEM systems can mitigate heat and manage energy, contributing to a habitable environment conducive to plant and animal growth, thereby promoting biodiversity conservation [24]. Moreover, UBEM assists in mitigating negative impacts on local ecosystems. Different property projects require efficient land use, which UBEM ensures, contributing to the preservation of natural habitats and meeting the goals of urban development and ecological conservation [8]. Thus, UBEM methodologies support environmental ecosystem preservation and urban development goals simultaneously, aiding biodiversity conservation and fostering environmental growth in urban areas (Fig. 3).

3.4 Resource Efficiency

Real estate properties face substantial energy consumption demands, which can be effectively managed and optimized through UBEM strategies [21]. By addressing energy consumption from fossil fuels and electricity, UBEM enhances energy resource efficiency and promotes the utilization of renewable energy sources to meet project requirements. In addition to electricity, real estate properties consume significant amounts of water and minerals, which can also be optimized with UBEM [27]. UBEM simulations can assess water and mineral usage in construction, maintenance, and regular operations, leading to reduced resource consumption and energy optimization. UBEM serves as a guiding strategy for developing water management and mineral preservation projects that align with user needs and sustainability goals. Furthermore, UBEM facilitates efficient layout development to optimize land usage in real estate projects [25]. This is particularly valuable for larger projects or properties with multiple land use cases, ensuring optimal energy flow and availability to meet diverse energy needs efficiently.

3.5 Social Inclusivity

The aim of social inclusivity is bolstered by UBEM practices, facilitating the development of affordable housing solutions by considering energy consumption needs

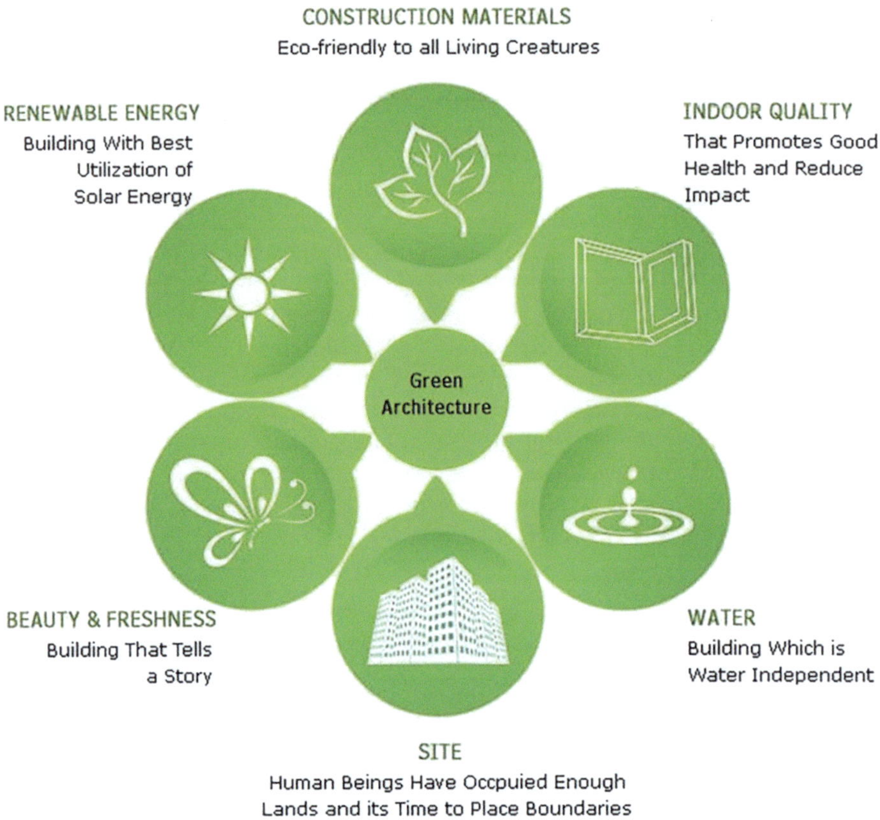

Fig. 3 UBEM and biodiversity benefits

and associated costs [23]. This approach fosters the creation of energy-efficient properties at accessible costs, expanding real estate projects' reach to a broader population and promoting equal housing opportunities in both urban and rural areas. Renewable energy generation, a key UBEM strategy, offers economic and social benefits for communities [15]. Economic gains include reduced utility bills and potential surplus energy sales, leading to lower maintenance costs. Social benefits manifest in improved environmental conditions and societal well-being, enhancing residents' quality of life. Residential projects in urban areas, as highlighted by [4], prioritize the development of recreational and communal services, with UBEM addressing energy needs. Incorporating green spaces further enhances urban well-being and residents' overall welfare, fostering a sense of belonging and social cohesion. These integrated efforts contribute to the creation of livable, affordable communities tailored to residents' needs (Fig. 4).

Fig. 4 Social Inclusivity and UBEM (Adapted from [28], November 20)

3.6 Research Gap

Analysis of work conducted by prior scholars has led to the identification of two major research gaps. The primary research gap which has been identified is the lack of comprehensive analysis which takes into account different aspects of UBEM ranging from energy management to social inclusion and sustainable building, and ensuring that all goals are achieved together. The second research gap which has been noted during the research is while prior researchers have established significant emphasis on different SDGs, they fail to take into account how UBEM can act as an effective measure to meet the need for sustainable building while also achieving SDGs. The current study identified 3 major SDGs namely SDG 9,11, and 12 which deal with innovation, sustainable infrastructure, inclusive and sustainable cities, and responsible consumption as well as production practices academic research has been done by prior scholars, but such research lacks depth and often fails to connect with the broader objective of ensuring energy management and optimisation. The current study has attempted to address these research gaps while at the same time, suggesting strategies which can lead to sustainable building, management of energy consumption in building projects, and also achieving sustainable development goals while not compromising on prospective buyer needs.

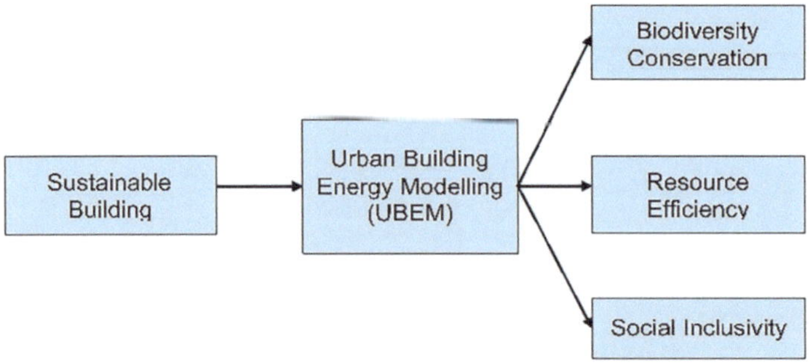

Fig. 5 Connection between sustainable building practices, UBEM, and outputs

4 Methods and Materials

The current study is based on secondary data which is collected from different journal articles published on relevant domains in 2020–2024 and is primarily openly accessible as affirmed by the study conducted by [5, 14], and [9]. A secondary data approach has been selected for the current study as such will allow an analysis of work conducted by primary scholars across different real estate projects and at the same time, who have adopted different strategies and frameworks under UBEM in order to meet energy management needs.

4.1 Data Analysis Tools

The current study is based on secondary data and hence content analysis method has been adopted for the study as affirmed by prior studies conducted by [10, 19], and [26]. All the articles which have been selected for the study have been analysed in order to derive major inferences as well as suggestions and limitations which have been noted by prior scholars during the research process which will aid in achieving all the objectives. Content analysis has been adopted for the current study as such allows extensive analysis of textual data to understand different aspects of UBEM (Urban Building Energy Modelling) and how such aids in different aspects ranging from resource efficiency to social inclusivity and biodiversity conservation (Fig. 5).

5 Results and Discussion

The study discusses various strategies in Urban Building Energy Modelling (UBEM) to ensure sustainable building practices. One primary strategy highlighted is integrating energy efficiency requirements during the initial design phase to optimize energy performance and manage thermal regulation. This approach aims to minimize energy loss and improve living conditions for residents, even during extreme weather events [13]. Another significant UBEM strategy involves integrating renewable energy systems, such as solar panels and wind turbines, into real estate projects to enhance environmental sustainability and reduce carbon footprint [6]. These strategies help determine resource requirements and facilitate systematic development stages. Moreover, the study mentions the adoption of smart building technologies, such as automated lighting and energy-efficient HVAC systems, to enhance user satisfaction and reduce utility costs based on occupancy and climate conditions [1]. Additionally, life cycle assessment, coupled with UBEM, aids in optimizing energy performance and minimizing costs associated with energy wastage or leakage [11]. However, challenges like implementation costs, legal regulations, and employee resistance need mitigation strategies for long-term goal achievement. Overall, UBEM plays a crucial role in optimizing energy performance, promoting environmental sustainability, and addressing challenges in sustainable building practices, aiming for long-term efficiency and affordability [1, 6, 11, 13].

Leveraging Urban Built Environment Models (UBEM) to attain Sustainable Development Goals (SDGs) 9, 12, biodiversity, and social inclusion involves incorporating injury prevention into the SDG framework. This integration is particularly evident in road safety targets like SDG 3.6 and SDG 11.2, complemented by global performance targets on road safety management, safer roads, and post-crash response. The SDGs stress the interconnectedness of social, economic, and environmental dimensions, underscoring the necessity for collaboration among international organizations to address these goals collectively. Monitoring SDG progress is a complex process, involving national statistical offices generating indicators at the national level for inclusion in a global database reported to the United Nations Secretariat and the Secretary-General. Despite recognizing the importance of embedding safety considerations across sectors for health equity within the SDGs, challenges persist in achieving policy integration within the United Nations system. Additionally, obtaining accurate statistics on UBEM adoption faces hurdles, hindering a comprehensive analysis due to these challenges.

6 Conclusion and Recommendations

Urban building energy models, or UBEM, serve as vital frameworks for managing energy performance in building and real estate projects, aiming to ensure resource efficiency and minimize wastage. These models align with Sustainable Development Goals (SDGs) 9, 11, and 12, focusing on industry innovation, sustainable cities, and responsible consumption. Despite their importance, UBEM implementation faces challenges such as high costs and public resistance, which require collaborative efforts and innovative solutions. The study recommends leveraging renewable energy sources and advanced artificial intelligence tools to track energy consumption and identify inefficiencies, facilitating effective energy management and meeting user demands. Regular system and operational audits are also advised to detect leakage and inform strategy development. Furthermore, involving all stakeholders in decision-making and collaborating with international peers in the real estate sector or research community can help adopt best practices and foster innovation.

7 Theoretical Implications or Managerial Implications

Certain technical as well as methodological limitations have been noted in the study which can negatively influence the validity and reliability of the results. One of the major limitations which has been noted in the study is the lack of primary data collection which has led to no statistical or empirical evidence in the research which can establish any relationship between different variables. Another major limitation which has been noted in the study is that all open-access journal articles as well as articles which have been published around 2020–2024 have been adopted, which leads to the exclusion of major inferences if such were in closed-access articles or have been published before 2020.

Future scholars shall consider this study as a foundational study and conduct primary data analysis which involves both quantitative as well as qualitative data. This will aid in analysing different aspects of urban building energy models from both objective as well as subjective perspectives. Further, future scholars shall consider adopting an experimental approach in which they engage with both real estate projects which are undergoing construction as well as real estate projects which has already been developed to understand how different aspects can be adopted and at the same time, what are the actual challenges faced and actual strategies which can be implemented.

8 Limitation of the Study and Any Future Research Direction

The present study holds significant implications for both academic researchers and project managers involved in construction and related fields. It synthesizes a literature review on urban building energy modelling (UBEM) solutions, focusing on innovation, smart technologies, and sustainability. Furthermore, it illustrates how UBEM can contribute to biodiversity conservation, resource efficiency, and social inclusivity, addressing the growing need for both rapid development and environmental responsibility in today's business landscape. Architects and designers can apply these findings to integrate UBEM systems into building designs, enhancing lighting while promoting biodiversity conservation, resource efficiency, and social inclusivity. This approach ensures the fulfillment of diverse stakeholder demands and balances economic and environmental considerations. The study also highlights UBEM's role in advancing Sustainable Development Goals (SDGs) 9, 11, and 12, benefiting various stakeholders including government officials, industry experts, and property buyers. Future research should delve deeper into UBEM adoption and its impact on achieving SDGs, particularly in optimizing energy usage, improving infrastructure sustainability, and fostering responsible consumption. Moreover, scholars are encouraged to explore UBEM's potential contributions to biodiversity conservation, resource efficiency, and social inclusion in urban development. These comprehensive studies will offer valuable insights to inform policies and practices for a more sustainable and inclusive future.

References

1. Akram MW, Mohd Zublie MF, Hasanuzzaman M, Rahim NA (2022) Global prospects, advance technologies and policies of energy-saving and sustainable building systems: A review. Sustain, 14(3), 1316
2. Ang YQ, Berzolla ZM, Reinhart CF (2020) From concept to application: A review of use cases in urban building energy modeling. Appl Energy 279:115738
3. Bardhan R, Debnath R, Gama J, Vijay U (2020) REST framework: A modelling approach towards cooling energy stress mitigation plans for future cities in warming Global South. Sustain Cities Soc 61:102315
4. Buckley N, Mills G, Letellier-Duchesne S, Benis K (2021) Designing an energy-resilient neighbourhood using an urban building energy model. Energs 14(15):4445
5. Budianto A (2020) Legal research methodology reposition in research on social science. Int J Criminol Sociol 9(1):1339–1346
6. Chen WH, You F (2022) Sustainable building climate control with renewable energy sources using nonlinear model predictive control. Renew Sustain Energy Rev 168:112830
7. Chung MA, Wang TH, Lin CW (2023) Advancing ESG and SDGs Goal 11: Enhanced YOLOv7-Based UAV Detection for Sustainable Transportation in Cities and Communities. Urban Sci 7(4):108
8. Dahlström L, Broström T, Widén J (2022) Advancing urban building energy modelling through new model components and applications: A review. Energy Build 266:112099
9. Dina Diatta I, Berchtold A (2023) Impact of missing information on day-to-day research based on secondary data. Int J Soc Res Methodol 26(6):759–772
10. Duan J, Zhang C, Gong Y, Brown S, Li Z (2020) A content-analysis based literature review in blockchain adoption within food supply chain. Int J Environ Res Public Health 17(5):1784
11. Estokova A, Samesova D (2021) Sustainable building materials and life cycle assessment. Sustainability 13(4):2012
12. Gao G, Yang S (2023). Construction and research of a data-driven energy consumption evaluation model for urban building operation. IEEE Access.
13. Hafez FS, Sa'di B, Safa-Gamal M, Taufiq-Yap YH, Alrifaey M, Seyedmahmoudian M, Mekhilef S (2023) Energy efficiency in sustainable buildings: a systematic review with taxonomy, challenges, motivations, methodological aspects, recommendations, and pathways for future research. Energ Strat Rev 45:101013
14. Hayat K, Yaqub K, Aslam MA, Shabbir MS (2022) Impact of societal and economic development on academic performance: a literature review. iRASD J. Econ., **4**(1), 98–106.
15. Hayes B, Kamrowska-Zaluska D, Petrovski A, Jiménez-Pulido C (2021) State of the art in open platforms for collaborative Urban Design and sharing of resources in districts and cities. Sustainability 13(9):4875
16. Hong T, Chen Y, Luo X, Luo N, Lee SH (2020) Ten questions on urban building energy modeling. Build Environ 168:106508

17. Johari F, Peronato G, Sadeghian P, Zhao X, Widén J (2020) Urban building energy modeling: State of the art and future prospects. Renew Sustain Energy Rev 128:109902
18. Kamel E (2022) A systematic literature review of physics-based urban building energy modeling (UBEM) tools, data sources, and challenges for energy conservation. Energies 15(22):8649
19. Kleinheksel AJ, Rockich-Winston N, Tawfik H, Wyatt TR (2020) Demystifying content analysis. Am J Pharm Educ 84(1):7113
20. Mutani G, Alehasin M, Usta Y, Fiermonte F, Mariano A (2023) Statistical building energy model from data collection, place-based assessment to sustainable scenarios for the city of milan. Sustainability 15(20):14921
21. Oraiopoulos A, Howard B (2022) On the accuracy of urban building energy modelling. Renew Sustain Energy Rev 158:111976
22. Pacific Northwest National Laboratory (2024) PNNL. https://www.pnnl.gov/
23. Rodríguez-Álvarez J (2023) Urban Building Energy Modelling for the Renovation Wave: A Bespoke Approach Based on EPC Databases. Buildings 13(7):1636
24. Romano P, Prataviera E, Carnieletto L, Vivian J, Zinzi M, Zarrella A (2021) Assessment of the urban heat island impact on building energy performance at district level with the eureca platform. Climate 9(3):48
25. Salter J, Lu Y, Kim JC, Kellett. R, Girling C, Inomata F, Krahn A (2020) Iterative 'what-if'neighborhood simulation: energy and emissions impacts. Build Cities, 1(1)
26. Vears DF, Gillam L (2022) Inductive content analysis: A guide for beginning qualitative researchers. Focus Health Prof Educ: Multi-Discipl J 23(1):111–127
27. Wong CHH, Cai M, Ren C, Huang Y, Liao C, Yin S (2021) Modelling building energy use at urban scale: A review on their account for the urban environment. Build Environ 205:108235
28. World Green Building Council (2023) World green building council. https://worldgbc.org/

Understanding Local Level Perception of Risk Dynamics: A Study of Gurugram's Urban Villages in Pursuit of Community Resilience

Pooja Lalit Kumar, Qamar Irshad, and Ila Gupta

Abstract India's metropolitan areas, marked by their intricate interdependence, expanding urban population, and high density, face significant disaster risks due to climate change. These challenges often impact neighboring urban villages, triggering cascading effects. This paper explores risk perception dynamics at the local level through a comprehensive study in Gurugram's urban villages. Moving beyond traditional risk assessments, the study highlights the importance of understanding subjective local risk perceptions. It investigates how climate change impacts are perceived within community dynamics, emphasizing inclusive development and community resilience. Using narratives as the main data collection tool, the research captures community members' lived experiences, providing a qualitative analysis of risk perceptions. The findings reveal disparities between top-down adaptation policies and grassroots-level coping strategies shaped by unique local perceptions. The study also underscores the limited acknowledgment of community-based approaches and shared resources in mainstream Disaster Risk Reduction (DRR) practices. Practical implications call for integrating local risk perceptions into planning and decision-making for effective risk management. The paper advocates adopting a Community Resilience framework tailored to Gurugram's urban villages. Ultimately, this paper contributes to enhancing community resilience against various risks by integrating climate change, inclusive development, and narrative data collection, promoting community-centric approaches in urban village contexts.

Keywords Changing perception · Community resilience · Disaster risks · Inclusive development · Ecology · Sustainable development goals

---

P. Lalit Kumar (✉) · Q. Irshad
Faculty of Architecture and Ekistics, Jamia Millia Islamia, New Delhi, India
e-mail: poojakumar@sushantuniveristy.edu.in

I. Gupta
Amity School of Architecture and Planning, Gurgaon, India

A. K. Sharma and A. Joshi (eds.), *Proceedings of the 3rd International Conference on Opportunities and Challenges for a Resilient Future*, Advances in 21st Century Human Settlements, https://doi.org/10.1007/978-981-96-0931-4_2

1 Introduction

Human beings inhabit a world where disaster risks manifests systematically, causing harm to the critical systems and infrastructure that serve as the foundation for human societies and economies. Never before in modern history have people encountered such a diverse array of both familiar and unfamiliar risks and hazards, interacting within a hyper-connected and swiftly evolving global environment. Reports from the World Bank [1], and the United Nations Office for Disaster Risk Reduction [2] highlight a concerning increase in both the severity and widespread impact of disaster risks. Over the last five years, casualties and affected individuals have exceeded the figures from the previous five years, indicating an unprecedented escalation in the challenges of disasters. This highlights the urgent need for concerted efforts in disaster risk reduction and resilience.

Human action is creating greater and more dangerous risk. Disasters have increasing impacts on communities and whole systems as risk multiplies. It has been seen that everyone is living downstream of something else and thus share risks to a certain extents and in no time global impacts become local, and vice versa. Impacts also cascade across sectors, creating new challenges [2]. Rapid urbanization is making people more vulnerable [3]. Population growth and expanded settlements put more people and infrastructure in the path of existing hazards, and there is an increase in frequency and intensity of climatic hazards due to climate change [4, 5] It is projected that by 2100, 200 million people in the world will be affected by sea-level rise, with most of those in Asia, in particular China (43 million), Bangladesh (32 million) and India (27 million) [6]. The United Nations Office for Disaster Risk Reduction (UNDRR) defines a disaster as a "serious disruption of the functioning of a community or a society at any scale due to hazardous events interacting with conditions of exposure, vulnerability and capacity, leading to one or more of the following: human, material, economic and environmental losses and impacts" Disasters branch from a combination of hazards with vulnerability and exposure of people and assets. Under this scenario, a hazard is a "process, phenomenon or human activity that may cause loss of life, injury or other health impacts, property damage, social and economic disruption or environmental degradation" [2, 7] while vulnerability describes "the conditions determined by physical, social, economic and environmental factors or processes which increase the susceptibility of an individual, a community, assets or systems to the impacts of hazards"[8] and Exposure is the "situation of people, infrastructure, housing, production capacities and other tangible human assets located in hazard prone areas" [2, 9] When hazards intersect with vulnerability and exposure, the likelihood of disasters significantly rises. This is because exposure amplifies the consequences, and vulnerability diminishes the ability to cope effectively.

The study examines the risk dynamics of hazard prone area, the nuanced experience, the adaptive capacity, the trade-offs among various potential risk The study uses *Narrative as a tool* to capture the nuances of risk perception in the urban villages of the millennium city, Gurugram (characterized by their juxtaposition of rural heritage against an urban backdrop, serve as microcosms where the impacts of global phenomena intimately intersect with local realities) which of late has been under the threat of disaster risks. It further comprehends how the perceived degree of risk influences or is influenced by the capacity to tolerate risk over time. While each of these components can be evaluated at the individual level, the study emphasizes the significance of community-level dynamics, which are generally ignored but play a crucial role. Ultimately, the research establishes the importance of risk perception as a tool for identifying indicators for assessing community resilience.

2 Community Resilience

'Community' and 'community resilience' are widely used concepts in literature addressing policies and strategies to prevent and reduce disaster risks and manage residual risks with preparedness, response and recovery activities [10]. Drawing on terminology from ecology and sociology, resilience is defined as "the capacity of a system, enterprise, or person to maintain its core purpose and integrity in the face of dramatically changed circumstances" [73–75]. Thus, is a dynamic process that is aimed at achieving a desired outcome, and "Community resilience is the existence, development, and engagement of community resources by community members to thrive in an environment characterized by change, uncertainty, unpredictability, and surprise" [11–14] showing various intricate layers involved.

Community on a general note refers to a group/groups of people linked by common identity, geography, commitment, interest, or concern. (Kirmayer & Whitley, 2009) though it has been often seen to be reduced to just denoting a specific geographical location at a local scale [15–17]. Scholars have explored the concept of "community" through various lenses. It has been conceptualized as a local scale of analysis, [15, 18] representing a network of actors [19–21] engaged in interactions within a specific locale [22]. Additionally, it has been viewed as a totality of social structures within a particular place, [21, 23] embodying unstructured interrelations among individuals [24, 25]. Furthermore, community has been described as an area nurturing shared identity and a sense of belonging [26]. Some scholars talk about community as networks involving specific types of actors, such as professional groups, [22, 27] while others highlight it as groups of individuals sharing a strong attachment to a particular place [28]. The existing literature indicates that the concept of community is often associated with a particular location, actor interactions, or a combination of both. Some interpretations encompass multiple layers or aspects of the community concept, highlighting the intricate nature of its definition. This complexity can be classified into three dimensions: ecological i.e. space and time, social structural i.e. networks and interactions, and symbolic cultural involving identities, norms, and values [21, 27, 28]. The dynamic and intricate nature of communities has been a subject of discussion in the social sciences for several decades. Surprisingly, these discussions are not prominently reflected in the literature on risk management or resilience frameworks.

Most of these frameworks prioritize the empowerment of local institutions and bottom-up, community-driven approaches for Risk Assessment. Yet, to incorporate such approaches into the decision-making process, it is imperative to investigate the ways that community members understand and perceive risks [19, 26]. People's perception depends on a number of factors, including risk communication among local communities, authorities, and decision-makers; current and future adaptive capacities; and, most importantly, the direct experiences, individually and collectively, of community residents [29].

3 Risk Dynamics

Climate change amplifies the risk of disasters through various mechanisms. It increases the probability, frequency, and intensity of climatic hazard events, impacting vulnerability to all hazards due to prolonged socioeconomic stresses [3, 76]. Climate change also leads to long-term consequences such as displacement and alters exposure patterns as climatic conditions evolve, causing hazards to emerge at new areas. As per the UNDRR analysis based on the International Disaster Database [77]. The number of reported disaster events per year has witnessed a significant increase over the past two decades. Although there were more peak disaster years in the 2000–2009 decade compared to 2010–2019, the overall frequency remains at an unprecedented level. In the period from 1970 to 2000, reports of medium- and large-scale disasters averaged around 90–100 per year. However, from 2001 to 2020, the reported number of such events surged to a range of 350–500 per year. These events encompassed geophysical disasters like earthquakes, tsunamis, and volcanoes, climate- and weather-related disasters, as well as outbreaks of biological hazards, including crop pests and epidemics. If current trends continue, the number of disasters per year globally may increase from around 400 in 2015 to 560 per year by 2030—a projected increase of 40% [78] which is a matter of concern.

The data clearly indicates a rising level of risk. However, this trend is not mirrored in surveys gauging individual risk perception. The dominant perception of risk, especially concerning long-term threats, tends to be characterized by optimism, underestimation, and a sense of invincibility. For example, the findings of a 2020 World Risk Poll suggest that, while the risks from climate change are generally understood and acknowledged, a significant proportion of people underestimate, remain skeptical or have no opinion on the issue (Lloyd's Register Foundation.2020. The Majority of People around the World Are Concerned about Climate Change. The Lloyd's Register Foundation World Risk Poll., n.d.) The surveys have shown that the average ***"risk perception gap"*** [2] between worrying about and having experienced hazards varies pointing towards experiencing as one of the most important aspect that affects people's risk perception. However multiple cognitive, behavioral, and sociocultural factors contribute to the complex landscape of disaster risk [79]. Yet, among these factors, ***risk perception & tolerance*** emerges as a critical aspect influencing how individuals/group prepares for, mitigate, and respond to hazards and overall the ***risk dynamics*** [2, 3, 30] of the area.

Risk dynamics refers to the ever-changing and evolving nature of risks over time-*the ways of perception* [31–33]—*the ways of tolerance* [34, 35]. It encompasses the attributes [30] that contribute to the fluctuations, uncertainties, and complexities associated with risk perception and tolerance. Understanding risk dynamics involves recognizing that risks are not static; they can manifest differently, intensify or diminish, and interact with each other in dynamic ways and thus forms an imperative contribution for creating an assessment framework for community resilience [6, 36, 37].

3.1 Risk Perception

Another important aspect of the challenge involves considering how individuals approach risks: do people/community identify risks in consistent patterns, do they understand and endorse expert opinion and recommendations regarding risk reduction strategies, and ultimately respond by adopting measures to reduce risks. Evidence confirms that challenges to risk assessment arise when disaster events coincide with the socio-economic vulnerability of geographically-defined communities in developing countries [3]. Due to inadequate information available regarding the changing impacts of disasters at the local level, decision-makers and professionals often rely on a predict-and-prevent approach based on historical climatic data, best practices, and recommendations by international networks, such as ICLEI and 100 Resilient Cities [38] the daily life of an individual is impacted by the overarching climate policies introduced at the city level. These policies and actions introduce both benefits and burdens, which are not always evenly distributed among their citizens [39, 40].

Scholars in Psychology have studied the theory of perceived risk under the head of social and behavior construct, differentiating it where it is considered the formal influencing the latter. While when it comes to implementation or management/ planning, risk has been studied under the aspects of ethnographic and social construct. In social aspects, as mentioned in Table 1, risk perception is often viewed as an *"experiential attitude,"* including the emotions provoked by an immediate response to a specific hazard [41]. According to this perspective, one's direct personal experiences play a pivotal role and are considered the most significant factor shaping risk perception. For instance, if individuals witness a natural hazard like flooding but do not undergo a substantial impact, their perceived risk of the hazard tends to be low. Social psychology approaches have evolved to include indirect dynamics as well. Scholars identify three indirect contributors [29] to risk perception: preparedness, worry, and awareness. From this viewpoint, the sense of risk transforms into fearfulness [22] Essentially, the social psychology literature suggests that the affective reaction of individuals to a particular risk is dominant, influencing human behavior. However, emotional reactions to significant disasters and their widespread impacts may not always align proportionally with the actual numbers of individuals harmed, unless the emotions are triggered by a particular event, such as mass killing [28, 42].

Behavioural scholars [43] also define risk perception as a subjective feeling emanating from a psychological process that involves the cognition and emotional changes that an individual associates with external risks [21, 44]. Thus Risk perception, (refer Table 1) as viewed in behavioural psychology, [45] represents individual attitudes and responses to a risk understood objectively and analytically [27]. In such an approach, people are theorized to be judgmental [46, 47] towards the world; they moderate their behaviors as per their perception. The argument in the literature shows that an individual's response to disaster risk is not solely determined by their individual experiences and related perceptions. Rather, responses are intricately shaped by complex social systems. Within these systems, various influences work together to persuade or, at times, compel individuals to embrace political biases and undergo social marginalization, factors that can magnify the risks associated with disasters [42].

These systemic influences, [44] encompassing shared beliefs, cultural practices, media exposure, levels of solidarity, and access to local institutions, play a significant role in shaping collective evaluations of exposure, significance, and trade-offs associated with various risks [44]. It is in this context that we can discern a collective or community-level perception of risks, which differs from individual perspectives.

The Ethnographic literature on disaster risk reduction risk assessment examines perceptions of risk within disaster planning at varying scales from local to global level collectively or separately depending on the type of framework. As we are now aware that perception of risk is driven by the available knowledge and information on natural hazards (e.g., estimated damage, probability of an event, hazard locations, etc.), [43, 51, 52] the community ability to comprehend that information, and propensity to accept it as valid remains arguable. As it has been seen at times being infused by the kind of information made available to them. As [54] says, "politicians and technocrats, through regulations and planning, seek to control people's behaviour to enforce order."

Planning studies consider risk perception as a dynamic process of constructing human (individual) behaviour vis-'a-vis a specific disaster as well as a community's socio-political system [43]. Collectively, an individual's perception of risks depends on many individual and communal factors, such as sociocultural beliefs and norms, sense of place, social cohesion, power structure, education, and disaster experience [34, 51, 53, 55]. Scholars have identified a consistent gap in government responses to risk: the neglect of community risk perception in risk assessment policies and frameworks [18, 56–58]. The isolation of risk assessment frameworks from community risk perceptions is evident from the study of various existing frameworks. Thus in order to have a framework which actually let the authorities know where to invest for resilience it's imperative to include a community level risk perception. "Place" and "Context" is an inseparable element of risk perception. Individuals learn from experiences of dealing with disasters that have occurred in a particular place based on a long process of evaluation and cognition [29, 42]. Thus, the character of risk

Table 1 Risk perception as in literature

Literature	Aspect	Keywords	Description	Source
Psychology	Risk as a social construct	Cognitive information (Feeling, Attitude)	"An individual's subjective feelings are based on prior experience, relevant knowledge, and psychological reactions." "As a psychological category, risk perception refers to individual understanding of external risks, emphasizing the impact of intuition and subjective feelings on personal cognition." "Intuitive feelings are still the predominant method by which human beings evaluate risk." "All current theories of choice under risk or uncertainty are cognitive and consequentialist."	[43, 47, 48]
	Risk as a behavior construct	Behavior, attitude, and individual practice (Individual and collective perception)	"An individual's understanding of external risks is composed of the impact of intuition and subjective feelings on personal cognition." "Cognitive maps of risk attitudes" and "judgments about the current and desired riskiness of diverse hazards" are relevant." A "subjective-construction process consisting of multiple dimensions, such as judgments on the severity and controllability of risks" that negatively correlates with mental health Risk perceptions "refer to whatever dimensions of the risk influence behaviour"; they may be products of the wider, complex social system	[35, 45–47, 49, 50]

(continued)

Table 1 (continued)

Literature	Aspect	Keywords	Description	Source
Planning	Ethnographic construct	Experience, Judgment, (Collective perception)	The local populations' risk perception corresponds to a specific rationality that does not comply with scientific rationality Risk perception is a perceived probability (i.e., a judgment of an events likely occurrence) and a perceived severity (i.e., a judgment of the likelihood of suffering death or injury, property damage, severe disruptions to daily routines, and financial losses)	[43, 51, 52]
	Spatial construct	Local knowledge Structures of privilege and power (Collective perception)	Disasters are socially constructed such that exposure to risks is uneven and societally shaped, reflecting prevailing patterns of inequality, marginalization, and exclusion Perceptions of risks are socially constructed and individually mediated, with dynamics of collective support, household dynamics, and individual voice, among others, at play. There may be collective understandings of what constitutes a risk, how risk is (fairly or unfairly) distributed, who or what is to blame, etc	[34, 51, 53– 55]

perception could vary among individuals with different life experiences, psychological conditions, knowledge, and modes of communications. Thus, it is studied further under its attributes *(i) Sense of belonging; (ii) Indigenous knowledge and adaptability; (iii) Information & communication.* The above is explored with respect to context and place based perception of an individual and/or a community.

Sense of Belonging

The degree of attachment an individual has to a particular place is closely knit with their experiences and perceptions of risk, elevating the "*sense of belonging*" [48]. Individual risk perception is constructed based on successive experiences in managing disasters, taking into account factors such as their occurrence, frequency, and intensity [23, 50]. People learn from experiences in a location, predicting future disasters and planning coping strategies like evacuation [59]. Risk perception differs for someone consistently affected versus those facing a disaster once. Personal experiences and traits, including location and cues, shape risk perception. Households in vulnerable areas may have distinct views on disasters. For example, when anticipating a significant storm, evacuation decisions depend on an individual's judgment. Some prioritize property protection, while others focus on securing food and shelter away from home. Expanding on the current body of literature, it is contended that "cultural characteristics" [34] like attachment to home or place (sense of belonging), cultural beliefs, and religious faith play a significant role in shaping individuals' perceptions and decision-making processes.

Indigenous Knowledge and Adaptive Capability

Research on both natural and human disasters has revealed a positive correlation between risk perceptions at the household or neighborhood level and the adoption of protective and adaptive measures by the local community [52]. These measures also reflect how people assess or value the adaptive capacity needed to respond to a specific disaster [39]. The indigenous knowledge of the area enhances the likelihood of residents choosing to stay in the area [62]. Community resilience is influenced by two factors: first, the individuals (or community) directly or indirectly affected, and second, the community's location, encompassing both its geographical and social risk profiles [61]. Local coping strategies and indigenous knowledge mirror the perceived risks, with residents evaluating their capacity based on how effectively they manage risks in comparison to others [60, 61]. Scholars like [8, 49, 62] evidently highlight a discrepancy in the coastal community's preferences and expected responses to climate risks. In this vulnerable coastal community, individuals perceive the risks related to rising sea levels as less concerning than the risks associated with relocating their homes, thus showing confidence in their coping ability.

Information and Communication

Access to disaster information, drawn from both observed and projected data, along with participation in disaster learning programs, enriches a community's knowledge and awareness, enabling them to better cope with disasters and shaping their risk perception [51]. In the face of an imminent environmental hazard, such as a hurricane in a hazard-prone area, individuals receive risk information through formal media channels and social media, guiding them to follow safety guidelines. This aids in perceiving emerging risks and recognizing environmental hazards. Risk perception is a negotiated stance based on perceived risk and expert-recommended benefits of disaster planning. Information empowers individuals to proactively protect themselves and access resources in advance.

Risk communication centers on how individuals and communities acquire knowledge, foresee, either directly or indirectly, a hazard, and anticipate its potential impacts on their lives. As aptly confirmed by [63] *"knowledge is and should be important in risk perception; if not, humans would have died out long ago."* He emphasizes three crucial concepts and the impact of "meaningful information" on risk perception. The first is production, addressing how knowledge is generated. Lay people's hazard knowledge is just as vital as expert-based disaster knowledge. The second concept is dissemination, where different channels, whether personal or media-based, may miscommunicate hazard information in terms of precision and accuracy. The third concept is information processing, highlighting the importance of interpreting data and behavioral modifications. This aspect, aiding individuals in understanding scientific conclusions, is often overlooked. Therefore, the experts must assess the community's knowledge level and determine the most effective means to disseminate information to individuals and groups, ensuring effective communication of messages about disasters and associated risks.

4 Method and Material

4.1 Case Study: Urban Villages of Gurugram

Gurgaon, in its rush towards becoming a Millennium City, has forgotten its roots. This erstwhile 'gaon' has 40 villages within it today—with 30 more within its 2031 Master Plan. Be it the villages in 'New' Gurgaon, like Sikanderpur, Nathupur, and Mollaheda, or those on the fringes of the City, like Ghata, Wazirabad etc. the fate is almost the same [64]. Gurugram's (refer to Fig. 1: Gurugram: Area of Study) present status as a metropolitan area includes surrounding settlements and has expanded through the establishment of new neighborhoods and districts.

The Haryana government's influential tax policy and improved infrastructure for the IT sector acted as a significant attraction. Gurgaon became a prominent outsourcing hub in India and Asia due to its proximity to the Indira Gandhi International Airport [65]. However, mentioned by [66] and [67] the city is currently facing challenges such as deteriorating infrastructure, a sudden decline in available productive agricultural land, loss of urban water bodies, and lakes. These issues contribute to air pollution, water contamination, health risks, and micro-climate fluctuations, all exacerbated by the growing population in urbanized areas.

Since the 2000s, Gurugram has seen a booming real estate sector with a focus on high-quality residential and commercial buildings, attracting multinational corporations and skilled professionals in IT, finance, and business process outsourcing.

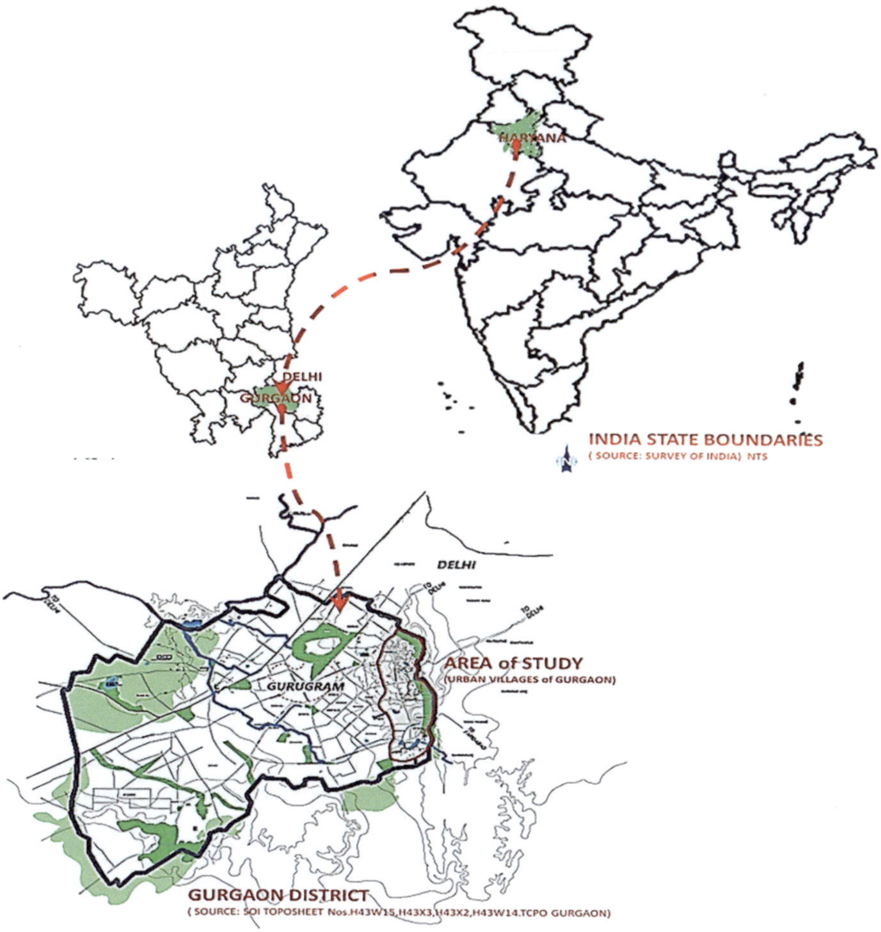

Fig. 1 Gurugram: area of study

The city's rapid growth has also created informal sector employment opportunities for low-income migrants, spanning industries, construction, and various services. Urban villages in Gurugram serve as crucial "arrival cities" for migrants, addressing housing and infrastructure needs for this workforce, although straining existing green and blue resources and exposing urban villages to environmental risks. The millennium city of late has witnessed a series of disasters in the form of floods, earthquakes and now heat waves. The intensity has been seen increasing and the impact was seen most on these urban villages, these instead of gearing up have seen dilapidating day by day, the influx of migrants and the idea of " it doesn't belong to me" has given way to un-thoughtful and un-conducive development in the urban villages.

4.2 Data Collection

We used Narrative as a tool to capture Risk Dynamics in terms of risk perception and ultimately risk tolerance of urban Villages of Gurugram. The villages falling under administrative zone 3 of Gurugram (refer to Fig. 1: Gurugram: Area of Study) Fig. 1: Gurugram: Area of Study were considered for further understanding of Risk Perception as they have been under scan for sustenance and have shown vulnerability to climate prone risks in the past [60, 68, 69]. To be more specific, our focus was on gaining understanding into three key elements of risk perception as mentioned earlier: the *sense of belonging, indigenous knowledge & adaptive capacity and information & communication.* We posed a set of questions in the local dialect to local stakeholders in risk-exposed urban villages of Gurugram to get an answer for the open-ended questions: In the era of climate change, what factors influence the perception of risk at both the individual and community levels?

The responses to our open-ended questions provided valuable information, experiences, and narrative frames. This documentation enabled us to capture the community's first hand experiences regarding the challenges posed by current and future disaster risks. Additionally, it shed light on crucial aspects that may have been previously overlooked, particularly in hazard-prone areas of Gurugram. We simultaneously conducted a desktop analysis to establish the context, reviewing scientific literature and policy documents related to climate change, resilience, and adaptation in Gurugram, with a specific emphasis on urban villages. Urban villages present an intriguing case due to their high climate vulnerability resulting from rapid land use changes in their spatial context. Despite active efforts by authorities to make the city resilient to climate changes, these areas, particularly urban villages, often remain overlooked.

We utilized websites, local knowledge, and established networks to identify suitable agencies and potential respondents, employing a snowball approach by seeking suggestions for additional contacts in each interview. Our selection criteria for further interviews included institutional stakeholders with influence over the issue, whether through political or social status, specific knowledge, formal legal power or responsibility, and the ability to shape concrete adaptation activities. In total, 51 individuals working in various spatial areas under investigation agreed to participate in the study and were interviewed. Our goal extended beyond capturing individuals' sentiments about the region; we aimed to assess specific risks, adaptive capacities, and imagery evoked by their narratives, representing, to some extent, the perspectives of numerous residents at risk with whom they collaborate. The interviews focused on eliciting perceptual images (memories and narrative explanations of circumstances) and symbolic representations (words and stories). Additionally, observational visits to the study areas were conducted to document risk scenarios through photos and casual interactions with residents. Media documents, grey literature, and scientific articles were also analysed to validate and complement interview data, particularly regarding relevant citizens and groups in the urban villages.

4.3 Data Interpretation

Narrative research as per [70, 71] has assumed diverse meanings across various social science and humanities disciplines, contributing to a dynamic field characterized by a plethora of methods and frameworks. This richness has led to both collaboration and occasional conflicts, as researchers seek to decipher the stories that imbue meaning into social life [72]. Here we employ narrative as a social science concept for interpreting how people – as 'storied animals'—cognitively make sense of their complex experience of the neighborhood where they live, and invest it with values and meaning [70]. The narratives played a crucial role in comprehending established facets of a sense of belonging, indigenous knowledge and adaptive capacity, and information and communication. They offered insights into community-level fears by examining internal and external factors associated with risk. Transcribed interviews were organized by focusing on spatial, socio-cultural, and psychological dynamics. The study documented individuals' experiences, attitudes, and perceptions, shedding light on residents' actions. We interpreted the information across scenarios to understand risk dynamics and how exposure exceeds individual or community acceptance to adapt and tolerate. Our methodology involves connecting themes to experiential memories and local stories, offering a comprehensive understanding of behavioral aspects shaping risk perception and contributing to community resilience.

5 Results and Discussions

5.1 Narratives from the Villages

"Even though Gurugram has dry weather, the winters used to be longer than it was before. Now winters are short. There is no green cover anymore. Trees have been felled to construct high-rise buildings." As narrated by one of the interviewed, evidently show the impact of urbansiation on climate change in the millennium city. It is noticed the urban villages especially in the Zone 3 have seen the population influx and many have been staying here only for more than 10 years now. The urban villages, which previously thrived with a lush green landscape due to abundant water resources and lakes fed by Aravalli range streams, now face depletion and deterioration. The streams that supply water to the village lake are now blocked due to unplanned growth. Over the years, the lake site has become a dumping ground for waste. The older community members attribute this change to two factors: reduced rainfall and the city's expansion. On the other hand, the younger generation or those who recently migrated accept the current situation as is and continue to adapt and build upon it.

"Our room shakes and the windows make a lot of noise,I don't know what to do, at times I feel like I should run away back to our homes but then our livlihood keeps me here", said one of the inhabitants staying on rent on the fourth floor in the village of Sikaderpur Ghosi after expericing earthquake shocks last year in the

month of March. The village which was initially just one story has over a period become multi- story on the same structural foundation and narrow staircase to lead to the upper floors. The material used is low standard inorder to make it affordable for the migrant worker. It is further noticed that risk perceived by him was higher that the person living on the ground floor who is an original resident. He believes that the built has survived the previous earthquake (20 + years back) and will survive the others too. This scenario is replicated in the other villages too.

"Our needs for water are met in this settlement and not a day goes by when we don't get water." In villages like Ghata, women do not face water shortages as taps supply water when the motorized pumps are activated. Landlords install these pumps to extract groundwater for tenants, accommodating the growing migrant population. Despite concerns about high soil salinity from excessive groundwater extraction with pumps, women appreciate the luxury of on-demand water in their busy lives.

Climate change contributes to issues like soil infertility, drying traditional water bodies, vanishing streams, and groundwater pollution. While villagers acknowledge strain from urban expansion, groundwater extraction, and changing weather, they may not connect these observed changes directly to climate change. *"The area around my house is filled with water, and water also enters my room, I am unable to reach my work place leaving my children."* Narrated one of the women after experiencing the water inundation, urban floods after continuous shower for a day," at times it takes more than a week to get back to our normal routine, as we get dirty water and our children get sick". This is the sad scene of most of the areas. A day long precipitation is not generally considered severe but it is not the case here, the risk is much higher that perceived by the authorities. It has also been observed that a major of original residents have already moved out of the villages considering the areas inhabitable The majority of residents in our study areas show a strong inclination to stay in their home communities, driven by a sense of belonging and a recognition of familiar risks. This attachment not only increases their willingness to adopt coping measures but also reduces their perceived risk. However, it's essential to note that, as mentioned in the above example, place attachment can sometimes lead to reduced coping behavior, resulting in higher future risks. Our study highlights that individuals base their decision on their evaluation of their ability to cope with perceived risks and endure household-level suffering. There is a prevailing tendency for people not to consider relocation, emphasizing the significance of mutual community adaptation and support, especially among the economically disadvantaged. Additionally, familiarity with the area, indigenous knowledge, and understanding the local environment, including people, culture, religion, and prevalent types of disasters, contribute to a diminished perceived risk, influencing the decision to remain in their communities. In this context, the sense of belonging tends to overpower risk perception.

5.2 *Risk Dynamics: Risk Perception & Tolerance*

This study emphasizes that risk perception is intricately tied to the social, political, and cultural context of a community, challenging the notion of it being an independent concept. The subjective sense of risk is significantly influenced by individuals' past experiences and their personal judgment of a situation in comparison to alternative outcomes. To comprehend how these conditions shape risk perception and its characteristics, we have organized people's reactions into four scenarios, aiming to understand the dynamics of risk perception in the millennium city of Gurugram.

The conceptualization of four iterative scenarios —(i) low risk perception & low risk tolerance; (ii) low risk perception & high risk tolerance; (iii) high risk perception & high risk tolerance; (iv) high risk perception & low risk tolerance—is crucial for understanding the dynamics of risk perception. Two other critical factors were considered while discussing the above scenarios. Refer to Fig. 2, The first factor being "time," as prolonged exposure to a known risk shapes one's perception. Vulnerable individuals become accustomed to environmental hazards, viewing them as routine over generations. This influences their risk perception based on unique adaptive capacity and planned arrangements. The second factor involves "trade-offs" between potential outcomes resulting from available options. In the face of disasters, individuals progress from the first scenario to the second, facing heightened risks and adaptation challenges. Over time, adaptive behavior allows them to better cope with or tolerate risks to some extent. Adaptation strategies involve navigating trade-offs among various available situations, such as reinforcing housing structures, creating alternative income sources during disasters, or seeking temporary or permanent relocation. Advancing to scenario three, individuals become accustomed to living with a known risk, using similar coping mechanisms repeatedly to accept even higher risks. Eventually, in scenario four, adaptive behavior shifts, and leading individuals to believe they can or must cope with risks considered unacceptable in scenario two. This trend continues over time.In our case studies, people tend to assess risks not primarily based on fearing the unpredictability of nature but rather by comparing them to past experiences. Particularly in scenarios two and three, vulnerable groups with limited coping capacity often opt for relocation as a strategy, accepting a different trade-off to mitigate potential adverse outcomes of risk acceptance. The gathering of data using narrative as a tool revealed various 'blind spots' in risk perceptions. The interpretation of terms such as "adaptive" or "at-risk" is subjective and varies among communities globally. Nevertheless, our analysis highlights that inadequate disaster risk management leads residents to consider specific risks as acceptable, even when they would typically be deemed unacceptable under more effective risk management practices.

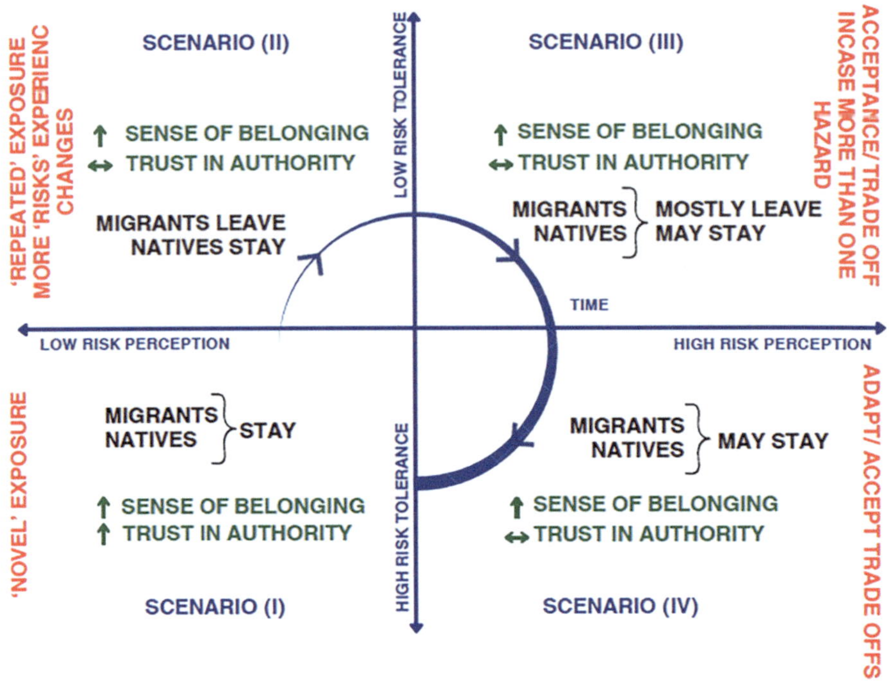

Fig. 2 Scenarios of risk dynamics

6 Conclusion

A critical aspect explored in the study is the disconnect between mainstream adaptation policies and the lived experiences of community members. The research illuminates the conflict between top-down decision-making processes influenced by social hierarchies and the grassroots-level coping strategies guided by the unique risk perceptions of the local population. Despite commitments to raise resilience, address climate change, and pursue sustainable development, current societal, political, and economic choices are moving in the opposite direction.

It is also observed that; Firstly, inadequate outcomes from institutional disaster responses and the escalation of disaster impacts, like waterlogging, compel local residents to coexist with high, and at times, unacceptable risks, resulting in social inequity. Secondly, community members tend to rely solely on their direct experiences and knowledge when gauging their perception of disaster risks, rejecting or not trusting institutional sources of risk communication. Institutional failures to acknowledge local challenges and aspirations intensify mistrust between residents and local governments, leading impoverished residents to adopt unique coping mechanisms. Thirdly, the risk perception of vulnerable communities often revolves around their financial capacity and disaster preparedness, relying on shared community resources. Unfortunately, institutional Risk Reduction practices seldom recognize these community-based approaches and shared resources. Our study further delved into risk perception as a lens to comprehend how urban village residents in Gurugram experience and adapt to disasters and their socio-economic impacts. This perspective challenges the traditional framing of risk as an individual calculation, proposing an alternative approach to understanding risk and risk perception within communities with limited resources. Future studies focusing on those directly affected by disasters could build upon our insights into risk perception dynamics. A key finding is the apparent conflict between mainstream adaptation policies influenced by social supremacy and local experiences and coping strategies guided by community members' risk perception. Practically, this study highlights the importance of incorporating local risk perception into planning and decision-making processes for Gurugram through a Community Resilience framework. Theoretically, it contributes to a broader understanding of the dynamics of risk perception. The outlined dynamics of risk perception can serve as a framework to evaluate local adaptations to risk and enhance community resilience. Knowledge, experience, and psychological reactions, both individual and collective, shape risk perception and lead to distinctive coping attitudes and behaviors in hazard-prone areas. Therefore, local disaster risk management must acknowledge and value such risk perception, recognizing the benefits of assessing local narratives of disaster experiences.

Acknowledgements This paper is a part of the on-going doctoral research on 'Exploring *and Assessing the Spatial and Socio-Ecological Resilience of Urban Communities for Disaster Risks. Study Area: Gurugram.*'at Faculty of Architecture & Ekistics in Jamia Millia Islamia, New Delhi. Guided by Dr. Qamar Irshad & Dr. Ila Gupta.

References

1. World Bank (2017) An evaluation of the world bank group's evolving approach background and context. In 2017. pp 1–28. Available from: http://documents.worldbank.org/curated/en/609841516898374978/pdf/122900-WP-PUBLIC.pdf
2. UNDRR (2022) Global assessment report on disaster risk reduction 2022: Our world at risk: transforming governance for a resilient future. GLobal Assessment Report on Disaster Risk Reduction. 2022. Available from: https://www.undrr.org/gar2022-our-world-risk-gar
3. Pörtner H-O, Roberts DC, Tignor MMB, Poloczanska E, Mintenbeck K, Alegría A, et al. (2022) IPCC: climate change 2022: Impacts, adaptation, and vulnerability. Working Group II Contribution to the IPCC sixth assessment report of the intergovernmental panel on climate change. Cambridge University Press, Cambridge, UK and New York, NY, USA, 3056
4. Van Aalst MK (2006) The impacts of climate change on the risk of natural disasters. Disasters 30(1):5–18
5. IPCC (2012) (Intergovernmental Panel on Climate Change). Climate change: New dimensions in disaster risk, exposure, vulnerability, and resilience. In Managing the Risks of Extreme Events and Disasters to Advance Climate Change Adaptation. A Special Report of Working [Internet]. Available from: https://www.ipcc.ch/report/managing-the-risks-of-extreme-events-and-disasters-to-advance-climate-change-adaptation/
6. Kulp SA, Strauss BH (2019) New elevation data triple estimates of global vulnerability to sea-level rise and coastal flooding. Nat Commun. 10(1). Available from: https://doi.org/10.1038/s41467-019-12808-z
7. Norris FH, Stevens SP, Pfefferbaum B, Wyche ÆKF, Pfefferbaum RL (2008) Community resilience as a metaphor, theory, set of capacities, and strategy for disaster readiness community

resilience as a metaphor, theory, set of capacities, and strategy for disaster readiness. Am J Community Psycology. 2014
8. Cutter SL, Boruff BJ, Shirley WL (2003) Social vulnerability to environmental hazards. Soc Sci Q
9. Kirmayer LJ, Whitley R (2009) Community resilience: models. Metaphors and Int J Indig Heal. 5(December):62–117
10. Rasanen A, Lein H, Bird D, Seeten G (2020) Conceptualizing community in disaster risk management. Int J Disaster Risk Reduct. 45
11. Al Rifat SA, Liu W (2020) Measuring community disaster resilience in the conterminous coastal United States. ISPRS Int J Geo-Information. 9(8)
12. Wilding N (2011) Exploring community resilience in times of rapid change. Trust; CU, editor. Carnegie UK Trust; 2011
13. Ostadtaghizadeh A, Ardalan A, Paton D, Jabbari H (2015) Community disaster resilience: a systematic review on assessment models and tools. PLOS. 0–16
14. Paton D, Millar M, Johnston D (2001) Community resilience to volcanic hazard consequences. Natural Hazards. Nat Hazards Earth Syst Sci
15. Kruse S, Abeling T, Deeming H, Fordham M, Forrester J, Jülich S, et al. (2017) Conceptualizing community resilience to natural hazards – the emBRACE framework. Nat Hazards Earth Syst Sci. 17(12)
16. Maclean K, Ross H, Cuthill M, Witt B (2017) Converging disciplinary understandings of social aspects of resilience. J Environ Plan Manag. 60(3)
17. Cutter SL (2016) The landscape of disaster resilience indicators in the USA. Nat Hazards 80:741–758
18. Pfefferbaum B, Pfefferbaum RL (2015) Community Resilience Interventions: Participatory,. Am Behav Sci. 2015
19. Clark A (2009) From neighbourhood to network: a review of the significance of neighbourhood in studies of social relations. Geogr Compass 3(4):1559–1578
20. Pauwelussen A (2016) Community as network: exploring a relational approach to social resilience in coastal Indonesia. Marit Stud [Internet]. 15(1):1–19. Available from: https://doi.org/10.1186/s40152-016-0041-5
21. Wilkinson KP (1970) The community as a social field. Soc Forces 48(3):311–322
22. Cox A (2005) What are communities of practice? A comparative review of four seminal works. J Inf Sci. 31:527–40. Available from: https://doi.org/10.1177/0165551505057016
23. Theodori GL (2005) Community and community development in resource-based areas: Operational definitions rooted in an interactional perspective. Soc Nat Resour 18(7):661–669
24. Beaumont E, Brown D (2018) 'It's the sea and the beach more than anything for me': Local surfer's and the construction of community and communitas in a rural Cornish seaside village. J Rural Stud. 59:58–66. Available from: https://doi.org/10.1016/j.jrurstud.2018.02.003
25. Kapferer B (2019) Victor Turner and The ritual process. Anthropol Today. 35(3)
26. Kuecker G, Mulligan M, Nadarajah Y (2011) Turning to community in times of crisis: globally derived insights on local community formation,. Community Dev J. 46(2):245–64. Available from: https://doi.org/10.1093/cdj/bsq002
27. Wenger-Trayner E (2000) Communities of practice and social learning systems. Organ
28. Gurney GG, Blythe J, Adams H, Adger WN, Curnock M, Faulkner L, et al. (2017) Redefining community based on place attachment in a connected world. Proc Natl Acad Sci. 114. Available from: https://doi.org/10.1073/pnas.1712125114.%0A
29. Barrett G (2015) Deconstructing Community. Sociol Ruralis. 55(2):182–204. Available from: https://doi.org/10.1111/soru.12057
30. Dhar T, Bornstein L, Lizarralde G, Nazimuddin SM, Risk perception—A lens for understanding adaptive behaviour in the age of climate change? Narratives from the Global South. Int J Disaster Risk Reduct [Internet]. 2023(95), (2022) 103886. Available from: https://doi.org/10.1016/j.ijdrr.2023.103886
31. Oliver-Smith A, Irasema A-A, Burton I, Lavell A (2017) The social construction of disaster risk: Seeking root causes. Int J disaster risk Reduct. 22:469–474
32. Burningham K, Fielding J, Thrush D (2008) 'It'll never happen to me': understanding public awareness of local flood risk. Disasters 32(2):216–238

33. Xu L, Li D, Peng L, Jing T (2019) Understanding sustainable disaster mitigation of stakeholder engagement: Risk perception, trust in public institutions, and disaster insurance. Sustain Dev 27:885–887
34. Billig M (2006) Is my home my castle? place attachment, risk perception, and religious faith. Environ Behav. 38(2):248–65. Available from: https://doi.org/10.1177/0013916505277608
35. Lindell MK, Prater CS, Wu HC, Huang S-K, Johnston DM, Becker JS et al (2016) Immediate behavioural responses to earthquakes in Christchurch, New Zealand, and Hitachi. Japan. Disasters. 40(1):85–111
36. Antronico L, Coscarelli R, De Pascale F, Condino F (2019) Social Perception of Geo-Hydrological risk in the context of urban disaster risk reduction: a comparison between experts and population in an area of southern Italy. Sustain 11(7):2061
37. Parsons M, Glavac S, Hastings P, Marshall G, McGregor J, McNeill J, et al. (2016) Top-down assessment of disaster resilience: A conceptual framework using coping and adaptive capacities. Int J Disaster Risk Reduct. 19:1–11. Available from: https://doi.org/10.1016/j.ijdrr.2016.07.005
38. Marschütz B, Bremer S, Runhaar H, Hegger D, Mees H, Vervoort J, et al. (2020) Local narratives of change as an entry point for building urban climate resilience. Clim Risk Manag. 100223. Available from: https://doi.org/10.1016/j.crm.2020.100223
39. Kates RW (2000) Cautionary tales: adaptation and the global poor. Clim Chang 45(1):5–17
40. Cowan T (2018) The Urban Village, Agrarian Transformation, and Rentier Capitalism in Gurgaon. India. Antipode. 50(5):1244–1266
41. Siegrist M, Gutscher H (2006) Flooding risk: a comparison of lay people's perceptions and expert's assessments in Switzerland. Risk Anal 26(4):971–979
42. Slovic P (2007) If I look at the mass I will never act: Psychic numbing and genocide. Judgm Decis Mak. 2(2):79–95. Available from: https://doi.org/10.1017/S1930297500000061
43. Snl MK (2008) Hwang. Households' perceived personal risk and responses in a multihazard environment, Risk Anal Int J. 28(2):539–556
44. Chelleri L, Waters JJ, Olazabal M, Minucci G (2015) Resilience trade-offs: addressing multiple scales and temporal aspects of urban resilience. Environ Urban 27(1):181–198
45. Bhatia S, Walasek L, Paul S (2021) The more who die, the less we care: evidence from natural language analysis of online news articles and social media posts. Risk Anal 41(1):179–203
46. Liu C, Huang N, Fu M, Zhang H, Feng XL, Guo J (2021) Relationship between risk perception, social support, and mental health among general chinese population during the covid-19 pandemic. Risk Manag Healthc Policy. 14:1843–1853
47. Peters E, MacGregor DG, Slovic P, Finucane ML (2010) Risk as analysis and risk as feelings: some thoughts about affect, reason, risk and rationality. In: The Feeling of Risk
48. Peng L, Tan J, Lin L, Xu D (2019) Understanding sustainable disaster mitigation of stakeholder engagement: Risk perception, trust in public institutions, and disaster insurance. Sustain Dev 27(5):885–897
49. Cannon T (1994) Vulnerability analysis and the explanation of "Natural" Disasters. Disasters, Dev Environ. 13–30. Available from: http://leeclarke.com/courses/disasters/cannon_vulnerability_analysis.pdf
50. Weinstein ND, Nicolich M (1993) Correct and incorrect interpretations of correlations between risk perceptions and risk behaviors. Heal Psychol. 12(3):235–45
51. Teka O, Vogt J (2010) Social perception of natural risks by local residents in developing countries—The example of the coastal area of Benin. Soc Sci Journal. 47:215–224
52. Buylova A, Chen C, Cramer LA, Wang H, Daniel T (2020) Cox. Household risk perceptions and evacuation intentions in earthquake and tsunami in a Cascadia Subduction Zone. Int J Disaster Risk Reduction. 44.
53. Páez H, Díaz J, Lizarralde G, Labbé D, Herazo B (2019) Coping with disasters in small municipalities: women ' s role in the reconstruction of Salgar, Colombia. 2019;(May):9–13.
54. Lizarralde G, Bornstein L, Robertson M, Gould K, Herazo B, Petter AM, et al. (2021) Does climate change cause disasters? How citizens, academics, and leaders explain climate-related risk and disasters in Latin America and the Caribbean. Int J Disaster Risk Reduct. 58
55. Brody SD, Walter WH, Peacock G(2005) Hurricane risk perceptions among Florida's single family homeowners. Landsc Urban Plan. 73(2)

56. Renschler CS, Frazier AE, Arendt L, Cimellaro GP (2010) Developing the ' PEOPLES ' resilience framework for defining and measuring disaster resilience at the community scale. 9th US 10th Can Conf Earthq Eng
57. Pfefferbaum B, Klomp RW, Services H, Reissman DB (2013) The Communities Advancing Resilience Toolkit (CART): An intervention to build community resilience to disasters. Public Heal Manag Pract
58. Sharifi A (2016) A critical review of selected tools for assessing community resilience. Ecol Indic. 69:629–47. Available from: https://doi.org/10.1016/j.ecolind.2016.05.023
59. Agrawal A, Gibson CC (1999) Enchantment and disenchantment: The role of community in natural resource conservation. World Dev 27(4):629–649
60. Kumar PL, Irshad Q (2020) Improvising vernacular ecological sustenance: an approach towards reviving the sinking cities. UGC Care J. 40(10):1–8
61. Alexander C, Bynum N, Johnson E, King U, Mustonen T, Neofotis P et al (2011) Linking indigenous and scientific knowledge of climate change. Bioscience 61(6):477–484
62. Krüger M (2019) Building instead of imposing resilience: revisiting the relationship between resilience and the State. Int Polit Sociol. 13(1):53–67. Available from: https://doi.org/10.1093/ips/oly025
63. Johnson BB (1993) Advancing understanding of knowledge's role in lay risk perception. In: Risk 4. p 189
64. Behl A (2013) Whose village is it? Friday Gurgaon
65. Narain V (2009) Growing city, shrinking hinterland: land acquisition, transition and conflict in peri-urban Gurgaon. India. Environ Urban. 21(2):501–512
66. Yadav K, Yadav V (2014) Tourism and eco development in aravalli ranges. Near Delhi. Deccan Geogr. 50(2):99–105
67. Gururani S (2013) Located roughly 30 kilometers south of the national capital of India, the city of Gurgaon,. Ecol Urban India. 119–43
68. Kumar PL (2017) Risk of floods in urban areas: a case of gurugram. Jamia Millia Islamia, New Delhi
69. Sharma S, Kumar PL (2023) Gaps in Emerging Techniques for Vulnerability Assessment of Urban Flooding. Int Conf Incl Resilient Cities Safe Sustain Futur. 1:296–303
70. Bruner J (1991) The narrative construction of reality. In: Critical inquiry. pp 1–21
71. Fina A De (2009) Narratives in interview—The case of accounts: For an interactional approach to narrative genres. Narrat Inq 19(2):233–258
72. Krauß W, Bremer S (2020) The role of place-based narratives of change in climate risk governance. Clim Risk Manag. 100221. Available from: https://doi.org/10.1016/j.crm.2020.100221
73. Langeland KS, Manheim D, McLeod G, Nacouzi G (2016) Definitions, characteristics, and assessments of resilience. In: How civil institutions build resilience. RAND Corporation. 4–10. Available from: http://www.jstor.org/stable/10.7249/j.ctt1btc0m7.8%0AJSTOR
74. Cutter SL, Christopher GB, Christopher TE (2010) Disaster resilience indicators for benchmarking baseline conditions. J Homel Secur Emerg Manag. 7(1).
75. Bhamra R, Dani S, Burnard K (2011) Resilience: the concept, a literature review and future directions. Int J Prod Res. 49(18):5375–93. Available from: https://doi.org/10.1080/00207543.2011.563826
76. Osuteye E, Johnson C, Brown D (2017) International journal of disaster risk reduction the data gap: an analysis of data availability on disaster losses in sub-saharan African cities. Int J Disaster Risk Reduct. 26:24–33. Available from: https://doi.org/10.1016/j.ijdrr.2017.09.026
77. Global Risk Report (2023) The global risks report 2023—18th Edition. 1–98 p. Available from: www.weforum.org
78. Ward PJ, Blauhut V, Bloemendaal N, Daniell EJ, De Ruiter CM, Duncan JM, et al (2020) Review article: natural hazard risk assessments at the global scale. Nat Hazards Earth Syst Sci. 20(4):1069–96
79. Haavik TK (2020) Societal resilience – Clarifying the concept and upscaling the scope. Saf Sci. 132:104964. Available from: https://doi.org/10.1016/j.ssci.2020.104964

Evaluating Land Use and Land Cover Transformations in Patna City, India, Through the Application of Remote Sensing and GIS Methods

Ruchita Rashmi, Shailendra K. Mandal, and Tarun Pant

Abstract India is experiencing accelerated urbanization, as is the case globally, offering Improved services in addition to enhanced social infrastructure, sanitary services, economic growth, technology, it is also the cause of much chaos such as environment degradation, quality of life, congestion, poor waste management, non-affordable housing, crime index etc. Considering all growing cities are expanding rapidly, it is estimated that 55% of the world's population lives in urban regions, with that number predicted to rise to 68% by 2050 (United Nations, Department of Economic and Social Affairs) [1]. Land Use and Land Cover Change (LULC) constitutes a significant process in the course of urbanization. This study examines at the changes in Land Use and Land Cover in the Patna Metropolitan Region during the last four decades (1980 to 2020) using GIS and remote sensing. The study area's satellite data was processed using ArcGIS software to detect noteworthy alterations in Land Use and Land Cover (LULC). The Image classification technique is used to create five signature categories: Built-up, Water-body, Fallow land, Forest, and Agricultural land. Each category generates around 40 unique signatures. To aid in image Interpretation, spectral signature plots are also generated. Following that, a complete examination of Land Use and Land Cover changes across multiple decades is carried out. The study shows that increased urbanization within the boundaries of the Patna Metropolitan Region has resulted in a significant change in spatial trends. This study's findings are distinct. The built-up area has almost doubled and vegetation has reduced. Fallow land is reduced as it was converted to settlement and built-up areas. Agricultural land has also decreased vastly for the same reason. These will

R. Rashmi (✉) · S. K. Mandal · T. Pant
Department of Architecture and Planning, National Institute of Technology Patna, Bihar 800005, India
e-mail: ruchitar.pg22.ar@nitp.ac.in

S. K. Mandal
e-mail: shailendra@nitp.ac.in

T. Pant
e-mail: tarun.ar18@nitp.ac.in; lncs@springer.com

A. K. Sharma and A. Joshi (eds.), *Proceedings of the 3rd International Conference on Opportunities and Challenges for a Resilient Future*, Advances in 21st Century Human Settlements, https://doi.org/10.1007/978-981-96-0931-4_3

be useful in anticipating the direction of urban expansion, with a focus on Patna's particular issues and potentials, to aid in future urban planning and development.

Keywords Urbanization · Land use land cover · Urban planning

1 Introduction

Patna Metropolitan Region, situated along the banks of the Ganges River in the state of Bihar, India, stands as a vibrant urban center experiencing rapid transformations in its land use and land cover patterns. As urbanization accelerates and socio-economic dynamics evolve, the cityscape undergoes continual changes, impacting its natural landscape and resource utilization. Understanding these shifts in land use and land cover is pivotal for informed urban planning, sustainable resource management, and environmental conservation. Land use refers to how humans utilize land and its resources, while land cover represents the physical features on the Earth's surface like water bodies, vegetation, forests, agriculture, and urban areas [2]. Land Use/Land Cover (LULC) analysis and change detection are vital for understanding the global environmental transform processes [3]. Changes in LULC are considered significant human-induced impacts on the local environment and can lead to various microclimatic alterations (Roberts et al. 1998) [4]. Land cover degradation signifies a decrease in land's capacity for profitable use and involves the characteristics of the land surfaces [5]. Understanding these aspects, is crucial for effectively managing natural resources, supporting sustainable practices, and assessing environmental changes.

LULC change is linked to the global climate change and urban/agricultural environment development [3, 6–8]. The complex relationship between LULC change and regional climate variations holds great importance for environmental adaptation and management. It is anticipated that LULC changes will alter land-surface-atmosphere interactions, consequently impacting weather and climate. The detection and monitoring of LULC changes using satellite multi-spectral image data has been a topic of research interest in remote sensing. Accurate and timely maps derived from remote sensing play a vital role in monitoring global and local climate changes, biodiversity conservation, and other environmental factors. Remote sensing, along with Geographical Information System (GIS), has been recognized as effective tools for managing land and natural resources, demonstrated in numerous studies analysing LULC changes [9–13].

Remote sensing data, especially from Landsat sensors, have been extensively used to map LULC changes over time, aiding in visually and quantitatively understanding these fluctuations in characteristics represented by different satellite datasets. Maximum Likelihood Classification (MLC) is a powerful and widely employed technique in the field of remote sensing and geospatial analysis for Land Use Land Cover (LULC) mapping. This statistical method leverages probability theory to assign pixels in remotely sensed imagery to predefined classes based on the likelihood that the observed spectral values belong to a particular land cover category. MLC operates on the assumption that each class has a distinct probability distribution for the spectral characteristics of its pixels. The classification process involves calculating the likelihood of a pixel belonging to each class and assigning the pixel to the class with the highest likelihood. This approach is founded on the principle of selecting the classification that maximizes the probability of observing the given set of spectral values, making it a robust and objective method for deriving land cover information from satellite or aerial imagery. The mathematical foundation of MLC provides a systematic and quantitative means of mapping and monitoring changes in land use and land cover over time, contributing to informed decision-making processes in diverse fields such as environmental management, urban planning, and agriculture.

2 Materials and Methods

2.1 Study Area

The metropolitan area encircling Patna, the capital city of Bihar, India, is officially defined as the Patna Metropolitan Region. The coordinates are 25.611°N latitude and 85.144°E longitude. It is often referred to as the Patna Planning Area or the Patna Metropolitan Area. As of the 2011 Census of India [14], the population of the Patna Metropolitan Region was approximately 3.62 million. (Since the Census of India has not yet released the data for the 2021 census, 2011 data is used in the study.) The study area constitutes of five municipal councils and one municipal corporation for the Patna district. The total area of the region is 1,156 km2, encircling 13 blocks: Patna Rural, Phulwari, Punpun, Sampatchak, Bihta, Daniawan, Dhanarua, Danapur, Khusrupur, Maner, Masaurhi, Naubatpur and Fatwah (Fig. 1).

2.2 Materials and Methodology

The remote sensing data has become a major source for change detection studies because of its high temporal frequency, digital format suitable for computation, synoptic view, and wider selection of spatial and spectral resolutions (Chen et al. 2012; Coops et al. 2006; Lunetta et al. 2004) [15]. In this study, multi-decadal digital remote sensing data from Landsat series satellites are used to map various LULC classes in the Study Area. The temporal information on land use and land cover helps identify the areas of change in a region. Geoinformatics allow to assign spatial connotations to land use land cover changes, namely, population pressure, climate, terrain, etc. which drive these changes [16]. Landsat series satellite images are among the most widely used satellite remote sensing data, and their high spectral, geographical,

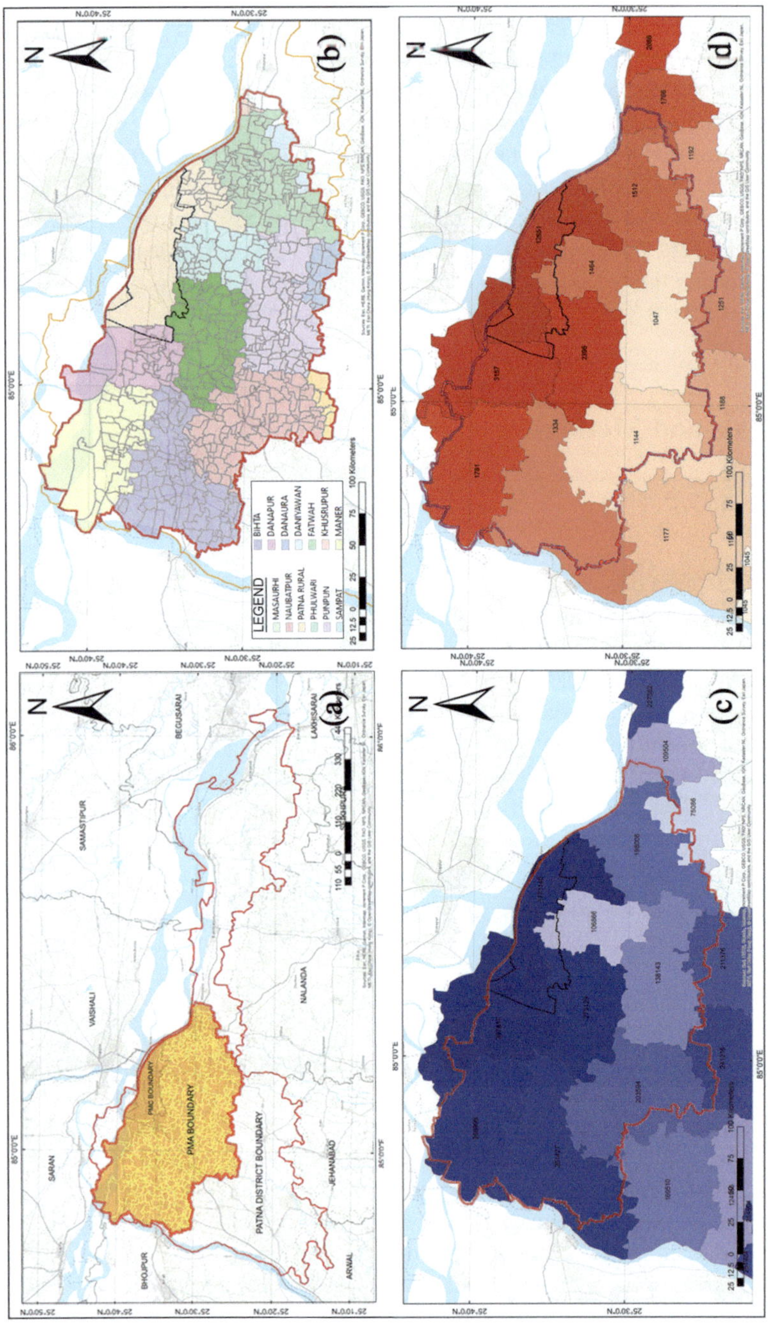

Fig. 1 **a** Administrative Boundaries of Patna District, Patna Metropolitan Area, Patna Municipal Corporation, **b** Block Map of Patna Metropolitan Area, **c** Block Population Map of Patna Metropolitan Area, **d** Block Population Density Map of Patna Metropolitan Area

and temporal resolutions make them useful for mapping and planning [17]. For the selection of the training and validation sets of pixels from the Landsat imagery, it is suggested that a minimum of 10–30 band cases per class can be used for training [18]. In this study, 30–35 representative training pixels per class are selected from the image based on stratified random sampling for multi decadal imagery. Remote sensing satellite data from Landsat-5 Thematic Mapper (TM) are acquired on April 6,1989, April 18, 1999 & April 29,2009 and Landsat-8 Operational Land Imager (OLI) image is acquired on April 25,2019. These are obtained from the US Geological Survey's official website (https://earthexplorer.usgs.gov). Landsat 4–5 Thematic Mapper (TM) images consist of seven spectral bands with a spatial resolution of 30 m for Bands 1 to 5 and 7. Spatial resolution for Band 6 (thermal infrared) is 120 m, but is resampled to 30-m pixels. Approximate scene size is 170 km north–south by 183 km east–west. Landsat 8 Operational Land Imager (OLI) and Thermal Infrared Sensor (TIRS) images consist of nine spectral bands with a spatial resolution of 30 m for Bands 1 to 7 and 9. The resolution for Band 8 (panchromatic) is 15 m. Thermal bands 10 and 11 are useful in providing more accurate surface temperatures and are collected at 100 m. Approximate scene size is 170 km north–south by 183 km east–west. The details of the satellite data used in the present study is shown in tabular form in Table 1. The Change Analysis is done in TerrSet software which enables users to detect, quantify, and visualize changes in land cover over time (Figs. 4 and 5).

Table 1 Details of satellite data used in the study

Sl. No.	Satellite sensor	Path/ Row	Band names	Spatial resolution (Meters)	Wavelength (Micrometres)	Month acquired	Cloud cover
1	Landsat-5 TM	141/42	Band 2-Green	30	0.52–0.60	March, April, May, 1989	0–10%
			Band 3-Red	30	0.63–0.69	March, April, May, 1999	
			Band 4-Near Infrared (NIR)	30	0.76–0.90	March, April, May, 2009	
2	Landsat-8 OLI	141/42	Band 3-Green	30	0.53–0.59	March, April, May, 2019	0–10%
			Band 4-Red	30	0.64–0.67		
			Band 5-Near Infrared (NIR)	30	0.85–0.88		

Flowchart showing Methodology of study:

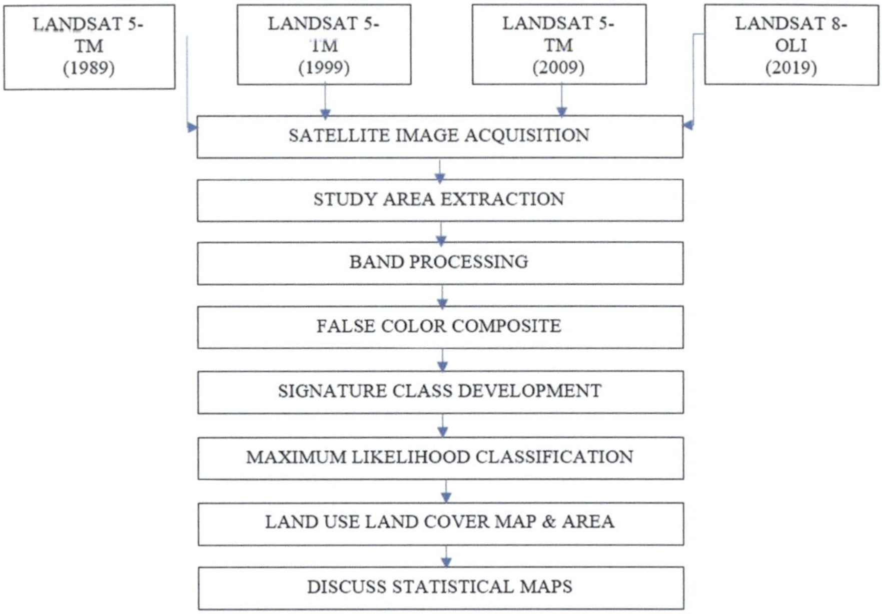

LULC Classification: The LULC maps for 1989, 1999, 2009 and 2019 are generated by classifying Landsat imagery from these four decades (Fig. 2). The images are classified using Arc GIS software. Each image is classified individually using a supervised maximum likelihood classifier. They are classified into five key LULC classifications, including built-up areas, agricultural land, vegetation, barrenland, and water bodies. Following is an overview of the key steps involved in LULC classification:

- Data Acquisition- We acquire satellite or aerial images with appropriate spatial and spectral resolution. Various sensors detect electromagnetic radiation in different bands, including visible, infrared, and thermal. We select images corresponding to the study's relevant time period, taking into account seasonality and the specific traits of the land cover types.
- Pre-processing- Adjusting the imagery for atmospheric conditions, sensor distortions, and geometric inaccuracies. The satellite imagery obtained is spatially referenced in the Universal Transverse Mercator (UTM) projection system within zone 45 north, using the World Geodetic System (WGS) 1984 datum, with a pixel resolution of 30 m. False color composites (FCC) for all images were generated using appropriate band combinations. These false color composite images were utilized to select training samples for each Land Use and Land Cover (LULC) class and to analyze them.

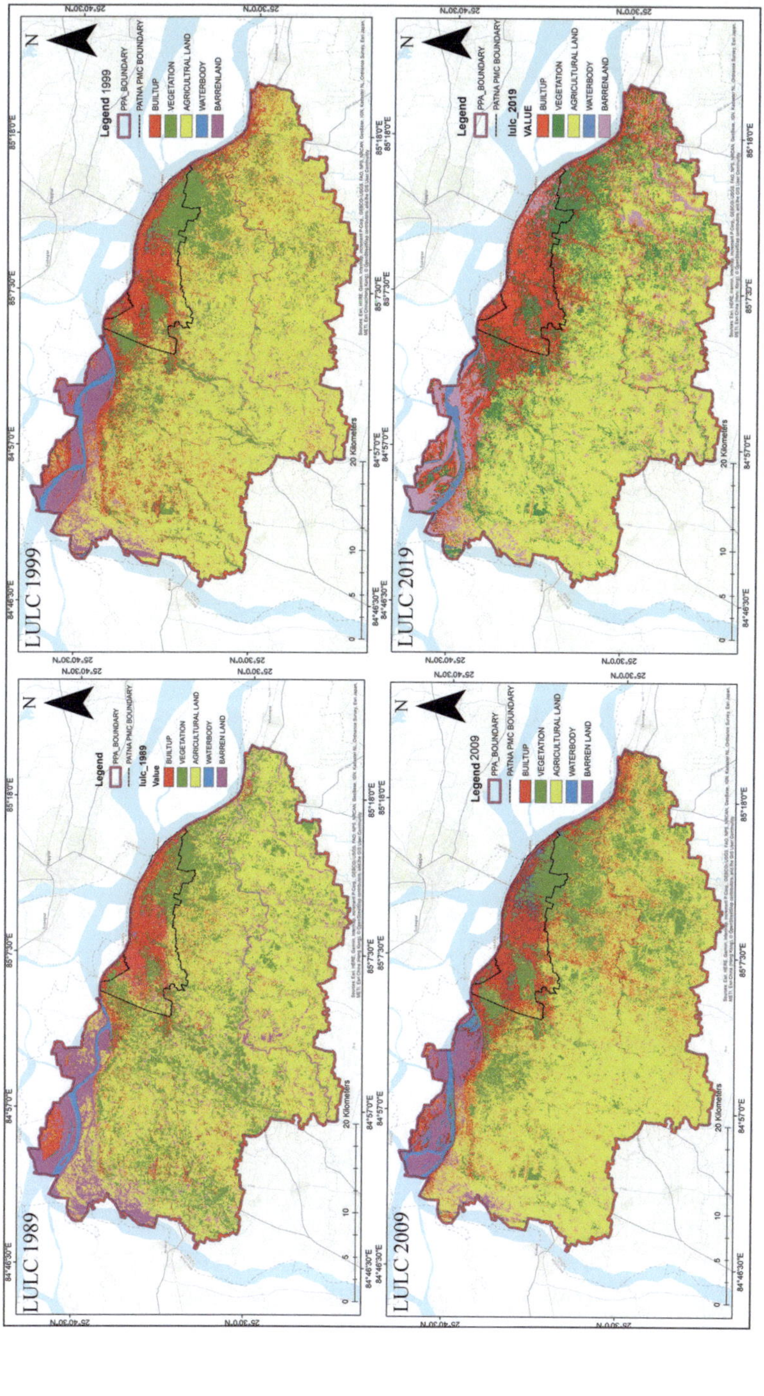

Fig. 2 LULC maps generated for the years 1989, 1999, 2009 and 2019

Table 2 Class Areas in Square Kilometers for the years 1989–2019

Classes	Area In Sq.Km-1989	Area In Sq.Km-1999	Area In Sq.Km-2009	Area In Sq.Km-2019
1. Builtup	91.329	134.533	142.142	199.002
2. Vegetation	225.442	146.516	194.366	177.094
3. Agricultural Land	689.672	756.628	732.114	646.779
4. Waterbody	13.135	18.581	26.874	20.228
5. Barrenland	132.699	96.020	56.782	109.174

- Training Inputs- For supervised classification, we gather training samples that represent each land cover class. These samples are used to train the classification algorithm. We use classification algorithms to assign pixels to various land cover categories. Popular algorithms include Maximum Likelihood, Support Vector Machines (SVM), Random Forest, and Neural Networks. In supervised classification, pixels are assigned to classes based on the training data, whereas in unsupervised classification, pixels are grouped into clusters without any prior training.

LULC Change Analysis:

See Figs. 2, 3.

Gains, Losses & Persistence Analysis of Classes:

See Figs. 4, 5.

3 Result and Discussion

3.1 Result

The LULC maps for year 1989, 1999, 2009, and 2019 are derived by classification of Landsat TM/ OLI images of respective years using the maximum likelihood classifier (Fig. 2). From the above study, we can see the following changes occurred during the four decades (1989–2019) (Table 2 and Fig. 3): Builtup almost doubled from 91 square kilometers to 200 square kilometers, while Barrenland decreased from 133 square kilometers to 109 square kilometers due to conversion to settlement and built-up areas. For the same reason, vegetation and agricultural land declined dramatically, from 225 sq.km to 177 sq.km and 690 sq.km to 647 sq.km. Waterbody area rose from 13 square kilometers to 20 square kilometers. Changes in precipitation patterns, temperature, and extreme weather events, as well as changes in agricultural methods, deforestation, and land use for industrial or residential purposes, can all

(a)

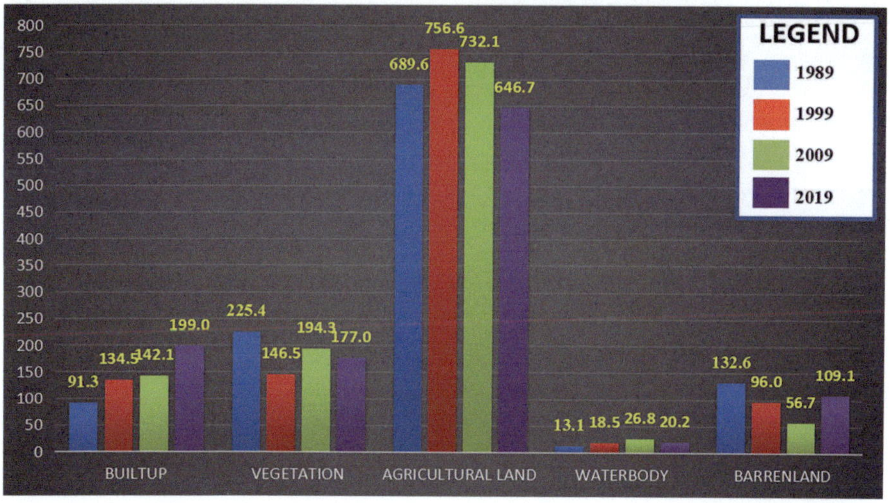

(b)

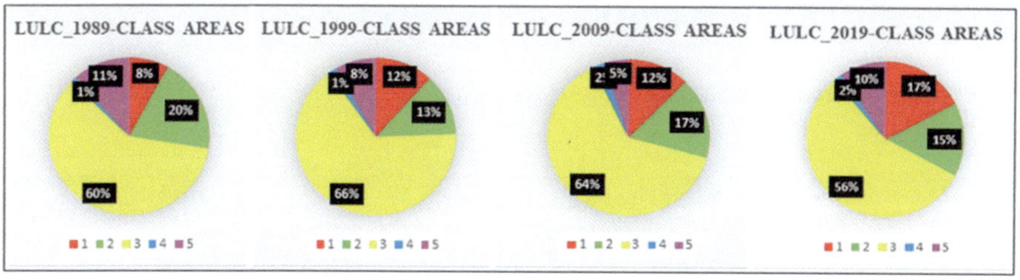

Fig. 3 **a** Graphical representation of Distribution of Class Areas in sq. Km (1989–2019), **b** Graphical representation of Percentage Distribution of Class Areas (1989–2019)

have an impact on hydrological cycles and contribute to water body formation or expansion.

The LULC change analysis during the period of 1989 to 2019 is also carried out using LULC maps with the help of change analysis tool available in Land Change Modeler in Terrset software. The LULC changes are evaluated by gains, losses and persistence experienced by different classes (Figs. 4 and 5). Most of the classes are showing losses (Unit- square kilometers of areas & percentage change in areas). During 1989 to 2019, Builtup has lost 48.08% and gained 76.17%, with net gain of 28.09%. Vegetation has lost 72.17% and gained 64.58%, with net loss of 7.59%. Agricultural Land has lost 31.44% and gained 26.89%, with net loss of 4.55%. Waterbody has lost 75.04% and gained 83.79%, with net gain of 8.39%. Barrenland has lost 69.35% and gained 62.75%, with net loss of 6.6%. So, it is clearly demonstrated that during the period 1989–2019, considerable changes and transitions were observed among various LULC classes. Contributions to net change in % area of

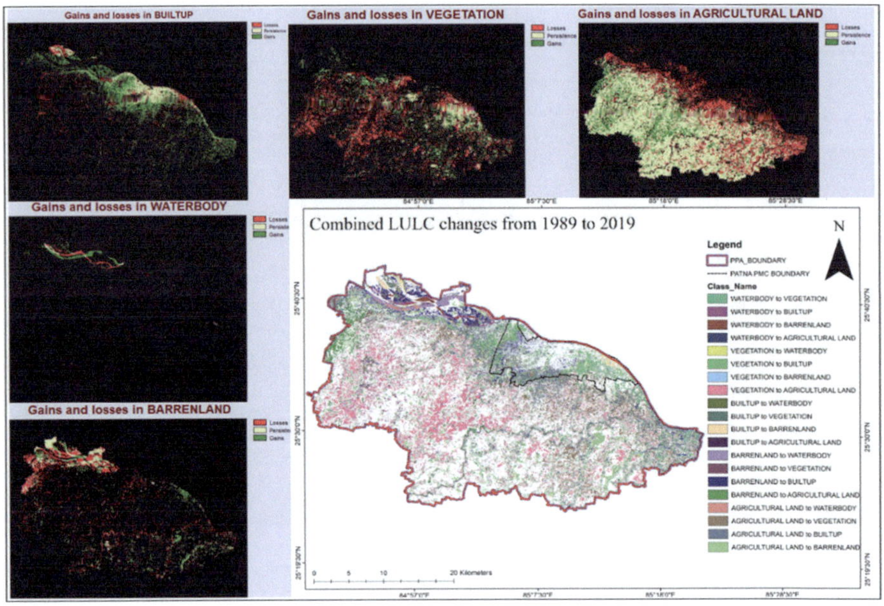

Fig. 4 Gains, losses and persistence analysis for each class and combined LULC change (1989–2019)

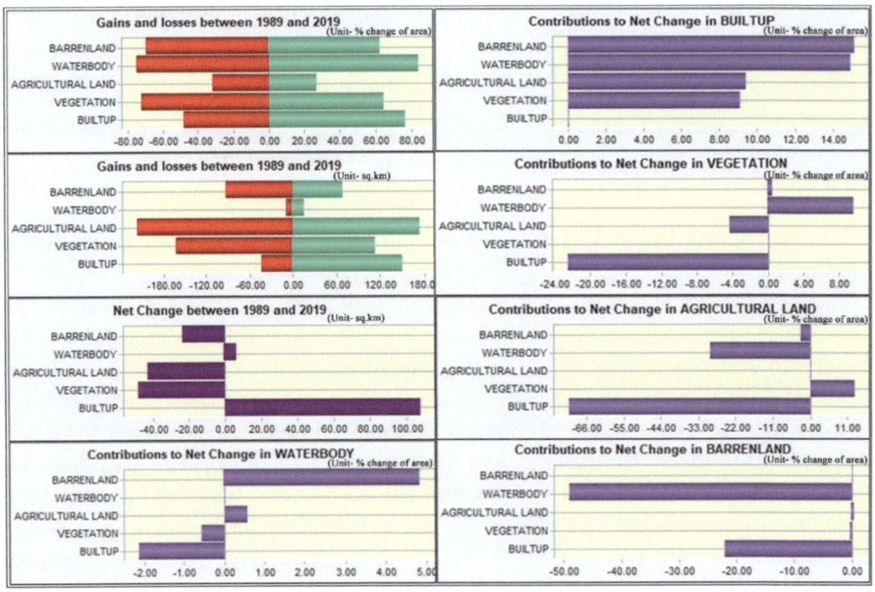

Fig. 5 Change Analysis-Gains and losses by category and classes which contributed to net loss/gain for each class (1989–2019)

each class with respect to other classes (in % change of area) during the given period are represented as graphical form (Fig. 5).

3.2 Discussion

In a similar study of LULC analysis, it is found that in the period of 25 years (1989–2014) the urban built-up area has nearly doubled in Patna Municipal Corporation boundary. In 1989 it was 31.343 sq.km which became 61.859 sq.km that is 61.521% of the total area of Patna Municipal Corporation. The water body made up of ponds, lakes and canal nearly halved, fallow land became nearly 1/5th in 2014 of what it was in 1989, agriculture land was reduced to 32.274 sq.km in 2014 from 53.708 sq.km in 1989. Vegetation increased more than 1.6 times in this period [19, 20].

In another similar study of LULC analysis of Patna District of time period 1988–2013, it was demonstrated that there is an enormous increase in built up area while decrease in agricultural land and vegetation covered area. It was observed that in 1988, built up covered an area of 31.85 sq.km, and it increased to 185.36 sq.km in 2013. The agricultural land in 1988 covered an area of 621.11 sq.km and it decreased to 555.06 sq.km in 2013. The water bodies in the study area have shown an enormous decrease from 11.12 to 5.44% during 1988–2013 [13, 14].

The results of the above examples [13, 14, 19, 20] show similarities with our findings of LULC change analysis for Patna Metropolitan Area for the years 1989–2019. The builtup area of waterbody, however was slightly increased in our study, from 13 sq.km to 20 sq.km. Hydrological cycles can be affected by variations in precipitation patterns, temperature, and extreme weather events. They can also be influenced by changes in agricultural practices, deforestation, and the usage of land for residential or commercial purposes. Spatial distribution of Vegetation also decreased in overall Patna Metropolitan Area from 225 sq.km to 177 sq.km for the years 1989–2019. The rapid growth of urban areas often leads to the conversion of natural landscapes, including vegetation-covered areas, into built-up environments.

4 Conclusion

The present study aims to demonstrate changes in LULC between 1989 and 2019, utilizing a combination of remote sensing imagery, GIS, and change analysis methods. To create LULC maps, a supervised maximum likelihood method of classification is employed with Landsat TM/OLI satellite images from 1989, 1999, 2009, and 2019. The LULC change analysis for the given period revealed a significant rise in built-up area while decreasing agricultural land and vegetation covered area. The changes in land use and land cover in Patna over the past four decades can be attributed to a complex interplay of various factors, including urbanization, population growth, economic development, infrastructure projects, agricultural practices, and environmental changes. The analysis of LULC change serves as a crucial tool for understanding the dynamics of urbanization and environmental transformations. This study not only contributes to the existing body of knowledge on urban growth but also highlights the significance of integrating technology to address the evolving challenges associated with rapid urbanization. By utilizing Remote Sensing and GIS methods to track LULC changes over time, researchers and policymakers gain a comprehensive view of how land is being utilized and modified. This knowledge forms the basis for informed decision-making in urban planning, resource allocation, and infrastructure development. Furthermore, LULC change analysis is instrumental in assessing the impact of human activities on ecosystems and natural habitats. It allows for the identification of areas undergoing deforestation, urban expansion, or changes in agricultural practices. Such insights are essential for crafting effective conservation strategies, mitigating the loss of biodiversity, and promoting ecologically sustainable practices. Moreover, the analysed data facilitates the identification of areas prone to environmental degradation or vulnerability to climate change. This foresight is crucial for implementing adaptive strategies and resilience-building measures in the face of evolving environmental challenges.

References

1. United Nations (2023) https://www.un.org/en/global-issues/climate-change
2. Di Gregorio A, Jansen LJM (2005) Land Cover Classification System (LCCS): Classification Concepts and User Manual http://www.fao.org/docrep/003/x0596e/x0596e00.HTM
3. Srivastava PK, Han D, Rico-Ramirez MA, Bray M, Islam T (2012) Selection of classification techniques for land use/land cover change investigation. Adv Space Res 50(9):1250–1265
4. Meyer WB, Turner BL (1992) Human population growth and global land-use/cover change. Annual Review of Ecology and Systematics, 23, 39–61. https://doi.org/10.1146/annurev.es.23.110192.000351
5. Roberts Dar, Batista, Getulio Pereira, Jorge Waller, Eric Nelson, Bruce (1999) Change identification using multitemporal spectral mixture analysis: applications in Eastern Amazonia.
6. Lambin EF, Geist HJ, Lepers E (2003) Dynamics of land-use and land-cover change in tropical regions. Annu Rev Environ Resour 28(1):205–241
7. Riebsame WE, Meyer WB, Turner BL (1994) Modeling land use and cover as part of global environmental change. Clim Chang 28(1):45–64
8. Walker RT, Solecki WD (1999) Managing land use and land-cover change: the New Jersey Pinelands Biosphere Reserve. Ann Assoc Am Geogr 89(2):220–237
9. Dhinwa PS, Pathan SK, Sastry SVC, Rao M, Majumder KL, Chotani ML, Sinha RLP et al (1992) Land use change analysis of Bharatpur district using GIS. J Indian Soc Remote Sens 20:237–250
10. Ghosh S, Sen KK, Rana U, Rao KS, Saxena KG (1996) Application of GIS for land-use/land-cover change analysis in a mountainous terrain. J Indian Soc Remote Sens 24:193–202
11. Kilic S, Evrendilek F, Berberoglu S, Demirkesen AC (2006). Environmental monitoring of land-use and land-cover changes in a Mediterranean region of Turkey. Environ Monit Assess 114:157–168
12. Raju K, Kumar RA (2006) Land use changes in Udumbanchola taluk, Idukki district—Kerala: an analysis with the application of remote sensing data. J Indian Soc Remote Sens 34:161–169
13. Majeed M, Tariq A, Anwar MM, Khan AM, Arshad F, Mumtaz F, Farhan M, Zhang L, Zafar A, Aziz M et al (2021) Monitoring of land use-land cover change and potential causal factors of

climate change in Jhelum District, Punjab, Pakistan, through GIS and Multi-Temporal Satellite Data. Land 10(10):1026. https://doi.org/10.3390/land10101026
14. Mishra VN, Rai PK (2016) A remote sensing aided multi-layer perceptron-Markov chain analysis for land use and land cover change prediction in Patna district (Bihar), India. Arab J Geosci 9:1–18
15. Census of India 2011—Bihar—Series 11—Part XII B—District Census Handbook, Patna https://censusindia.gov.in/nada/index.php/catalog/299
16. Attri P, Chaudhry S, Sharma S (2015) Remote sensing & GIS based approaches for LULC change detection–a review. Int J Curr Eng Technol 5(5):3126–3137
17. Roy P, Roy A (2010) Land use and land cover change: a remote sensing AND GIS perspective. J Indian Inst Sci 90:489–502
18. Sadidy J, Firouzabadi PZ, Entezari A (2009) The use of radarsat and landsat image fusion algorithms and different supervised classification methods to improve landuse map accuracy–case study: sari plain–Iran. Dep Geogr, Tarbiat Moallem Sabzevar Univ p 7
19. Mohammad A (2014) An assessment of land use land cover change pattern in patna municipal corporation over a period of 25 Years (1989–2014) using remote sensing and GIS techniques. Int J Innov Res Sci, Eng Technol. 03. 16782–16791. https://doi.org/10.15680/IJIRSET.2014.0310053
20. Petropoulos GP, Kontoes C, Keramitsoglou I (2011) Burnt area delineation from a uni-temporal perspective based on Landsat TM imagery classification using Support Vector Machines. Int J Appl Earth Obs Geoinf 13(1):70–80

Climate Resilience as the Future Trend of City Planning, Case of Indian Cities

Anisha Deb and Seema Praliya

Abstract Climate-resilient city planning and infrastructure has the potential to improve the reliability of service provision, increase city lifespan and protect their people. City planning and climate resilience, both are interrelated and play a significant role in overall development of a settlement. Climate resilience is the preparedness capacity of a society to cope with hazardous and vulnerable impact of climate change. Urban development frequently substitutes concrete, impermeable surfaces for natural components, which can destroy and damage ecosystems, natural flows and processes (such as the cycles of water and nutrients), and biodiversity. Haphazard planning interventions such as large scale deforestation projects, illegal construction activities, hydropower projects along with change in land use (CLU) are few significant causes of climate change and urban disasters. The effects of climate change have highlighted the need for reinforced and integrated infrastructure system, particularly for those supporting vital city services related to ICTs, health, education, and transportation among others. For instance in case of Kedarnath and Joshimath, where both are hill towns of Uttarakhand, India which recently suffered climate crisis as cloudbursts, floods and land subsidence. As a result, anthropogenic causes are the major driver for all such incidents for which authorities and individuals at their own stage need to align their developmental activities towards climate resilience as a whole. For which globally organisations such as UN Climate Change Conference (UNCCC), UN Environments Programme (UNEP) and Inter-governmental panel of climate change (IPCC) are working to reduce the emissions and atmospheric factors causing climate change. The paper focuses on climate resilient planning taking the case of two vulnerable Indian cities affected with past tragedies. The research is carried out mainly based on secondary study and to understand the gaps in the

A. Deb (✉)
Student of Masters in Urban and Regional Planning, School of Architecture and Planning, DIT University, Dehradun, Uttarakhand 248009, India
e-mail: 1000017748@dit.edu.in; debanisha07@gmail.com

S. Praliya
Assistant Professor, School of Architecture and Planning, DIT University, Dehradun, Uttarakhand 248009, India
e-mail: seema.praliya@dituniversity.edu.in

A. K. Sharma and A. Joshi (eds.), *Proceedings of the 3rd International Conference on Opportunities and Challenges for a Resilient Future*, Advances in 21st Century Human Settlements, https://doi.org/10.1007/978-981-96-0931-4_4

climate resilient planning for cities that are impacted by catastrophes. The study approach includes prevention, government initiatives for disasters, causes, damage, impact and planning strategies to mitigate the damage caused by them. This paper will assist researchers and authorities in field of disaster and allied works.

Keywords Climate · City planning · Disaster · Resilience · Hazard · Ecological and inclusive development · Inclusive development

1 Introduction

Climate resilience is essential in the context of global climate change, as it enables societies to minimize the adverse impacts of climatic shifts, protect vulnerable populations, and promote sustainable development in the face of a changing climate. The significance of climate resilience lies in its capacity to address and mitigate the adverse impacts of climate change on ecosystems, communities, economies, and infrastructure. As the Earth's climate undergoes changes, including rising temperatures, extreme weather events, sea-level rise, and shifting precipitation patterns, building climate resilience becomes increasingly crucial. The ability of a system, group, or civilization to foresee, plan for, respond to, recover from, and adapt to the effects of climate change is referred to as climate resilience. It entails developing the capacities to both long-term adapt to changes in climatic patterns and to endure and recover from the negative consequences of climate-related events. The idea of climate resilience is broad and involves several elements, such as social, economic, cultural and environmental issues. One prevalent human-induced danger impacting many big Asian cities, such as Tokyo, Osaka, Shanghai, Taipei, Bangkok, Jakarta, Calcutta, Manila, and Hanoi, is land subsidence from excessive deep well pumping. The conditions and impacts of the subsidence in some of these cities, namely Tokyo, Osaka, Taipei and Shanghai, were quite severe but the subsidence was soon brought under control by implementing effective groundwater management and control schemes along with the timely development of adequate tap water supply facilities from surface water sources [1]. City planning plays a critical role in enhancing climate resilience by shaping the physical, social, and economic aspects of urban areas to withstand and adapt to the impacts of climate change. Effective city planning strategies integrate climate resilience measures into various aspects of urban development, infrastructure, and community life. Here are key ways in which city planning contributes to climate resilience:

Green Infrastructure and Urban Design Incorporating green infrastructure elements, such as parks, green roofs, and permeable surfaces, to enhance the absorption of rainfall, reduce heat island effects, and mitigate flooding. Thoughtful urban design can optimize the use of open spaces and promote sustainable land use.

Resilient Transportation Systems Designing transportation systems that are resilient to extreme weather events and can adapt to changing conditions. This includes incorporating climate-resilient infrastructure for roads, bridges, public transit, and bicycle networks.

Integrated Water Management Implementing integrated water management strategies that address challenges related to water scarcity, flooding, and water quality. This may involve sustainable storm water management, rainwater harvesting, and the protection of water bodies.

Climate-Resilient Buildings and Infrastructure Enforcing building codes and standards that enhance the resilience of structures to withstand climate-related hazards such as storms, floods, and heatwaves. Incorporating climate-resilient infrastructure design for energy, water, and waste management systems.

Zoning and Land Use Planning Implementing zoning and land use policies that consider climate vulnerabilities and promote sustainable development practices. Avoiding construction in high-risk areas and preserving natural buffers can reduce the impact of climate-related disasters.

Community Engagement and Social Resilience Engaging local communities in the planning process to understand their needs, vulnerabilities, and traditional knowledge. Fostering community resilience through social cohesion, education, and the inclusion of vulnerable populations in decision-making processes.

Emergency Preparedness and Response Developing and implementing emergency preparedness plans that account for climate-related risks. This includes establishing early warning systems, evacuation plans, and coordination mechanisms among various stakeholders.

Economic Diversification and Innovation Encouraging economic diversification and innovation to reduce dependence on climate-sensitive industries. Supporting the development of a diverse and resilient economy helps communities adapt to changing climate conditions.

Green Building Standards Promoting and enforcing green building standards to improve energy efficiency, reduce carbon emissions, and enhance the sustainability of urban structures. Green building practices contribute to both climate mitigation and adaptation efforts.

Preservation of Natural Ecosystems Preserving and restoring natural ecosystems within urban areas, including wetlands, forests, and coastal areas. Healthy ecosystems provide valuable services, such as flood regulation, water purification, and habitat for biodiversity.

Climate-Resilient Energy Systems Planning for climate-resilient energy systems, including the incorporation of renewable energy sources, distributed energy generation, and measures to protect energy infrastructure from extreme weather events.

Data-Driven Decision-Making Utilizing data and technology for informed decision-making. Climate modeling, risk assessments, and Geographic Information System (GIS) tools help planners identify vulnerable areas and prioritize interventions for maximum impact.

International Collaboration Engaging in international collaboration and knowledge-sharing to learn from best practices and innovative solutions implemented in other cities facing similar climate challenges.

2 Methodology

The study is conducted through understanding the background of hill towns in general, recent disaster trends in the country, causes, damage occurred, impact in and around the site, prevention, government initiatives for recovery, strategies for mitigation and planning proposals. It is carried out broadly in local level to analyze and correlate the data and impact. The author studied thoroughly more than sixty articles, media reports and government guiding documents to understand the causes, impact and planning interventions which will lead us to a new era and direction in the field of urban planning.

3 Climate Resilience and City Planning

The City planning for climate resilience involves a holistic and integrated approach that considers the interconnectedness of urban systems. By addressing vulnerabilities, incorporating sustainable practices, and fostering community engagement, city planners contribute to the development of resilient and adaptive urban environments capable of withstanding the impacts of a changing climate. Acute occurrences, such as heat waves, torrential downpours, hurricanes, or wildfires, are frequently linked to climate resilience and will likely increase in frequency or intensity as a result of climate change. Climate-resilient infrastructure reduces, but may not fully eliminate, the risk of climate-related disruptions [2]. In terms of planning aspect, efficient and adaptive interventions can mitigate the adverse affects of climate change.

Today's human societies are being impacted by the consequences of climate change such as disasters. A catastrophe is an abrupt, catastrophic occurrence that significantly disrupts, destroys, and distresses the affected community, frequently beyond its capacity to recover with its regular resources. Natural events, human activity, or a mix of the two can cause disasters. Flooding is a temporary overflow of water onto land that is normally dry. In case of floods, it can result from rain, snow, coastal storms, storm surges, overflows of rivers, and dam failure. United Nations (UN): 43% of natural disasters worldwide, from 1995 to 2015, were caused by floods, which impacted approximately 56% of people affected by natural disasters and caused casualties for approximately 26% of them. A gradual settling or sudden sinking of the Earth's surface due to removal or displacement of subsurface earth materials is known as land subsidence. Approximately 60% of all disasters are caused by such incidents of land subsidence caused by groundwater extraction globally.

4 Case of Kedarnath, Uttarakhand, India

Kedarnath is located in Rudraprayag district which is 86 km from the town known for its Kedarnath temple. Kedarnath is surrounded by two holy rivers Mandakini and Saraswati. It lies in elevation of 3583 m (11,755 ft). The tragic incident occurred in 16th June 2013 while it was there in small scale from a number of days. It is one of the most remote of 'chota char dham' of the state. The town is in proximity to Chorabari glacier which is just above the temple. The incident recorded continuous heavy rains, cloudbursts which resulted in flash floods and landslides. The eruption of the Mandakini River caused heavy floods in the Rudraprayag district and adjacent areas. The enormous volume of water also induced to erosion along all the river valleys, which in turn, triggered landslides at a number of places. As per the State Government, a total of 169 people died and 4021 people were reported missing (presumed to be dead) (09 May 2014). 2,845 death certificates were issued by the government. About 10,000 people died [3] (Figs. 1, 2, 3 and 4).

Damage through disasters is not limited to a single aspect. Material damage and loss of life; Interruption of economic activity in the flooded areas; Infection by water-borne diseases such as leptospirosis, cholera, etc.; Water pollution by flooding of dumps of toxic materials, treatment plants, etc. As per the study a number of reasons are there behind the tragedy such as heavy rains and cloudburst, Overflow of Chorabari lake (glacier), Overflow of river waters, Poor planning of space, No building codes, guidelines, Lack of local people's awareness, Overflow and collapse of the Chorabari Lake which released large volume of water, Poor crowd management system, new construction projects in eco-sensitive zone, Deforestation of forests, Rapid Construction of dams disturbed the water bodies, Poor communication and coordination among the authorities; after warning from IMD and GSI, no restriction was imposed in tourist movements; Cities' urban development master plans generally impose no restrictions regarding the settlement of areas at risk of flooding; Invasion of government-owned areas near rivers, by low-income population; Settlement of medium-risk areas, less often flooded, but when they are they suffer significant

Fig. 1 Location map of Kedarnath town showing terrain (author, Google earth)

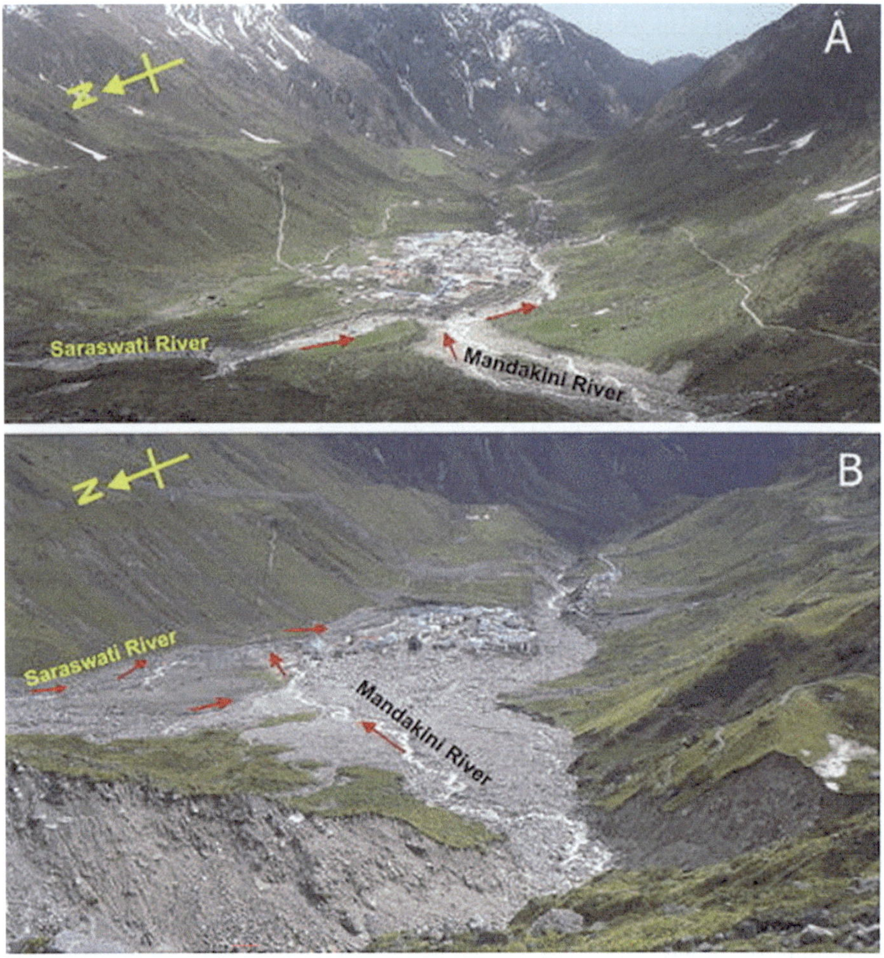

Fig. 2 River flow along Kedarnath before and after (Deen, 2020)

damage. Also unregulated construction and congested street pattern made it easier for destruction, there are no building guidelines, No proper planning of settlement, construction in floodplain area, contamination of water sources are indirect causes of the catastrophe. As per the Indian Meteorological Department (IMD), the rainfall in the State was measured at 385.1 mm, against the normal rainfall of 71.3 mm, which was in excess by 440%. As per the Geological Survey of India (GSI), the melting of Chorabari Glacier at the height of 3800 m was responsible for the floods.

Post disaster, 105 satellite phones were distributed by the Government of India to various Central and state agencies (also BSNL). Army, Air Force and Central Paramilitary Forces (ITBP, NDRF) on June 16. A Coordination Committee was formed under

Table 2: The extent of damage as reported by the State Government

S. No	Item	Details
1.	Number of affected districts	13
3.	Number of dead and missing persons (Presumed to be dead)	(169 + 4021)
4.	**No. of houses damaged :-**	
	(i) Fully damaged pucca houses	2,119
	(ii) Severely damaged pucca houses	3,001
	(iii) Partially damaged pucca houses	11,759
	(iv) Fully damaged kuchha houses	394
	(v) Severely damaged kuchha houses	360
	(vi) Number of kuchha houses partially damaged	1,676
	(vii) Number of huts damaged	471
	(viii) Cowsheds damaged	361
5	**Animals lost :-**	
	(a) No. of big animals lost	3,280
	(b) No. of small animals lost	7,811

Fig. 3 Damage caused report by state government (SDMA UTTARAKHAND, 2016)

Fig. 4 Rescue operations by armed forces ((Kedarnath Redevelopment @ Www.Inidesignstudio.Com, 2023)

the leadership of the Chief Secretary for conducting daily reviews. Government initiatives include Flood Plain Zoning Act regulating construction within the flood plain of a river should be implemented strictly. For clearance of all hydro-power and other mega projects in ecologically sensitive regions like Uttarakhand, the Disaster Impact Assessment (DIA) should also be made compulsory besides Environmental Impact Assessment (EIA). Landslide risk zonation mapping be completed on priority. Development and enforcement of guidelines, regulations and codes for landslides is critical. Effective stabilization of slopes in shear and weak zones is undertaken using scientific techniques available at national/international levels. A Special Central Programme is undertaken for construction of new roads and renovation of existing roads in a scientific manner.

5 Case of Joshimath, Uttarakhand, India

Joshimath is a hilly town located on the Rishikesh-Badrinath National Highway (NH-7) in Chamoli district of Uttarakhand. City is an overnight halt for visitors travelling to Badrinath, Auli, the Valley of Flowers, and Hemkund Sahib. Joshimath is one of the Army's most significant cantonments, holds tremendous strategic importance to the Indian armed forces. Running' streams with a high gradient from Vishnuprayag, a confluence of the Dhauliganga and the Alaknanda rivers, through the town (fall in high-risk seismic Zone-V). It is located on an elevation of 6,150 feet (1,875 m). Town covers an area of 2458 sq km and population of 16,709 (2011). Joshimath lies on an ancient landslide, resting on a deposit of sand and stone, not rock. The debris has angular sediments, which are worse than river-deposited sediments. These sediments have voids, making them extremely unstable, geologically. Critical infrastructure such as buildings and roads, changes drainage patterns and increases the risk of flooding. Poor groundwater availability with significant environmental implications for cities and their inhabitants is seen. 41 lakh pilgrims visited Badrinath in 2021, 2006, Times of India (TOI) report: Joshimath sinks by 1 cm every year; At least 66 families have fled the town while 561 houses have reported cracks (Fig. 5 and Table 1).

The town sits on the intersection of the Main Central Thrust 1, 2, and 3, the intra-crustal fault lines, where the Indian Plate has pushed under the Eurasian Plate along the Himalayas. The reactivation of these fault lines nearly 50–60 kms under the surface remains a big mystery. It is triggered by an incessant spell of rain in the region in the last few years, which deposited more water on the surface. Due to the unavailability of solid rocks underneath, the water seeped into the soil and loosened it from within. The glacial and cloud outbursts brought debris worth 10,000 houses in one day. Similar cities under threat: Nainital, Champawat, and Uttarkashi. Urban sinking or land subsidence are refers to the gradual sinking of the land surface in urban areas. Natural and human-made factors including groundwater extraction, construction works, soil consolidation and geological processes are major causes. A-seismic ground failure (AGF) is mostly linked to land subsidence brought on by compaction

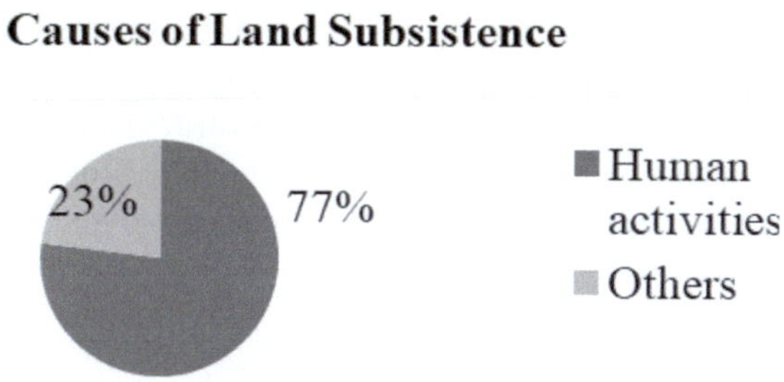

Fig. 5 Location map (Google earth, author)

Table 1 Condition of buildings in Joshimath (SDMA UTTARAKHAND, 2016)

Ward name	No. of houses cracked
Gandhi nagar	134
Palika marwari	28
Lower bazaar	24
Singdar	56
Mohanbagh	80
Upper bazaar	31
Sunil	38
Parsari	51
Ravigram	161

of underlying unconsolidated sediment. AGF is generated by human-induced drops in groundwater levels [4]. Some of the identified causes of the incident are excessive extraction of groundwater in urban areas, Excessive construction of heavy buildings and infrastructure and changes in land use cause soil compression. "Joshimath is a deposit of sand and stone—it is not the main rock—hence it was not suitable for a township. Vibrations produced by blasting, heavy traffic, etc., will lead to disequilibrium in natural factors..." (1976 government report by Garhwal collector). Lack of proper drainage facilities; The existence of soak pits, which allow water to slowly soak into the ground, is responsible for the creation of cavities between the soil and the boulders; Increased construction activity and growing population have contributed to frequent landslides in the area. Locals: NTPC Limited's 4 × 130 megawatt Tapovan Vishnugad hydel project for the situation–a 12-km tunnel has been carved into the hill. NTPC: denied allegations. Government initiatives include houses in the area with the maximum damage to be demolished, areas that have become inhabitable be identified, and people relocated on priority. High court: asked independent experts from the fields of hydrology, geology, glaciology, disaster management, geomorphology and landslide to study the land subsidence issue. Joshimath Bachao Sangharsh Samiti: first to draw the attention of authorities to the land-subsidence crisis in Joshimath. Eight central technical and scientific institutions, including the Central Building Research Institute, Geological Survey of India, Wadia Institute of Himalayan Geology, National Geophysical Research Institute, Central Ground Water Board, Indian Institute of Remote Sensing, National Institute of Hydrology and IIT, Roorkee were engaged to study the problem and ascertain its causes.

6 Results and Discussion

Under prevention measures for Kedarnath incident, preparation and implementation of early disaster management and mitigation plan, Awareness and training of public and rescue forces, Emergency disaster fund stock, Research and technology in field of disaster and related pursuits, Capacity building programmes for people and relief/rescue communities can be considered. Structural measures such as watershed plan, surface runoff, embankments and dykes can be carried out. Non-Structural measures such as city sanitation plan/drainage, risk assessment, early warning systems: local meteorological departments and authorities, Monitoring programmes: environment, climate/rainfall, water cycle etc., vulnerability mapping: identify flood prone areas, Education and awareness programmes can be considered. Strategies to mitigate such incidents include large scale plantations for resisting soil erosions during floods, Generation of technological and research funds through policies/schemes, Spatial planning and zoning, preparation and implementation of early disaster management and mitigation plan, Awareness and training of public and rescue forces, Emergency disaster fund stock, Research and technology in field of disaster and related pursuits, Capacity building programmes for people and relief/rescue communities, Construction of Dams and Reservoirs, Redirecting the excess water to canals and flood ways, excess water can be used for groundwater replenishment. Public finance can be used to mobilize private finance for climate-resilient infrastructure [2]. Structural adaptation measures: e.g., changing the composition of road surfaces so that they do not deform in high temperatures, building seawalls or using permeable paving surfaces to reduce run-off during heavy rain falls. Ecosystem-based approaches using natural infrastructure to design adaptation measures are also key alternatives to be considered alongside structural adaptation measures [2] (Fig. 6).

When surface water supplies are no longer sufficient, groundwater resources must be used. The major rivers flow into the seas quickly due to the steep topography. Few methods to control subsidence include reduction of pumping draft, artificial recharge of aquifers from the land surface, and re-pressuring of aquifers through wells, or any combination of these methods [6]. Prevention for land subsidence are reduction of pumping draft: Import of substitute surface water, Through improvement of irrigation methods, from ditch and furrow or flood irrigation to overhead sprinkler irrigation or to drip irrigation, For overdrawn ground-water basins, adjudication (equitable distribution) of available supply, In urban areas, by recirculation and reuse of treated water by industrial plants, By changing the depth range of perforated intervals in well casings or screens to tap less compressible deposits, Imposition of restrictions on heavy construction, Artificial recharge of aquifers from the land surface, and re-pressuring of aquifers through wells, Stop trees cutting and running plantation drives, Agriculture on the slopes must be avoided. In order to investigate the mechanism of land subsidence in central Taiwan, a multi-sensor monitoring

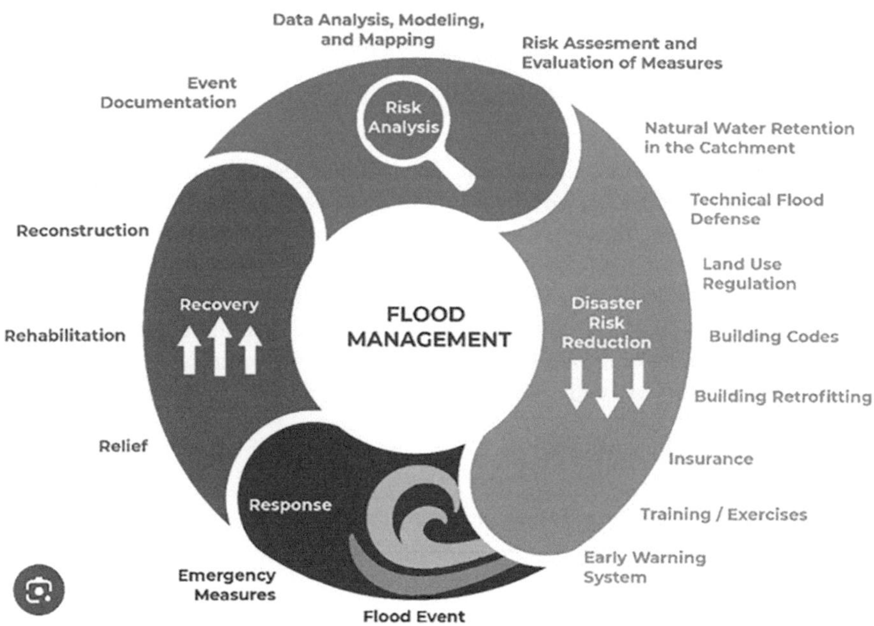

Fig. 6 Flood management measures [5]

system is installed. The system includes InSAR, GPS, leveling, multi-layer compaction monitoring well and Piezometer [7]. Strategies for land subsidence includes Multi-sensor land subsidence monitoring system, Spatial planning and zoning with strict planning byelaws, Injection of river water through production wells, deep soil mixing, artificial recharge, Considering carrying capacity for planning cities/towns, Using alternative construction techniques such as lightweight building materials and foundation systems, examining the load-bearing capacity of the soil and the stability of the site, River training is the construction of structures to guide the river's flow., Cracks which have developed on the slopes should be sealed. The toe of a landslide is its bottom-most point, coordination of civic bodies and military forces.

7 Conclusion

Climate-resilient city planning is an imperative and proactive approach that holds profound significance in addressing the challenges posed by a changing climate. As urbanization continues to grow and climate-related risks intensify, the need for resilient cities becomes increasingly critical. The following points encapsulate the key considerations and outcomes of climate-resilient city planning:

Holistic Approach:

Climate-resilient city planning requires a holistic and integrated approach that considers the interconnectedness of urban systems. Addressing vulnerabilities in infrastructure, communities, and ecosystems simultaneously is essential for building comprehensive resilience.

Adaptive Urban Design:

Incorporating adaptive urban design principles allows cities to respond effectively to changing climate conditions. This involves green infrastructure, sustainable building practices, and the incorporation of nature-based solutions to enhance resilience.

Risk Reduction Strategies:

Implementing risk reduction strategies at various levels', including zoning and land use planning, ensures that urban areas are less susceptible to climate-related hazards. Identifying and protecting vulnerable areas, as well as preserving natural buffers, are crucial components.

Community Engagement and Social Equity:

Prioritizing community engagement fosters resilience from the ground up. Inclusive planning processes that consider the needs and knowledge of local communities contribute to social equity and enhance the overall adaptive capacity of the city.

Flexible and Adaptive Infrastructure:

Climate-resilient cities invest in infrastructure that is flexible and adaptive to changing climate conditions. This includes robust transportation networks, resilient buildings, and smart technologies that can withstand and recover from extreme weather events.

Economic Diversification and Innovation:

Promoting economic diversification and innovation reduces dependence on climate-sensitive industries. A diverse and resilient economy is better equipped to absorb shocks and continue functioning during and after climate-related disruptions.

Policy Integration:

Integrating climate resilience considerations into urban policies and regulations ensures that long-term sustainability goals are aligned with development objectives. This involves updating building codes, land-use policies, and zoning regulations to reflect climate considerations.

Data-Driven Decision-Making:

Utilizing data and advanced technologies for informed decision-making enhances the effectiveness of climate-resilient city planning. Climate modeling, risk assessments, and data analytics contribute to evidence-based strategies and targeted interventions.

International Collaboration:

Engaging in international collaboration and knowledge-sharing allows cities to learn from best practices and innovative solutions implemented in other urban centers facing similar climate challenges. This global cooperation fosters a collective response to climate resilience.

Public Awareness and Education:

Public awareness and education campaigns play a crucial role in fostering a culture of resilience. Educating citizens about climate risks, preparedness measures, and sustainable practices empowers communities to actively participate in building a climate-resilient future.

Climate-resilient city planning is a dynamic and ongoing process that requires adaptive management, collaboration across sectors, and a commitment to sustainability. By incorporating resilience measures into the fabric of urban development, cities can navigate the uncertainties of a changing climate, protect their residents and assets, and strive for a more sustainable and liveable future.

References

1. Phien-wej N, Giao PH, P N (2006) Land subsidence in Bangkok, Thailand. Eng Geol 82(4):187–201. ISSN 0013–7952. https://doi.org/10.1016/j.enggeo.2005.10.004
2. OECD (2018) Climate-resilient infrastructure. Policy Perspectives OECD Environ Policy Pap 14:1–46

3. Fatewar M, Kaur M (2020) Meet, AFM, Article. Low Version.pdf
4. Hernández-Madrigal VM, Muñiz-Jáuregui JA, Garduño-Monroy VH, Flores-Lázaro N, Figueroa-Miranda S (2014) Depreciation factor equation to evaluate the economic losses from ground failure due to subsidence related to groundwater withdrawal. Nat Sci 06(03):108–113. https://doi.org/10.4236/ns.2014.63015
5. Conitz F, Zingraff-Hamed A, Lupp G, Pauleit S (2021) Non-structural flood management in European rural mountain areas—are scientists supporting implementation? Hydrol 8(4):1–37. https://doi.org/10.3390/hydrology8040167
6. Poland S (1985) 7 review of methods to control or arrest subsidence. In Joseph F (ed) Guidebook to studies of land subsidence due to ground-water withdrawal, 127–130.
7. Hung WC, Chen YA, Hwang C (2020) IoT technology and big data processing for monitoring and analysing land subsidence in Central Taiwan. Proc Int Assoc Hydrol Sci 382:103–109. https://doi.org/10.5194/piahs-382-103-2020

Shifting Sands of Design: Investigating the Evolution of Mosque Architecture in Islamic Contexts

Shubham Jaiswal, Mohd Arif, and Akhilesh Kumar

Abstract The Islamic architecture is known for their unique style and their elements. Throughout the history, it is seen that elements such as minarets, domes, jaali and arches of Islamic architecture were recognized and appreciated by Muslims and non-Muslims scholars. Elements of Islamic architecture is not limited to only religious structure, but it is richly used in all types of structure religious as well as secular. The aim of this study to examine the changes in application of elements of Islamic architecture over the periods from Historic times to modern times. It seeks to make the readers appreciate the Islamic architecture and clear the misconception about the application of elements of Islamic architecture in historic and modern period. It highlights the characteristics of elements of Islamic architecture that make it unique from the other forms of architecture. This research examined the various type of Islamic architecture and their elements, application of elements in historic as well as in modern period, to draw comparative analysis study of elements of Islamic architecture in historic and modern period. The study concludes that Islamic architectural elements were applied in various contexts across time. Hence the elements of Islamic architecture have direct links to the religion's divine principles and its aspirations.

Keywords Islamic architecture · Muslims · Elements of Islamic architecture · Mosque · Evolution

S. Jaiswal (✉)
Graduate School of Advanced Science and Engineering, Hiroshima University, Hiroshima, Japan
e-mail: ar.shubhamjaiswal@gmail.com

M. Arif
Jamia Millia Islamia, Delhi, India

A. Kumar
School of Architecture, Planning and Design, DIT University, Dehradun, India
e-mail: akhilesh.kumar@dituniversity.edu.in

A. K. Sharma and A. Joshi (eds.), *Proceedings of the 3rd International Conference on Opportunities and Challenges for a Resilient Future*, Advances in 21st Century Human Settlements, https://doi.org/10.1007/978-981-96-0931-4_5

1 Introduction

Architecture is indeed a unique art form that can be used to identify people, cultures, and civilizations. The design, construction, and style of buildings and structures can provide valuable insights into the history, traditions, beliefs, and technological advancements of a society [2]. Indeed, architecture plays a significant role in shaping the culture and image of Islamic societies. Islamic architecture is regarded as one of the most prominent forms of Islamic art, encompassing a wide range of structures that are both religious and secular. From mosques and palaces to bazaars and public buildings, Islamic architecture reflects the principles, values, and aesthetics of the Islamic faith and has had a profound influence on construction works in Muslim culture [2].

Islamic architecture provides valuable insight into the beliefs and practices of Muslims throughout history. It serves as a tangible manifestation of Islamic principles, adapting and responding to different cultures and traditional practices while preserving the spirituality of its creators. Also, the architects who have focused on Islamic architecture have showcased their creativity by incorporating abstract forms and unique designs that result in magnificent works of art. One of the defining characteristics of Islamic architecture is its ability to transmit Islamic messages and ideas. The primary aim of these architectural works is not merely to provide aesthetic pleasure but rather to serve as a means of communication, expressing and reinforcing Islamic beliefs. Using intricate patterns, calligraphy, geometric shapes, and other symbolic elements, Islamic architecture conveys spiritual and religious concepts [5, 6].

Islamic architecture is often associated with mosques, palaces, and castles, as these structures have been historically significant and have received significant attention from scholars and researchers. Early publications, such as those by [6], have provided valuable insights into Islamic architecture, but they may not cover the topic extensively due to the limited knowledge and resources available at the time. In recent years, there has been an increased interest in Islamic architecture, leading to a growing body of literature on the subject. While it is true that some modern publications may provide a shallow or superficial coverage of the various aspects of Islamic architectural designs, signs, and significations, there are also many comprehensive and in-depth resources available.

2 Literature Review

2.1 *Islamic Architecture*

Islamic architecture represents Islamic culture and civilization and the faith of Islam, going beyond simple creative expression. It embodies the culture and aesthetic sensibility of Muslim societies while carrying their identity. Islamic architecture has changed over time to accommodate Muslim communities' concerns and requirements. It has been used to meet the community's functional, social, cultural, and religious needs. Islamic architecture aims to create spaces that are both practical and uplifting spiritually, whether it be a mosque, a palace, a market, or a residential building. The idea of sustainability and sustainable development is central to authentic Islamic architecture. Islam places a strong focus on caring for the environment and its resources. Islamic architecture thus seeks to complement nature by considering elements like climate, materials, energy effectiveness, and environmental impact [12].

Islamic architecture does result from the cultural, ideological, and value-based facets of Islamic teachings. Islamic architecture may not have a single canonical manuscript devoted to it, but its concepts are based on Islamic precepts like modesty, sustainability, and humanistic design. Mosques, palaces, and private homes were among the many structures with Islamic architectural features that were common during the Golden Age of Islam. These architectural elements were created and improved over time with the help of numerous regional architectural styles, including Moorish, Fatimid, Persian, and Timurid architecture [5]. In Islamic Architecture, special attention was given for crucial feature of Islamic construction. It provides a framework for putting Islam into practice and embodies the essence of Islamic culture and civilization [7, 12]. The influence of Islamic culture on building design and construction leads to the development of unique architectural movements, regional styles, and structural components within Islamic architecture [5, 9].

2.2 Elements of Islamic Architecture

The distinctive characteristics of Islamic architecture can be found in mosques, palaces, tombs, and other buildings, among other places. These components, which are rooted in Islamic precepts, add to Islamic architecture's distinctive aesthetic and practicality. Domes, Minarets, Jaalis (lattice screens), and arches are examples of Islamic architectural features that are still used and cherished today. These architectural characteristics are more than just ornamental additions; they also have profound symbolic and cultural importance, defining Islamic architecture. The spread of Islamic architecture was considerable throughout numerous historical eras and locales. The building of mosques, madrasas (Islamic study institutions), and shrines were essential in the dissemination of Islamic thought and culture. Various dynasties, including the Timurids, Safavids, and Ottomans, each of which left a distinctive mark on Islamic architecture, ruled during the period of this architectural growth [4].

The growth and spread of Islamic architectural forms have mostly been attributed to Iran in particular. It is a focal point for the study of Islamic architecture because of its rich architectural heritage and the abundance of information available. The architectural concepts and forms that were developed in Iran throughout the Seljuk era and thereafter have had a significant impact, influencing the following architectural styles in various regions. Every Islamic building, like a mosque, is made up of several spaces, and it is occasionally feasible to create additional spaces as an extension of existing structures [7]. Muslims place a great deal of value on the masjid, which is regarded as the beating centre of Islam. It serves as both a place of worship for Muslims and a hub for knowledge and instruction. They are also recognised as centres for social assistance and dispute resolution [1].

Dome: The dome is a key architectural element in Islamic architecture with many unique qualities and purposes. The use of domes to cover holy places like mosques, cemeteries, and mausoleums is one of its famous uses. Some of the significant features and purposes of domes in Islamic architecture are Firstly, Domes in Islamic architecture stands for the celestial world and the link between the material world and the divine. They arouse feelings of transcendence and spirituality. Secondly, Domes add to the overall aesthetic attractiveness of Islamic architecture in terms of design. They frequently incorporate elaborate calligraphy, geometric patterns, and decorative accents, producing aesthetically attractive focal points. Thirdly, Large open spaces below the main structure are made possible by the structural support provided by domes, which are built to equally disperse the weight of the roof. They give the building structural stability and support. Fourth, Enhancing Acoustics-The curved shape of domes contributes to improving the acoustics of the enclosed space. It enhances sound dispersal and amplification, making sermons and prayers easier to hear [5, 13].

Minarets: A tall tower-like structure called a minaret is frequently found close to mosques. In Islamic architecture, it has both religious and utilitarian values. Some significant facts about minarets are First, Function, and Symbolism-Minarets are used to identify mosques and represent the Islamic call to prayer (adhan). The call to prayer was delivered by the muezzin, who would ascend to the top of the minaret. Second, Construction Details-Calligraphy, geometric patterns, and elaborate decorations are frequently used to embellish minarets, varied geographical areas and historical eras have varied minaret designs and styles, which represent regional cultural influences. Also, the Minaret design and construction have changed over time because of improvements in architecture and regional fashions. Early Islamic minarets had simpler designs and frequently borrowed features from traditional architecture [6, 13].

Jaalis: Islamic architecture frequently uses the jaali, a distinguishing architectural feature. It is a delicately carved lattice screen or grill, usually constructed of stone, wood, or plaster. The jaali is created in a way that offers privacy and adornment while allowing for the dispersion of light and air. The term "jaali" comes from the Arabic word "jali" or "jal," which means "net" or "mesh,". Jaalis are frequently seen in mosques, palaces, tombs, and residential buildings, among other Islamic architectural constructions. They can also be found in various regions where Islamic architecture has had an influence, including the Indian subcontinent, Iran, and Central Asia's traditional architecture [7, 13].

Arches: Arches play a key role in Islamic architecture both in terms of practical use and symbolic value. Following are some crucial details regarding the significance of arches in Islamic architecture. Firstly, Arches' ability to distribute the weight of the structure or roof above makes the structure stable. The excellent force transfer provided by the arch's curved shape enables it to traverse greater distances without the aid of additional supporting columns or walls. This structural element made it possible to design mosques and other Islamic structures with opulent interiors. Second, Arches are significant from a symbolic standpoint in Islamic architecture. They inspire a sense of transcendence and spirituality because they are viewed as images of the skies or the holy realm. Ascension and the link between the earthly and celestial are frequently symbolized by the arch's upward curvature. Arches are also excellent examples of the Islamic concepts of unity and harmony. They provide the architectural space with a sense of cohesiveness and movement while expressing God's unity and the interdependence of all things in the cosmos. Arches can be used repeatedly in a structure to produce a rhythmic pattern that improves the design's overall harmony and balance. Third, Light and Illumination-Windows or latticework are common features included in Islamic arches, allowing natural light to reach the interior chambers. This emphasis on light represents the divine presence and spiritual enlightenment. The arches' creation of a dance of light and shadow improves the building's aesthetic appeal [8, 13].

Overall, arches in Islamic architecture have both practical and symbolic functions, enabling the creation of aesthetically magnificent and spiritually significant places that represent the tenets and ideals of the Muslim faith [3]. A crucial component of Islamic architecture is Arches. Various architectural traditions, including Roman and Greek, served as inspiration for Islamic monarchs and architects. They modified Byzantine styles to fit their own artistic and cultural inclinations. As a result, Islamic architecture employed a variety of arches, each with its special traits. Hence, In Islamic architecture, the following prominent arch types are common such as Multifoil Arch, Horseshoe Arch, Point Arch, Ogee Arch, Stalactite Arch, etc. In Islamic architecture, arches are used not only for practical reasons but also to provide aesthetic beauty and symbolic significance to the overall composition [7, 13].

3 Evolution of Islamic Architecture and Empires

After Mohammed died in 632, Islam grew under the Umayyad (661–750) and Abbasid (750–1258) Dynasties to encompass a large portion of the Near and Middle East, as well as the coast of North Africa and much of Spain. By the ninth century, however, the Baghdad-based Caliphate had started to lose control of some of its peripheral territories to the east. From the early eleventh century, a group of tribes of Turkic origin, the Seljuks, started to establish a stable Islamic social order that eventually controlled Iran and Asia Minor. Then, Under Fatimid dominance (969–1171), a separate Caliphate was founded in Egypt. Egypt continued to be the dominant force in the Near East under the Ayyubid (1171–1250) and Mamluk (1250–15I7) dynasties. By the thirteenth century, the Mongols, a nomadic race from China's steppes, had repeatedly invaded, upending the stability of Islamic Middle Eastern society. They overthrew Baghdad in 1258 and established Iran's Ilkhanid dynasty. They converted to Islam themselves in the year 1295 and then were immediately assimilated into the Muslim community [5, 7].

3.1 Islamic Dynasties and Their Application of Islamic Architectural Elements

1. Umayyad Dynasty (661–750 CE)
2. Abbasid Dynasty (750–1258 CE)
3. Fatimid dynasty (909–1171 CE)
4. Seljuk Empire (1037–1194 CE)
5. Mamluk Dynasty (1250–1517 CE)
6. Ottomans Empire (1299–1923 CE)
7. Mughal Empire (1526–1858 CE)

Umayyad Dynasty (661–750 CE): In the year 661, upon Ali's death in Kufa (Muhammad's son-in-law), the Umayyad caliphate was established. As the first Umayyad caliph, Muawiyah I, the governor of Syria, assumed power. Damascus served as the Umayyads' capital. The Arab empire continued to grow under the Umayyads, eventually reaching Central Asia, Yemen, India's boundaries in the east, Morocco's Atlantic coast in the south, and the Iberian Peninsula in the west. The Umayyad Caliphate (661–750) made substantial contributions to the region's architectural landscape with the development of Umayyad architecture. It included novel features while also incorporating elements from numerous civilizations and cultures. Some essentials of Umayyad architecture are Firstly, Affectations of Other Civilizations-Umayyad architecture was influenced by pre-Islamic Arab and Nabatean cultures, the Byzantine Empire, Sassanian Persia, and other Middle Eastern civilizations. This mix of components and styles gave the architecture a unique identity. Making use of Byzantine methods, Byzantine architectural components, and styles, such as the use of marble, mosaics, and complex ornamental details, were borrowed and changed by Umayyad architecture. The Dome of the Rock in Jerusalem, which was built during the Umayyad era, bears witness to this impact. Secondly, New Building Types- During the Umayyad era's architecture gave rise to new building types such as minaret- and mihrab-adorned mosques [8]. Third, Innovative Decoration-The architecture of the Umayyad period featured novel decorative elements, including sophisticated geometric and floral patterns, calligraphy, and ornamental motifs. Minarets served as lofty towers from which the call to prayer (adhan) was made, while the mihrab served as a focal point for prayer and indicated the direction of Mecca. These decorative elements were typically made of stucco, mosaic, and carved stone.

Along with other religious themes, Islamic imagery was incorporated into Umayyad architectural designs. Quranic inscriptions and Islamic calligraphy were commonly used as building decorations, emphasizing the connection between the constructed world and the Muslim faith. The Umayyad architectural style, which had a lasting impact on succeeding Islamic architectural traditions, influenced the evolution of architectural forms and ornamental elements in the countries under Islamic authority. It left a distinctive architectural legacy by reflecting a mix of aesthetic and cultural influences [10].

After the Arab capture of Jerusalem in 639, Caliph Omar designated the Dome of the Rock (Qubbat al-Sakhrah). The dome was erected in 691. Grecian-style columns and mosaic decorations are just a couple of the characteristics shared by many Byzantine Christian churches. When designing the Dome of the Rock, Muslim architects used the church's dome as an inspiration. The dome's proportions and increased level of impressiveness were achieved by covering it in gold, which are only design changes that have been altered from the original design of the dome. They constructed domes and arches for both structural and aesthetic. The Dome of the Rock's architectural characteristics does show certain Byzantine influences, especially in the way it uses columns and mosaic ornamentation [8]. This is explained by the existence of talented Byzantine artisans and the region's exposure to Byzantine architecture. Therefore, it is conceivable that the architecture was influenced by features of earlier Christian buildings [10].

Abbasid Dynasty (750–1258 CE): The centre of Islamic political and cultural life moved from Syria to Iraq during the Abbasid caliphate (750–1258), which succeeded the Umayyads (661–750) in 750. In 762, Baghdad, the round City of Peace (Madinat al–salam), was established as the new capital. Persian, Mesopotamian, and Byzantine cultures all had an impact on the architecture of the Abbasid dynasty, resulting in a distinctive fusion of architectural forms. Some significant architectural details and constructions from the Abbasid era is throughout their reign, the Abbasids built many impressive mosques. The Great Mosque of Samarra in present-day Iraq is one of the most well-known instances. The enormous hypostyle prayer hall in this mosque has spiral minarets, emphasizing the building's verticality and majesty.

Fatimid Dynasty (909–1171 CE): In terms of adornment, the Fatimid dynasty (909–1171 CE) did have a unique aesthetic. The calligraphy, brilliant colours, and elaborate geometric patterns of Fatimid decoration are well known. These components were evident in textiles, metalwork, ceramics, and architectural ornamentation, among other forms. Protruding gateways from the wall, domes atop mihrabs and qiblas, porches and arcades with keel-shaped arches supported by a series of columns, façade embellishment with iconographic inscriptions, and stucco decorations are only a few of the architectural features of Fatimid mosques. The mosques were constructed using the hypostyle architectural design, which had a central courtyard surrounded by arcades and a roof supported by keel arches, which were initially supported by columns with Corinthian capitals. Example of Fatimid architecture is Great Mosque of Mahdiya were Horseshoe arches and domes were used for both symbolic and structural purposes [3, 6].

Seljuk Empire (1037–1194 CE): The Seljuk Empire dominated the majority of the Middle East, Central Asia, Iran, and Anatolia. Seljuk architecture introduced many innovations like the symmetrical four-Iwan pattern and the first widespread construction of state-sponsored madrasas. Seljuk dynasty has also made a significant contribution to the architecture of Iran and the surrounding areas. Their houses were mostly made of brick, with stucco carvings, tiles, and brickwork used for decoration. During the Seljuk Empire, a Turkic monarchy developed a distinctive architectural style called Seljuk architecture. Different types of buildings that fall under the umbrella of Seljuk architecture are mosques, madrasas, caravanserais, mausoleums, and palaces. The style was a mix of Persian, Byzantine, and Islamic architectural influences that is evident throughout the period [13].

Mamluk Dynasty (1250–1517 CE): Mamluk architecture was the flourishing architectural style that was popular in Egypt and Syria from 1250 to 1517. A caste of slave soldiers known as the Mamluks governed the area as sultans and funded the creation of numerous works of art in architecture. Mamluk architecture created an original and distinctive architectural legacy by combining aspects and architectural elements of Islamic, Byzantine, Persian, and native Egyptian styles. Cairo's Sultan Hassan Mosque and the Mosque of Ibn Tulun, both of which have intricate stone carvings, domes, and minarets, are magnificent examples of Mamluk architecture. The aspects of structure, symbolism, and aesthetic purposes were served by architectural elements like domes, minarets, and arches.

Ottomans Empire (1299–1923 CE): Ottoman Empire was a large and strong state between the end of the 13th and the beginning of the twentieth centuries. It was established by Osman in 1299, and during Suleiman the Magnificent's rule in the sixteenth century, it was at its height. The empire was a significant world force since it spanned numerous continents in Europe, Asia, and Africa. At its height, the Ottoman Empire ruled over regions that included what is now Turkey, Greece, Bulgaria, and parts of the Balkans in southeast Europe, as well as parts of North Africa and the Arabian Peninsula, Iraq, Syria, Jordan, Lebanon, Israel, and Palestine in the Middle East. The Ottoman Empire left behind a considerable legacy since it had a long-lasting effect on the areas it formerly controlled. Its accomplishments in the arts, architecture, and culture are still admired, and many nations in the Middle East and south-eastern Europe still reflect its influence in their government and traditional buildings [8].

The Ottoman Empire's architecture is a complex synthesis of many influences, with components drawn from Byzantine, Islamic, Persian, and European architectural traditions. Over the years, Ottoman architecture changed to reflect the empire's growth and connections with other cultures. The widespread use of domes and semi-domes is one of Ottoman architecture's most defining characteristics. Particularly Ottoman mosques are renowned for their imposing domes, which are frequently supported by enormous columns or piers. The Hagia Sophia, The Süleymaniye Mosque and the Blue Mosque (Sultan Ahmed Mosque), of Istanbul, are well-known examples of Ottoman mosques. There were many other magnificent buildings in the Ottoman Empire besides mosques. Palaces, like the Topkapi Palace in Istanbul, displayed lavish architecture and minute decorations. Large courtyards, lovely gardens, and elaborate rooms embellished with delicate wood carvings, sophisticated tile work, and calligraphy were the distinguishing features of these palaces.

Throughout the Ottoman Empire, geometric designs, complex calligraphy, and ornamental tiles were frequently used in building architecture. Mosques, palaces, and other buildings featured these decorations on their walls, domes, and mihrabs (prayer niches). While Many Ottoman-era architectural marvels are still standing today, especially in Istanbul and other important Ottoman cities. They continue to be revered for their beauty and historical value, serving as reminders of the empire's opulence [11].

Mughal Empire (1526–1858 CE): During the Mughal Empire's dominion over the Indian subcontinent from the 16th to the nineteenth centuries, a particular architectural style called "Mughal architecture" developed and flourished there. An original style of architecture developed from the blending of Islamic, Persian, Indian, and Central Asian elements (particularly Timurid architecture). Mughal architecture is recognised for its splendour, symmetry, and high-quality work. The Indo-Islamic style known as Mughal architecture was established by the Mughals throughout the ever-changing extent of their empire in the Indian subcontinent. A uniform pattern of structure and character such as large bulbous domes, slender minarets at the corners, massive halls, large vaulted gateways, and delicate ornamentation are all hallmarks of Mughal architecture. Afghanistan, Bangladesh, India, and Pakistan all include examples of the style in their contemporary architecture.

Mughal architecture blends Hindu elements with Persian and Islamic elements. Large, onion-shaped domes that are often flanked by four smaller domes are a recurring design element. The use of intricate embellishment techniques, such as pachin kari and jali-latticed screens, calligraphic inscriptions in Persian and Arabic (containing Quranic verses), use of decorative chhatris, jalis and jharokhas, and grand entrances leading up to the main structure are only a few examples. Mughal architecture has also influenced later Indian architectural styles such as Indo-Saracenic style popular under the British Raj, the Rajput style, and the Sikh style are only a few of the subsequent Indian architectural traditions that owe a debt to Mughal architecture [10].

Modern Islamic Architecture: Modern Islamic architecture refers to structures built using contemporary building materials and techniques yet inspired by Islamic principles, aesthetics, and cultural traditions. This architectural style emerged in the late twentieth century and is continuing to evolve today. Modern mosques blend expressive contemporary architectural styles with an eye toward meeting the needs of the Muslim community. Some of the modern Islamic Buildings with minimum symbolic elements are Dandaji Mosque in Niger, Valiasr Mosque in Tehran, and Bait ur Rouf mosque in Bangladesh. All of these Islamic buildings that have the minimum Islamic architectural elements used in the structures and are provided only to give the essence of Islamic architecture. Elements are only used as an identity so that one can recognise the building. The structures of both Dandaji Mosque in Niger and Bait ur Rouf mosque in Bangladesh, have flat roof while Valiasr Mosque in Tehran pitched roofs, but they use arches are there but no domes in their designs, so the use of minimum design elements are just for the identification. And for the Contemporary Mosque in the countryside the buildings and structures are designed to meet the needs of rural people while combining Islamic principles and aesthetics is the main goal of contemporary Islamic architecture in rural areas. It considers the cultural setting, weather, and resources accessible in rural locations. The visual and symbolic value of minarets, arches, jaalis, and domes in contemporary mosques is mostly ornamental and are used for aesthetic and symbolic purposes.

4 Conclusion

Islamic architecture is influenced by Roman and Byzantine architecture because they were earlier civilizations concerning Islamic architecture. In the early Islamic architecture period like the Ummayad dynasty architecture, they constructed domes, arches, and minarets which could be seen in the Great Mosque in Damascus. After that, all Islamic dynasties such as Abbasid, Fatimid, Seljuk, Mamluk, Ottoman, and Mughals also used domes, arches, minarets, and jaalis in their design.

In early Islamic architecture, simple domes, arches, and minarets were used but with the passing years the architectural style was nourished with add on to new and more complex types of arches, domes, and minarets, and later jaalis in their design; still in modern Islamic architecture, domes, arches, minarets, and jaalis are used. Since the construction techniques, technology, and materials were not so much developed during the early Islamic architecture as it is developed in the present time, Hence in early Islamic architecture style simple design elements and construction techniques were used for construction of domes, minarets, and arches but since the techniques and technology have been developed in present time these elements are now being used in a variety of styles and for a wide range of purposes although even in present time these elements are mainly used for functional as well as aesthetic purposes.

As per the Umayyad Dynasty in 661 AD to Ottoman Empire in 1923 AD and Mughal Empire in 1858 AD Muslim power rose steadily as they expanded their territories and built impressive structures and buildings featuring elements like arches, domes, jaalis, and minarets and they eventually invented some construction techniques and new types of elements. Because of the widespread usage of these elements (arches, minarets, domes, and jaalis) in Islamic architecture, these features became symbols and identity of Islamic architecture. In today's world, technology, construction techniques, and materials are already advanced, and it is possible to construct Islamic structures without dome arches, minarets, and jaalis although they do not affect the functionality of these structures but still, these are in practice because they became the identity of Islamic architecture. For example, if we construct a mosque without domes, minarets, jaalis, and arches still that mosque plays the same function Hence there is no relation between slah and these Islamic architectural elements. Islamic architecture evolved with time, inventing new building methods, and nurturing primitive styles that had been inspired by Persian, Roman, and Byzantine architecture, etc.

References

1. Abd Rahim SA, Ismail NH, Abd Latip NS (2022) Analysis of current condition of heritage masjid in Negeri Sembilan. Plan Malays 20(2):184–195. https://doi.org/10.21837/pm.v20i21.1102
2. Ahmad SD (2010) Islam, science, and the challenge of history. https://www.jstor.org/stable/j.ctt1nq893
3. Ahmed HT (2023) The influence of fatimid architecture heritage on sustainable design of contemporary. Int Des J 13(1):381–392. https://doi.org/10.21608/IDJ.2023.280085
4. Baloch QB (n.d.) The Islamic civilization. In: Islamic civilisation, the dialogue. pp 37–56. https://www.qurtuba.edu.pk/thedialogue/The%20Dialogue/1_1/04--Islamic%20Civilisatino.pdf. Accessed 15 June 2023
5. Faiz Toorabally MFT, Hoe Sieng C, Binti Norman H, Binti Razalli Z (2016) Impact of modern technologies on Islamic architecture in Malaysia and middle east. Nova J Eng Appl Sci 5(1):1. https://doi.org/10.20286/nova-jeas-050101
6. Ghasemzadeh B, Fathebaghalli A, Tarvirdinassab A (2013) Symbols and signs in Islamic architecture. Rev Eur Estud Artist 4(3):62–78. https://doi.org/10.37334/eras.v4i3.86
7. Kaptan K (2013) Early Islamic architecture and structural configurations. Int J Arch Urban Dev 3(2):5–12
8. Omer S (2011) Islamic architecture and the prospects of its revival. J Islam Thought Civ 1(2):103–121. https://doi.org/10.32350/jitc.12.07
9. Rabah S (2002) Introduction to Islamic architecture-Muslim heritage. Muslim Herit: Discov Gold Age Muslim Civ 1–3. https://muslimheritage.com/introduction-to-islamic-architecture/
10. Saoud R, Al-Hassani S (ed) (2002) Muslim architecture under the Umayyad patronage (661–750AD). Found Sci Technol Civ 1(1):1–6. http://www.fstc.co.uk
11. Selçuk Akşin S (2010) The A to Z of the Ottoman empire. Scarecrow Press
12. Spahic O (2021) A conceptual framework for sustainability in Islamic architecture: the significance of the concepts of man and the environment. J Islam Contemp World 3:49–74
13. Yassin AA, Utaberta DN (2012) Architecture in the Islamic civilization: Muslim building or Islamic architecture. J Islam Arch 2(2):52–60. https://doi.org/10.18860/jia.v2i2.2202

Reinterpretation of Ghats: Exploring Cultural Narratives and Enhancing Thriving Urban Spaces (Cases of Religious Town as Per Hindu Beliefs)

Sanskriti Gupta, Shubham Jaiswal, and Akhilesh Kumar

Abstract *"Oh, Varanasi and Ujjain, cities of ancient lore, Where the sacred rivers Ganges and Kshipra flow, Your ghats are a sight to behold, A tapestry of culture, history, and the untold"* Varanasi and Ujjain, two ancient cities on the banks of the Ganges and Kshipra rivers, respectively, are renowned for their iconic riverfront spaces, or ghats. These ghats are steeped in history, culture, and religious significance. They serve as a place of spiritual solace for millions of pilgrims, as well as a vibrant hub for social, cultural, and economic activities. Varanasi, the city rises from the western bank of the river Ganga, where the river takes a broad crescent sweep towards north. Along the Ganges there are 84 bathing ghats, reaching from Assi Ghat to Adi Keshava in the north. The relationship between the sacred river and the city is the essence of Varanasi. Ujjain, one of the seven sacred cities of Hinduism, located on the eastern banks of the Kshipra River is deemed to be an agglomeration of history, traditional beliefs, myths and legends. This research project aims to reimagine these iconic riverfront spaces as vibrant cultural, recreational, and economic hubs, while preserving their cultural significance, developing sustainable urban spaces, and promoting tourism and culture by identifying the cultural significance of riverfronts in the context of ghats. It will then examine the current state of the ghats in Varanasi and Ujjain and identify the challenges and opportunities for their redevelopment. Drawing on best practices from around the world, the paper will develop recommendations and a framework for the redevelopment of the ghats. These recommendations will be guided by the principles of preservation and development of these ghats.

S. Gupta (✉)
Manipal School of Architecture and Planning, Manipal, Karnataka, India
e-mail: sanskritigupta66@gmail.com

S. Jaiswal
Graduate School of Advanced Science and Engineering, Hiroshima University, Higashihiroshima, Japan
e-mail: ar.shubhamjaiswal@gmail.com

A. Kumar
School of Architecture, Planning and Design, DIT University, Dehradun, India
e-mail: akhilesh.kumar@dituniversity.edu.in

A. K. Sharma and A. Joshi (eds.), *Proceedings of the 3rd International Conference on Opportunities and Challenges for a Resilient Future*, Advances in 21st Century Human Settlements, https://doi.org/10.1007/978-981-96-0931-4_6

Keywords Ghats · Riverfront · Culture · Development · Sacred landscape

1 Introduction

1.1 Kashi: A Microcosm

Kashi is the whole world, they say. Everything on earth that is powerful and auspicious is here, in this microcosm. All the sacred places in India and all her sacred waters are here. All of gods reside here, attracted by the brilliance of the city of lights. All eight directions of the compass originate here, receiving jurisdiction over the sectors of the universe. And all the time is here, they say, for the lords of the heavenly bodies which govern time are grounded in Kashi and have received their jurisdiction over the days and months right here. Thus, all organising forces of space and time begin here, and are present here, within the sacred boundaries of Kashi. A city with many names Like a loved one, the sacred city is called by many names which helps us to understand various aspects of Varanasi-*Kashi*: The city of lights, *Varanasi*: Between Varuna and Assi, *Avimukta*: The never forsaken, *Anandvana:* The forest of bless, *Rudravasa:* The city of Lord Shiva, *Mahashamshana*: The great cremation ground [1, 2].

1.2 Ujjain: A Microcosm

Ujjain, a miniature representation of the cosmos, is considered to be the navel of the world. It is said that the city is located at the exact centre of the earth, and that it is a place where the cosmic forces of creation, preservation, and destruction are all in perfect balance. Like Kashi, Ujjain has many names, each of which reflects a different aspect of the city's rich culture and heritage-*Avanti*: The ancient name of the region in which Ujjain is located. *Avantika*: The name of the city's namesake river. *Ujjaini*: The Sanskrit name for Ujjain, meaning "bright" or "illustrious." *Pratikalpa*: The city of equals, meaning that Ujjain is equal to any other city in the world in terms of its importance and significance. *Padmavati*: The city of the lotus, reflecting the city's beauty and serenity.

2 Literature Study

2.1 The Ghats of Varanasi and Ujjain

The ghats of Varanasi and Ujjain are two of the most sacred and significant riverfront spaces in Hinduism. They are steeped in history, culture, and religious significance. The ghats have been used for religious and ceremonial purposes for centuries, and they continue to be an important part of the lives of Hindus today. According to Hindu mythology, the ghats of Varanasi and Ujjain were created at the beginning of time. It is said that the god Brahma created the ghats as a place for people to come and bathe in the sacred rivers Ganges and Kshipra, respectively [3].

The ghats of Varanasi and Ujjain are of great cultural and historical significance. They have been associated with many important Hindu legends and traditions. For example, the Manikarnika Ghat in Varanasi is said to be the place where the goddess Sati immolated herself [4]. The Ram Ghat in Ujjain is said to be the place where Lord Rama performed his morning ablutions. The ghats of Varanasi and Ujjain are also important religious centres. Millions of pilgrims visit these cities every year to bathe in the sacred rivers and visit the many temples and shrines located along the ghats. The Kumbh Mela, one of the largest religious gatherings in the world, is held in Varanasi and Ujjain every 12 years.

2.2 Cultural Significance

From its source in the Himalaya to its mouth in the Bay of Bengal, covering a course of about 2,525km, only in Varanasi does the Ganga River flow in a crescent shape from south to north, converging into a meander. The 6.4km (4 miles) long riverfront of the river Ganga presents a specific vision of a magnificent architectural row of lofty buildings and holy sites [5, 6]. The stone stair ghats started construction in the fourteenth century, and as late as 1995 a few of the ghats were finally built into stones. There are 96 sacred waterfront sacred spots (jala tirthas) concentrated around the 84 ghats. Most of the Ghats have been active spots for festivities and rituals since the historical past [7]. The 84th ghat, Adi Keshava Ghat, is not taken into consideration in the heritage zone with an aim to avoid the discontinuity of settlement between the last two ghats [3].

Ujjain has been a major centre of astronomical studies since ancient times. Indian astronomy treats Ujjain as at 0° longitude. Vedic astronomy was based on the movements of the moon among the several starry constellations and as the moon completes her revolution in 28 days or a little less, the Vedic rishis divides the zodiac into 27 mansions. But, the Greeks following the Chaldeans who were first to mathematically study the movements of planet's among stars, divided the zodiac into 12 houses, corresponding to the sun's position among the stars during 12 months in which the sun makes a complete circuit of the heavens. This proves that later Indian astronomy borrowed from the Greeks these 12 houses and methods of calculations based upon them and amalgamated the two divisions, namely, 27 Nakshatras and 12 Rashis. This amalgamation must have taken place at Ujjain under Ashoka in whose time we read that there were astronomical schools at Ujjain.

2.3 Rituals and Festivals

Rituals and festivals form a significant component of the cultural life of this city. Numerous daily rituals, festivals and annual fairs like-Kumbh mela and deep Diwali bring in a large number of pilgrims and visitors to the ghats, especially to the sacred Ram Ghat, Assi Ghat and Dashashwamedh Ghat. Occupying the centre of the riverfront, at the heart of the old city, the *Dashashwamedh Ghat* is the focal point of all ghat activities. *Assi Ghat* in Varanasi is geographically at its lowest point. It is located in the southernmost part of the city and is where the Varuna River meets the Ganges River. The name *'Varanasi'* emerges from this ghat, where Varuna meets Assi. Assi Ghat is a popular spot for both locals and tourists alike and is known for its peaceful and serene atmosphere [8].

Located on the banks of the Shipra River, in the heart of the city is *Ram Ghat.* According to the Hindu epic Ramayana, Lord Rama and his wife Sita spent 14 years in exile in the forest after Rama's elder brother Bharata was tricked into believing that Rama had died. During their exile, Rama and Sita travelled extensively throughout India. It is believed that they stayed at Ram Ghat in Ujjain for some time. One legend associated with Ram Ghat says that Lord Vishnu dribbled some drops of amrita (the nectar of immortality) at the ghat. According to the legend, Lord Vishnu was carrying amrita in a pot when he was attacked by the demon Rahu. Vishnu accidentally dribbled a few drops of amrita at four different places on Earth, including Ram Ghat. These four places are now known as the "Char Dham" (four abodes) of Vishnu. Another legend associated with Ram Ghat says that Lord Shiva created the ghat to welcome Lord Rama to Ujjain. The legends associated with Ram Ghat are significant because they highlight the spiritual importance of the ghat. The Kumbh Mela takes place every 12 years on this ghat, the last recorded Kumbh Mela in Ujjain was 2016 and the next Kumbh Mela will take place in 2028 and the cycle of Kumbh Mela repeats every 12 years in Ujjain [9].

3 Need of the Study

This research paper aims to reimagine iconic riverfront spaces, that are historically and culturally significant areas; and to harness their potential as vibrant cultural, recreational, and economic hubs with the objectives of (1) To identify the cultural significance of riverfronts in context of ghats. (2) To prepare recommendations and framework for the development of ghats and cultural hubs. The paper aims at finding answers for the following research questions: How have ghats been used for religious, cultural, and social purposes throughout history? What are the different ways in which ghats are used today? What are the best practices and key objectives for developing ghats and cultural hubs?

Redeveloping riverbank areas, like ghats, can be a potent instrument for creating sustainable urban areas, encouraging tourism, and protecting culture, history, and legacy. This essay's focus is on the preservation of culture, history, and heritage, which includes defending and upholding the diversity of historical locations, customs, traditions, and assets [10]. It is necessary to create living surroundings that respect and accommodate different religious observances in order to construct urban spaces in line with religious practices. Redeveloping historical and cultural monuments to increase tourism and generate employment possibilities is part of the process of promoting tourism and culture [10]. Nonetheless, every task comes with its own limitations. The project's scope is confined to a defined area due to time constraints, limiting the exploration to select ghats. This spatial restriction acknowledges the vast expanse of ghats spread across different regions, necessitating a targeted scope of work to a specific subset of ghats. As a result, certain ghats may not be incorporated into the study. However, the project is still a significant undertaking that has the potential to make a real difference in the preservation and revitalization of ghats.

This research paper focuses on 2 major Hindu sites-Varanasi and Ujjain. Varanasi, has a total of 84 ghats located along the banks of the River Ganga. But, there are five ghats (panchtirth), which are consideredthe mostt sacred among all the ghats. This research paper focuses on 3 of these ghats-Assi Ghat, Dashashwamedh Ghat and Manikarnika Ghat for its sacred and cultural significance. Ujjain has a total of 13 ghats. However, this research paper focuses on 3 of these ghats-Ram Ghat, Mangalnath Ghat, Triveni Ghat.

3.1 Research Methodology

For the research cluster sampling was done for 2 places Varanasi in Uttar Pradesh and Ujjain in Madhya Pradesh and then the data collection was done by visiting these two places and collecting primary data by conducting offline surveys, personal interviews with the public. The secondary data was collected through books, journals, online documentaries and tourist feedback.

3.2 Scared Landscape

For those with a long-standing or ancestral connection to a particular region, certain geographic locations hold particular significance as sacred landscapes. They can be noteworthy for other reasons, such as their natural beauty, cultural value, or historical relevance, but they are frequently linked to religious or spiritual beliefs. Two of India's

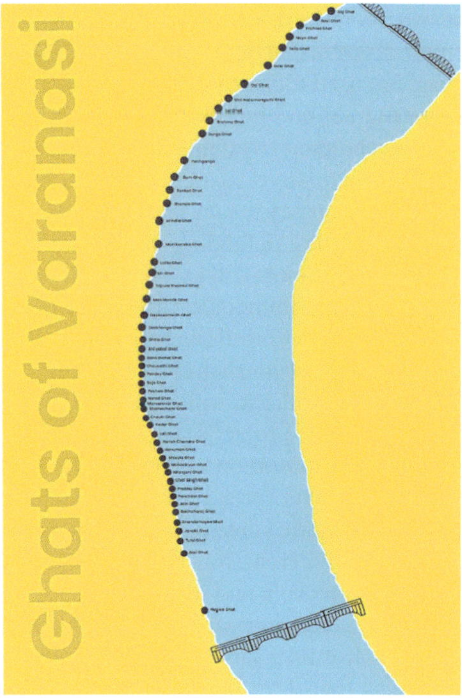

Fig. 1 A map showing all the ghats of Varanasi. *Source* Ghat of Varanasi map, Agni Amrita

holiest locations are the ghats of Varanasi and Ujjain. They are situated, respectively, on the banks of the sacred rivers Shipra and Ganges. They are a significant part of many people's lives. These are locations where people go to celebrate their culture, learn about their history, and establish a spiritual connection. Within the framework of Hinduism and Indian mysticism, they have great religious and cultural significance. At these ghats, devotees believe that taking a dip in the Ganges absolves them of their sins and opens the door to salvation [11, 12].

The cremation rites, which are thought to release the deceased soul from the cycle of reincarnation, are among the religious practices that take place at the ghats (Fig. 1).

4 Ghats of Varanasi

4.1 Assi Ghat

Assi Ghat, located along the banks of the River Ganges in the ancient city of Varanasi, holds significant historical and cultural importance in Hindu mythology and Indian spirituality. The name Assi comes from the Sanskrit word "ashti", which means "bone". It is believed that the name refers to the fact that Assi Ghat is a place where people come to cremate their loved ones but in recent times the cremation on Assi Ghat is limited and Manikarnika Ghat is preferred over Assi ghat for cremation [6]. Another theory is that the name Assi comes from the number 80. This is because Assi Ghat is one of the 80 ghats in Varanasi that are considered to be particularly sacred; There is a continuous semi-circular stretch of ghats along the river from the Assi in the south to the Raj Ghat in the north. Varanasi now has 84 ghats, 4 new ghats have been redeveloped and introduced in recent times [13].

According to Hindu mythology, it is believed that Goddess Durga, after defeating the demon Shumbha-Nishumbha, threw her sword into the river at this very spot. The ghat is also associated with the place where the revered poet Tulsidas is believed to have written parts of the famous Indian epic, the Ramcharitmanas. The ghat has a museum dedicated to Tulsi Das and has been attacked on multiple occasions, hence to protect the museum, it has been closed and can be accessed on limited days only. It has been home to many poets-Kabir Das, Tulsi das, Munshi Premchand, Philosopher Swami Vivekananda and the spiritual leader Maharishi Mahesh Yogi. The Assi ghat attracts several tourists and localities for its popular Ganga Aarti, which happens every morning and evening; while it is also popular for its boating activities. The ghat has a temple dedicated to Lord Shiva, The Asia Sangameshwar Temple, it is a place where they can come to worship Lord Shiva in a very sacred setting. The temple is dedicated to a lingam, which is a phallic symbol that represents Lord Shiva. The lingam at the Asia Sangameshwar Temple is said to be one of the oldest and most powerful lingams in Varanasi [6].

4.2 Dashashwamedh Ghat

The name Dashashwamedh Ghat is derived from the Sanskrit words "dasha" (ten) and "ashwamedha" (horse sacrifice), the ancient Vedic ritual known as the "Dashashwamedh Yajna," which translates to the "ten-horse sacrifice." Legend has it that Lord Brahma, the Hindu god of creation, is believed to have performed this intricate and elaborate sacrificial ritual at the very location where the ghat stands today, on the banks of the River Ganges in Varanasi. Another theory states that the name Dashashwamedh Ghat is derived from the fact that the ghat was rebuilt ten times over the centuries. Whatever its origin, the name Dashashwamedh Ghat is a reminder of the sacredness and antiquity of this ghat. The ghat houses the 'Sitala Mata Temple' located on Sitala ghat which is a part of Dashashwamedh ghat. The temple is dedicated to Shitala Mata, a goddess who is worshiped to protect people from diseases and 'Ahalyeshwar Mahadev Temple-Replica of Kashi Vishwanath' The Ahilyeshwar Mahadev Temple, located on the Sitala Ghat in Varanasi, is renowned for being a remarkable replica of the famous Kashi Vishwanath Temple. Dedicated to Lord Shiva, the temple is named after Queen Ahilyabai Holkar, the 18th-century ruler of the Malwa kingdom, who is known for her patronage of Hindu temples and her contributions to the restoration and construction of various religious sites across India [6, 8].

The ghat is extremely popular for its Ganga Aarti, devotees secure their places much before the evening aarti starts since seating is limited and the Ghat gets extremely crowded during the time of Aarti. The ghat is situated close to the Vishwanath Temple and is easily accessible from various parts of the city, making it a popular destination for pilgrims, tourists, and devotees seeking spiritual blessings and a profound connection with the sacred river [8].

4.3 Manikarnika Ghat

The name Manikarnika comes from the Sanskrit words "mani" (jewel) and "karna" (ear). It is believed that the name refers to the fact that an earring fell from Sati's ear, the first wife of Lord Shiva as she was immolating herself. It is located on the western bank of the Ganges River. It is believed that cremation at Manikarnika Ghat can help the soul to achieve moksha (liberation from the cycle of birth and death), the very significance of ghat is mentioned in the puranas. Today, Manikarnika Ghat is one of the busiest cremation sites in the world. Every day, hundreds of bodies are cremated at the ghat and the fire at Manikarnika Ghat never stops, the pyre of wood used to cremate the body runs 24*7 throughout the year [4].

The main temple at Manikarnika Ghat is the Ratneshwar Mahadev Temple. This temple is dedicated to Lord Shiva, and it is one of the most photographed temples in Varanasi. The temple is known for its leaning structure, which is believed to be due to the weight of the millions of Hindus who have been cremated at the ghat over the centuries. The temple inclines towards the river Ganga and is labeled as 'Leaning Tower of India' by the localities. The Shiva-Durga temple-Shri Bata Devi Shri Gayathri Devi Mandir, of Raja of Amethy (Oudh, presently in Uttar Pradesh) looms over the Manikarnika Ghat. With its five deep-red spires and gilded pinnacles, this Shiva-Durga temple, constructed in CE 1857 by Raja Lal Madhav Singh, the monarch of Amethy, is a masterful example of architectural and sculptural depiction [4] (Figs. 2 and 3).

Havell [1905: 169, 173] describes it: *"It is built on a terrace overlooking the river, and is approached by one of those steep, staircased streets, leading from the ghats up into the city, which suggest a town of southern Italy or Spain. Clambering up a side staircase, you pass under the Naubata Khana, where musicians are chanting praises of the goddess with strange but not unpleasing accompaniments. On the right side of the entrance is a fine bronze lion of Durga, and on the left Shiva's bull. The quiet and cleanliness inside are a relief from the bustle, sloppiness, and dirt, and the somewhat sordid atmosphere of more popular Benares shrines. … Its cusped arches and graceful stone tracery betray the Saracenic influence, which is very prominent in modern Hindu art in northern India."*

The ghat also is the home to Baba Mashan Nath Mandir and is one of the most photographed temples of Varanasi. The name of the temple comes from the Sanskrit word "masan," which means "death." The temple is dedicated to Baba Mashannath, an incarnation of Lord Shiva. Baba Mashan Nath is considered to be the god of death,

Fig. 2 The Baba Mashan Nath Mandir on Manikarnika Ghat in Varanasi

Fig. 3 Manikarnika Ghat on river Ganga-Varanasi, 1992. *Source* The empire of India, Bampfylde fuller

and his temple is an important pilgrimage site for those seeking moksha, or liberation from the cycle of rebirth. The temple was built in the seventeenth century and began as a small shrine. However, the temple has since been expanded and renovated, and it is now a large and grand temple. The temple has a main shrine that houses a statue of Baba Mashan Nath, and the temple complex also includes several other temples and shrines to other deities. The temple is the subject of many legends, and it is a mysterious and fascinating place. Pilgrims who visit the temple typically pray to Baba Mashannath for moksha. They also offer offerings to the temple, such as flowers, incense, and candles. Some pilgrims also offer copper vessels to the temple, which are believed to help them achieve moksha. Tourists who visit the temple are also interested in the temple's architecture and history. The temple is a beautiful and fascinating place, and it represents an important religious site in Varanasi [6].

The ghat is also a popular tourist destination, especially for foreign tourists since it is believed inauspicious for Hindus to visit a cremating ground unnecessarily, and it is known for its vibrant atmosphere and its unique culture among the foreign tourist. Aghori are a sect of Shaivite sadhus (ascetics) who are known for their extreme religious practices and their unusual lifestyle. They are often associated with death and cremation, and they are often seen as being outsiders to mainstream Hinduism. A bunch of Aghori activities take place on this ghat, which has been labeled as 'occult practices' because of the sudden publicity of this ghat but they are merely occult and just religious practices followed by Aghoris according to Mayur Kalbag who is the author of the book-'Aghori' The book highlights the importance of Manikarnika to Hindus and explains the Aghori practices beautifully explaining the importance of Aghori in Hinduism and how they are a part of the religion and not the 'outcast' as labeled by the mainstream media. The history of Manikarnika Ghat dates back centuries. The ghat is mentioned in the Puranas, which are ancient Hindu scriptures. The Puranas say that Manikarnika Ghat is one of the most sacred places on Earth, and that cremation at the ghat can help the soul to achieve moksha [4].

5 Ghats of Ujjain

5.1 Ram Ghat

Ram Ghat, situated along the Shipra River in the ancient city of Ujjain, Madhya Pradesh, India, holds a significant historical and cultural importance in Hindu tradition. The name "Ram Ghat" is believed to be derived from the legendary association of Lord Rama, an incarnation of the Hindu god Vishnu, with the sacred river. According to Hindu mythology, it is believed that Lord Rama visited this ghat during his exile and performed religious rituals here, sanctifying the site and bestowing it with divine significance. Another theory states that this is believed to be the place where Lord Rama performed the last rites of his father, King Dasharatha (Figs. 4, 5, 6, 7 and 8).

Fig. 4 The Mangalnath temple located on Mangalnath Ghat in Ujjain

Fig. 5 The Shri Bhuvaneshwar Prithvi Mata present inside the Mangalnath temple

Fig. 6 The lord Shiva Lingam at the Mangalnath temple

Fig. 7 Pooja performed for Mangal Dosha at the temple

Fig. 8 The river Shipra at Mangalnath Ghat during Monsoon season. *Source* Author

Ram Ghat is also the site of the annual Ram Ghat Mela, which is a religious festival that takes place during the month of November. The festival attracts millions of pilgrims from all over India. Devotees from all around the country come to Ujjain during the festival of Savan to complete their Kawan Yatra-a ritual where the holy Ganga water is offered to Lord Shiva Lingam present at the Mahakaleshwar Jyotirlinga. Its cultural significance is further accentuated by the numerous temples and religious institutions located in its vicinity, making it a vital religious and cultural landmark in Ujjain. Ram Ghat continues to be a spiritual haven for pilgrims and visitors seeking spiritual solace, blessings, and a deeper connection with the rich spiritual heritage of Ujjain and Hindu mythology.

Fig. 9 The entrance of Triveni temple located at Triveni ghat in Ujjain

5.2 *Mangalnath Ghat*

Mangalnath Ghat is a ghat on the banks of the Shipra River in Ujjain, India. It is dedicated to Lord Mangal, the god of Mars. The ghat is believed to be the birthplace of Lord Mangal, the God of Mars. The name Mangalnath Ghat is derived from the Sanskrit words "Mangal" (the god of Mars). The ghat is home to the Mangalnath Temple (Fig. 4). This temple is dedicated to Lord Mangal, Lord Shani and the nine planets of astrology. The main temple on the ghat (Mangalnath Temple) is dedicated to Lord Shiva and Shri Bhuvaneshwar Prithvi Mata-who is considered the mother of Lord Mangal. People often come here to do specific poojas and rituals to get rid of their Graha dosha (one of the most nefarious dosh that one could have in their planetary alignment) of the planet Mars (Mangal). It is a popular pilgrimage site for Hindus, especially during the Mangal Navami festival. Mangalnath Ghat continues to be a revered spiritual destination for pilgrims and devotees, drawing people from different parts of the country who seek to experience the sanctity and divine aura of this sacred site deeply entrenched in Hindu mythology and Vedic traditions (Figs. 9, 10, 11 and 12).

Fig. 10 The temple complex of Navagraha Shani Mandir

Fig. 11 The interiors of the temple

Fig. 12 The entrance of the Navagraha Shani mandir. *Source* Author

5.3 Triveni Ghat

The name Triveni Ghat is derived from the Sanskrit words "tri" (three) and "veni" (confluence). The ghat is named after the confluence of three rivers-the Saraswati, the Kshipra, and the Ganges-which meet at this sacred ghat, forming a significant tributary junction. Triveni Ghat is a popular pilgrimage site for Hindus. It is especially popular during the Simhastha Kumbh Mela, which is a religious festival that takes place every 12 years. The Simhastha Kumbh Mela is one of the largest religious gatherings in the world, and millions of pilgrims come to Triveni Ghat to bathe in the Kshipra River. The ghat is home to the Navagraha Shani (temple of nine planets and the God of Saturn and justice) Temple present at Triveni Ghat (Figs. 10, 11 and 12). A variety of pujas are performed at Triveni Ghat in Ujjain, depending on the devotee's needs and beliefs. Some of the most common pujas include

- Navagraha Pooja—The pooja of nine planets to get rid of Doshas.
- Shani Pooja—This pooja is performed to appease Lord Shani; the god of Saturn to remove obstacles from life
- Triveni Sangam Pooja—This pooja is performed to worship the confluence of the Kshipra River with the two invisible rivers, Saraswati and Ganga. This confluence is believed to be a very sacred place, and this pooja is performed to seek the blessings of the three rivers.

Overall, Triveni Ghat is a place of great historical and cultural significance. It is a place where people come to worship the nine planets of Hindu astrology, Lord Shani (the god of Saturn), and the confluence of the three rivers. It is also a place where people come to experience the unique culture and spirituality of Ujjain. Subsequent paragraphs, however, are indented.

6 Inferences, Recommendations and Framework

Sl. No	Criteria	Varanasi				Ujjain		
		Assi Ghat	Mani-karnika Ghat	Kedar Ghat	Da-shashwamedh Ghat	Ram Ghat	Man-galnath Ghat	Triveni Ghat
Landscape Elements								
1	**Outdoor Illumination**	✓	x	x	✓	✓	✓	✓
		Reorganizing public lightning and existing supply lines by providing new wiring, meters, etc. wherever necessary						
2	**Signage**	✓	✓	✓	✓	✓	x	✓
		Most of the ghats have very minimal signage and this creates confusion among tourists. Addition of signage to reduce confusion among tourists. The signage should be such that they bolster the site identity, highlight the key entry point, convey clear pedestrian wayfinding to facilities. Typography for signage : Font name - Ariel Bold (100mm Height) and Ariel Regular (30mm Height) for English texts, Surendra (100mm Height) for Hindi texts.						
	Colour palette for signage							
3	**Shading**	✓	x	x	✓	x	x	x
		Ghats have very limited Chattris for shade. Chattris can be provided on the octagonal seating and pattas available on ghats. Additionally interactive seating space with shading can be added wherever required.						
4	**Street Furniture**	✓	x	x	✓	✓	✓	✓
		Since most of the ghats have inadequate seating, people gather in the evening and secure their places before the Evening aarti starts. Benches with chattri can be added. More Octagonal platforms can be provided. Seats around religious trees can be provided.						
5	**Vegetation**	✓	x	x	x	✓	✓	✓
		Very little vegetation can be found on ghats and most of the vegetation seen is a religious tree where religious activities take place. Provision of more landscaping to reduce heat island effect on every						

		ghat in each zone. Removal of trees and saplings growing out of the existing infrastructure.						
6	**Dustbins**	✓	✓	✓	✓	✓	x	x
		Dustbins can be found at irregular intervals. Addition of colour coded dustbins at regular intervals of 10m - 15m distance.						
		Architectural Elements						
7	**Material**	Sand-stone	Sand-stone	Lime-stone	Sandstone	Sand-stone	Granite, Red Sand-stone	Black, Grey granite
8	**Colour**							
		Varanasi - The ghats in Varanasi exhibit a vibrant and diverse color palette. The architectural elements and structures on the ghats are adorned with various colors, creating a visually appealing environment. Ujjain - The overall color palette of the ghats in Ujjain is a beautiful and harmonious blend of warm and cool tones. It reflects the rich cultural heritage of the city and the beauty of the natural environment.						
9	**Ramp**	✓	x	x	x	✓	✓	✓
		Assi ghat has multiple ramps for easy accessibility whereas Ram Ghat and Triveni Ghat in Ujjain have only one ramp each located at the southern and western end of the ghat that leads down to the water. Slope: The slope of the ramp should not exceed 1:12. Width: The minimum width of the ramp should be 1800 mm. This will allow two wheelchairs to pass side by side. Handrails: The ramp should have handrails on both sides. The handrails should be 800 mm high and extend 300 mm beyond the top and bottom of the ramp. Landings: There should be a landing at the top and bottom of the ramp. The landings should be at least 1800 mm long and 2000 mm wide. This will allow wheelchairs to turn around easily. Surface: The surface of the ramp should be non-slip and durable with adequate lighting.						
10	**Handrails**	x	x	✓	x	x	x	x
		There are no handrails for elderly people on ghat which makes it difficult for them to access without any help. Provision of handrails near the ritual bathing area to prevent accidents and easy access to elderly people who often visit the ghats. Provision of handrails near the temples.						

11	**Ornamen-tation**	✓	✓	✓	✓	✓	✓	x
		Varanasi - Bastions, balconies, aedicules, portals, pavilions, and platforms, are combined in different sizes and materials to create a visually complex structure. Repetition of these elements is seen throughout the Ghats of Varanasi Additionally, the ghats are adorned with thatched umbrellas, which add a distinctive touch to the landscape. Ujjain - The ghat is lined with pillars that are decorated with intricate carvings. The steps leading down to the water are also decorated with balustrades that are carved with flowers, animals, and other motifs. Repair of Minor cracks, missing and broken elements, surface disfiguration caused by natural erosion. Replacement of missing or broken elements. Restore and preserve the existing ornamentation.						
Architectural Parameters								
12	**Parking**	✓	x	x	x	✓	✓	✓
		Varanasi - Assi Ghat has its own parking area for visitors. Whereas Manikarnika Ghat, Kedar Ghat, and Dashashwamedh Ghat do not have parking spaces. Ujjain - Ram Ghat has 2 unofficial parking areas which causes trouble to both the visitors and localities during peak season (Kumbh mela and Kawan Yatra). Mangalnath also has a unofficial parking space which is surrounded by street vendors and cafes. Triveni Ghat has a parking space right before the Temple for buses and 2 wheelers, whereas 4-wheeler parking is located 200m away from the temple. Open parking areas (avoid concrete and asphalt as it absorbs heat and contribute sustainably to heat island effect) but this can be mitigated by rendering these areas with white or light colours instead of black. Providing shade, open grid pavements for the sites with non- impervious surfaces, including parking lots, walkways, plazas, etc.						
13	**Pedestrian Pathway**	x	x	x	x	x	x	x
		None of the Ghats have a pedestrian pathway since entry of automobile is prohibited to these sites and people walk on feet while visiting these places. 2.2m wide lane with shallow, open drains connected to the city sewer lines and white glazed tiles to be installed for ease of access and movement						
14	**Waste Manage-ment**	✓	✓	✓	✓	✓	✓	✓
		The waste management strategies on Ghats focus on collecting and treating organic waste, using deployable structures for consolidated vending, and creating amphibian spaces to reduce congestion. Some						

		of the challenges faced by the authorities are Lack of awareness, High volume of waste and Lack of Infrastructure development. Waste Management system to be implemented and river polluting objects like Reusable plastics, Cigarette butts to be banned from the site. Solid waste should be collected and disposed of properly, and liquid waste should be treated before being discharged into the river. Regular monitoring of water quality.						
15	**Drinking Fountains**	x	x	x	✓	x	✓	x
		Varanasi - No drinking fountains are found on the Ghats of Varanasi, except for Dashashwamedh ghat located on the western end of the Ghat. Ujjain - Water fountains can be seen across Ujjain and in every public place but when it comes to ghats water fountains are found only near the temple vicinity. Drinking water facilities to be provided at a regular interval of 100-150m on every ghat.						
		Built Elements						
16	**Temple**	✓	✓	✓	✓	✓	✓	✓
		Temples play a major role in all of these ghats as it is primarily a religious space. The temples on the ghats can be maintained, preserved and restored to improve its longevity. Yearly check on temples for any defects and maintenance check. Yearly Restoration and Preservation is also recommended.						
17	**Museum**	x	x	x	✓	x	x	x
		Virtual Experiential Museum is a virtual reality museum that takes visitors on a journey through the history and culture of Varanasi. Jantar Mantar, Man Singh Observatory are also near Dashashwamedh Ghat. Government and NGO's can take initiative to build more such places near the ghats that highlight the religious and cultural influence of the ghat.						
18	**Restaurants/ Cafe**	✓	x	x	✓	✓	✓	x
		Cafes and Restaurants can be seen in the majority of the Ghats both in Varanasi and Ujjain. Manikarnika Ghat doesn't have any cafes or restaurants being a cremation ground, but some shops are available within the 1km radius. Triveni ghat is one of the most isolated Ghats of Ujjain and there are no eateries around. Cafes and restaurants can be provided secluded from the place of religious activities on ghat. The structure shouldn't be on water edge and sustainable materials to be used to build the new structures for cafes and restaurants.						
19	**Cremato-**	x	✓	x	x	x	x	x

	rium							
		Varanasi - Manikarnika Ghat is the main cremation ground in Varanasi. It is believed that a person who is cremated here will attain moksha, or liberation from the cycle of rebirth. Crematoria on ghats should be equipped with adequate pollution control measures to minimize air and water pollution. This may include measures such as installing electrostatic precipitators and chimneys of adequate height. Additionally, the design of the crematorium should minimize the amount of smoke and ash that escapes into the environment. Crematoria on ghats should be regularly maintained to ensure that they are in good condition and that the pollution control measures are functioning properly.						
20	**Tourist Information Centre**	✓	✓	✓	✓	x	x	x
		Varanasi - Each Ghat in Varanasi has a tourist information center located either at the entrance of the ghat or near the Police Station of the respective Ghats. Ujjain - There are no tourist information centers on Ghats of Ujjain. However, there is a tourist information center located at the Mahakaleshwar Temple, which is about 1 km from Ram Ghat. Tourist Information Kiosk to be set up at regular intervals on each ghat. Government can also take initiative to set up a All-in-One mobile Application for tourists, solving language barrier problems.						
21	**Public Toilets**	✓	✓	x	✓	x	x	x
		There are 24*7 SBM Toilets available on most of the Ghats in Varanasi which are in good condition. The toilets are maintained by Sulab International and users have to pay Rs. 5 per usage. If any ghats lack public toilet then one is available within a 2km radius. Area statement of proposed new toilets: Gents - 3 W.C, 2 urinals, 4wash basins. Women's toilet – 3W.C, 4 wash basins, a small storage and pump room. 1 small shop near the toilet, provision of water tank with a place where devotees can wash their feet and a water trough for animals outside. Total Area - 35sq.M for each male and female toilet respectively.						
		Religious Elements						
22	**Ritual Bathing**	✓	✓	✓	✓	✓	x	x
		Hindus believe that bathing in the sacred rivers of India can cleanse one's sins and bring spiritual enlightenment and this is done for spiritual purification. Provision of changing room near the ritual						

		bathing area for ease and comfort.						
23	**Religious Trees**	✓	x	✓	✓	✓	x	✓
		Preservation of religious trees						
24	**Lamp**	✓	✓	✓	✓	✓	x	✓
		Diyas or lamps are offered as prayers to River Ganga and Shipra as a form of prayer. Lamps are considered auspicious as they symbolize the absence of darkness and unhappiness. The lamps could be in the form of butter lamp which is often kept on a disposable plastic cup before offering it to the river. A separate area can be allotted for Diya offering to the river to maintain pollution of the Rivers. Disposable plastic can be banned on Ghats to preserve the river and the environment.						
25	**Religious Offering**	✓	x	x	x	✓	✓	✓
		Flowers are usually offered to the River during Aartis, religious rituals and cultural events.						
26	**Aarti**	✓	x	✓	✓	✓	x	x
Activities on Ghat								
27	**Boating**	✓	x	✓	✓	✓	x	x
28	**Economic hubs**	✓	✓	✓	✓	✓	x	✓
29	**Handicraft Store**	✓	x	✓	✓	✓	x	x

The Recommendations and Framework are designed to be universal and can be implemented on any ghat in the country. In other words, the Recommendations and Framework are a set of guidelines that can be used to improve the condition and management of any ghat in India. The Framework and Recommendations serve as a useful tool for anyone seeking to enhance the state of India's ghats. Government representatives, NGOs, community leaders, and people can all use them to change things.

7 Conclusion

A significant tourist destination with a unique appeal, a long history, and exceptional architectural and cultural legacy, Varanasi is a historic city that seems to have withstood the rigors of socioeconomic, commercial, and industrial progress. The stunning developments in Varanasi around the banks of the river Ganga and inside its water body are made up of a variety of unique buildings and areas that combine to form a unique architectural landscape alongside the river. Its soul lies in its history, the mind in learning and the body in Ganga. Varanasi would lose its very soul if Ganga disappeared. Both the mind and the soul are housed in the body. Regretfully, the Ganga River, whose waters cleanse the filthy, is about to lose its identity. One of the most polluted rivers in the world and the most contaminated main river in India is the Ganga. To quote Carl Jung: "The people of India will never find true peace until they can cultivate deeper feelings of reverence for and come into a harmonious relationship with the Ganga River, which is the cradle and identity of India's culture and civilization since time immemorial." As expressed by a pious traveler, "The Varanasi community and the global community must act to stop the pollution of the Ganga [1]". There needs to be a push to modernize the city while maintaining its cultural identity, and to return River Ganga to its original state.

It is evident that Varanasi has incorporated the sacred River Ganga into its architectural structure, with the river flowing through the city like a dreaming breath. On the sacred map of India, there is a city called Ujjain. It was a city woven from a tapestry of threads, each thread representing a different religion, culture, and tradition. The enchanting tapestry of India's sacred destinations, Ujjain emerges as a vibrant chapter adorned with the graceful dance of numerous temples and the serene embrace of the city's revered ghats, where the Kshipra River weaves its tranquil melody. The importance of Ujjain on the sacred map of India is irrevocably centered around its rich cultural heritage, religious fervor, and the spiritual significance of its sacred ghats. The ghats of Ujjain have served as pivotal centers for religious gatherings, cultural festivals, and sacred ceremonies, fostering a sense of unity and spiritual communion among its visitors. Despite the challenges posed by modern development and environmental concerns, the ghats remain steadfast in their role as spiritual sanctuaries, preserving the timeless essence of Ujjain's cultural and religious heritage for generations to come.

The ghats of Varanasi and Ujjain, steeped in centuries-old history and religious fervor, stand as timeless testaments to India's rich cultural heritage. From the sacred banks of the Ganges in Varanasi, where the spiritual aura is palpable in the echoes of ancient rituals and the flickering lamps of devotion, to the tranquil embrace of the ghats along the Kshipra River in Ujjain, where the confluence of faith and tradition weaves a serene tapestry, these sacred spaces embody the spiritual soul of the cities. Despite the challenges posed by modern development and environmental concerns, the ghats remain steadfast in their role as the spiritual sanctuaries, preserving the timeless essence of India's diverse cultural and religious tapestry for generations to come.

References

1. Basak S, Sarkar A, Ghosh R, Chaudhuri A (2015) Ghats of Varanasi—an emerging centre of pollution. IOSR Journal of Humanities and Social Science
2. Gurmeet SS (2007) Caste/d space: a reinterpretation of the movement and architecture of Dashashwamedh Ghat, Varanasi, India. Doctoral dissertation, University of London, University College London (United Kingdom)
3. Kumari Saha S (2019) Landscape development of the edge of river Ganga in Patna, Bihar. Doctoral dissertation, SPA Bhopal. http://14.139.241.254:8080/jspui/bitstream/123456789/725/1/2017MLA015%20%28%20TH001035%20%29.pdf
4. Video document. YouTube (n.d.) Riverfront development in India: an insight. Retrieved from https://youtu.be/31CyQx_xxAU
5. Video document. YouTube (n.d.) The makeover of riverfront has given a complete new experience for the residents: a video analysis. Retrieved from https://youtu.be/Tk6e5qCGo8s
6. Video document. YouTube (n.d.) Historic ghats of Varanasi: conservation and challenges. Retrieved from https://youtu.be/bNW8Ov08-m8
7. Singh RPB (2004) The Ganga riverfront in Varanasi: a heritage zone in contestation. Context: Built, Living and Natural, 1(1):25–30. DRONAH
8. Sinha A (2018) The performative landscape of Dashashwamedh Ghat, Varanasi. Retrieved from https://www.researchgate.net/publication/327418392_The_Performative_Landscape_of_Dashashwamedh_Ghat_Varanasi#full-text
9. Samant S (2004) Manifestation of the urban public realm at the water edges in India—a case study of the Ghats in Ujjain. Cities 21(3):233–253
10. Samant S (2010) Reflections on water: architectural manifestations in the historic and cultural quarter along Lake Pichola in Udaipur. J Sustain Dev 3(2):12–21. https://doi.org/10.5539/jsd.v3n2p12
11. World Bank (2015) Recommendations for Patna redevelopment. Retrieved from https://documents1.worldbank.org/curated/pt/510601468252652037/pdf/SFG1326-EA-P119085-PUBLIC-Disclosed-8-28-2015-Box393194B.pdf
12. Lall N (2019) Recommendation for redevelopment of ghats in Patna. Tekton 3(2):56–68. Retrieved from https://tekton.mes.ac.in/wp-content/uploads/2019/10/05-nishant-lall-volume-3-issue-2.pdf
13. Sinha A (2015) The enacted landscape of Varanasi Ghats: beyond the picturesque. Vis Arts Res 12(April 2013–March 2015)

Retrofitting for Resilience—A Case of Srirangam Temple Town

V. Vishaliny and Amanjeet Kaur

Abstract Heritage conservation is integral to inclusive development as it ensures the protection of cultural and historical assets for future generations. This paper is based on the case of Srirangam temple town, the world's largest operational temple, which embodies the complexity of preserving cultural heritage and achieving sustainable urban development. The paper employs exploratory study utilizing various primary and secondary data sources, including literature study, visual surveys, and interviews with a diverse array of stakeholders. The parameters for the study focus on identifying the historical and cultural significance of Srirangam, the challenges faced by the temple town and deriving strategies to mitigate them. Drawing daily crowds of thousands, and lakhs during the festive seasons, the temple's prominence has led to compelling urban issues. The sacred temple streets have now commercialized, inducing congestion in the temple town. The foundation of the ancient temple ramparts are now imperiled by improper sewage management. The paper identifies such critical challenges that have led to a change in the temple landscape and discusses retrofitting strategies for the same. These strategies will help conserve the cultural, ecological, and historical importance of Srirangam and simultaneously promote sustainable urban development. In conclusion, this research extends beyond analyzing Srirangam's challenges, serving as a valuable resource for urban planners, architects, authorities, and communities dealing with identical issues. Further, Srirangam depicts the similar challenges encountered by temple towns across India and highlights the need for innovative sustainable strategies that safeguard cultural heritage, and promote inclusive urban development.

Keywords Srirangam temple town · Sustainable urban development · Heritage conservation · Retrofitting strategies

V. Vishaliny (✉) · A. Kaur
National Institute of Technology Hamirpur, Hamirpur, India
e-mail: vvvishaliny@gmail.com

A. K. Sharma and A. Joshi (eds.), *Proceedings of the 3rd International Conference on Opportunities and Challenges for a Resilient Future*, Advances in 21st Century Human Settlements, https://doi.org/10.1007/978-981-96-0931-4_7

1 Introduction

Preserving the character of historic towns while addressing the challenges of new development remains a global spatial planning dilemma [1]. The study is centered on Srirangam and its vast Sri Ranganathaswamy temple complex, and explores the intricate links between the temple, community, and urbanization (Fig. 1).

There are four main periods that played a crucial role in the continuous development of the Srirangam Temple Town. The ancient period (third century BCE–fourteenth century CE) saw the birth of the Sri RanganathaSwamy Temple, marking the town as a religious and cultural center. Dravidian temple architecture, royal patronage from the Chola and Pandya dynasties, and the establishment of religious practices shaped the town's identity [2]. During the medieval period (fourteenth century CE–seventeenth century CE), the temple complex expanded with monumental gopurams, fostering trade activities and structured administrative systems. Controlled urbanization preserved natural surroundings, contributing to the town's prosperity [3]. The colonial period (eighteenth century CE–nineteenth century CE) witnessed colonial impact on the temple administration and subsequent efforts to preserve the cultural heritage of the temple town. Urban development initiatives, especially road connectivity, laid the groundwork for modern infrastructure [4, 5]. Post-independence (twentieth century CE–present), the contemporary period accentuated tourism, pilgrimage, and laid stress on heritage conservation. Technology integration has improved temple management, and ongoing heritage conservation efforts strive for sustainable development [6, 7] (Table 1).

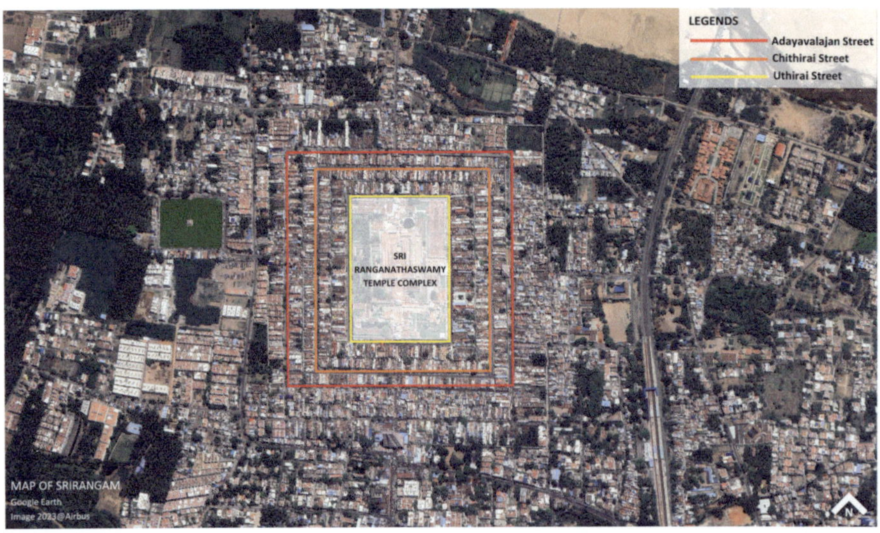

Fig. 1 Map of Srirangam town illustrating the temple complex and the adjacent streets (Image taken from google earth)

Table 1 Historical timeline of Srirangam's evolution tracing significant development over the centuries (Compiled by author)

Parameter	Ancient period (third century BCE–fourteenth century CE)	Medieval period (fourteenth century CE–seventeenth century CE)	Colonial period (eighteenth century CE–nineteenth century CE)	Contemporary period (twentieth century CE–present)
Key development	Birth of Dravidian temple architecture and construction of the Ranganatha Swamy temple. Support from Chola Pandya dynasties, leading to infrastructure development around the temple. Establishment of rituals and religious practices contributing to the cultural identity	Addition of monumental gopurams as architectural landmarks. Formation of trade guilds and marketplaces near the temple. Introduction of administrative systems for efficient temple management	Addition of monumental gopurams as architectural landmarks. Formation of trade guilds and marketplaces near the temple. Introduction of administrative systems for efficient temple management	Emphasis on tourism and pilgrimage as key economic drivers. Implementation of technology for temple management. Conservation efforts for heritage and environmental sustainability
Historical changes	The temple became a focal point for regional dynasties, shaping the historical narrative	Documentation of temple affairs and administrative records	Periodic reviews and documentation of the temple's historical significance	Ongoing documentation of contemporary developments for future records
Ecological changes	Early temple construction using locally available materials	Controlled urbanization with attention to preserving natural surroundings	Increased urbanization with attention to environmental conservation	Implementation of eco-friendly practices for sustainable tourism

(continued)

Table 1 (continued)

Parameter	Ancient period (third century BCE–fourteenth century CE)	Medieval period (fourteenth century CE–seventeenth century CE)	Colonial period (eighteenth century CE–nineteenth century CE)	Contemporary period (twentieth century CE–present)
Cultural changes	Formation of religious practices and cultural rituals	Expansion of religious festivities and cultural events	Preservation and documentation of cultural practices	Preservation of traditional rituals alongside modern practices
Contribution to further development	Establishment of the temple as a cultural and religious hub laid the foundation for subsequent expansions. Attraction of pilgrims and royal patronage contributed to economic growth	Architectural enhancements increased the temple's prominence, attracting more devotees. Economic activities stimulated by trade guilds contributed to the town's prosperity	Cultural preservation efforts provided a foundation for future heritage conservation. Urban development initiatives laid the groundwork for modern infrastructure	Integration of technology enhanced the visitor experience and facilitated efficient temple management. Heritage conservation efforts ensure the preservation of cultural and historical richness

This contextual overview of the spatial planning emphasizes the necessary balance between preserving the historic town character of Srirangam and accommodating new development.

2 Methodology

This paper follows an exploratory study approach, understanding the interplay between human experiences and the built environment in Srirangam. The study utilizes primary and secondary sources of data including literature study, visual survey, questionnaire surveys, semi-structured and focus group interviews. To have a holistic approach, a diverse spectrum of participants, ranging from residents, local vendors, temple priests, and city authorities to pilgrims, tourist guides, auto drivers, real estate agents, and hotel representatives were interviewed in the process. Further, heritage significance assessment, temple town spatial assessment and templex complex condition assessment were done to identify the historical significance, and challenges faced by the temple town. Analysis was done based on simple statistical tools, and qualitative data from interviews were analyzed to extract commonalities, variations and trends. Further successful case examples globally were studied to draw insights from their retrofitting strategies for resilience, and potential solutions for Srirangam.

3 Analysis and Assessment

Sri Ranganathaswamy temple has been a positive influence on the growth of Srirangam town, as stated by its residents. The local vendors mentioned the influx of tourists has significantly boosted their local economy. Furthermore, the temple's presence has led to an increase in real estate values in the temple town. To understand and identify the challenges posed by the continuous development of Srirangam, three levels of assessment were done.

3.1 Heritage Significance Assessment

According to the temple authorities, the temple experiences a daily influx of 3000 to 5000 visitors on regular days while this number escalates to 10,000 to 12,000 during weekends. The Vaikunda Ekdasi Festival, typically observed in December and January, witnesses approximately 1.5 lakh pilgrims. The Chithirai Ther festival that is held between March and April to coincide with the Tamil calendar's Panguni month draws more than 50,000 attendees during it's eleven-day festival (Fig. 2).

Fig. 2 Image showing the pilgrims gathered in large numbers within the temple premises during the Vaikunda Ekdasi festival, with police managing the crowd around the idol carried by the priests (Sriranga Pankajam (n.d.). Srirangam Temple Gallery. Sriranga Pankajam. https://www.srirangapankajam.in/srirangam-temple-gallery.php)

3.2 Temple Town Spatial Assessment

The temple premises are not only sacred spaces but also home to families dedicated to temple seva (religious service). The streets once solely designated as processional routes or avenues for movement have undergone major transformations. According to a temple authority, the streets have now evolved into areas of duplex houses, apartments, and commercial zones.

In particular, urban development has outpaced the provision of adequate parking facilities, leading to congestion issues. The expanding commercial sector is witnessing an increase of tourists and pilgrims visiting the temple town, leading to a rise in the number of hotels and lodges. Moreover, overcrowding is a prevailing issue, as individuals are drawn to settle in this sacred town post-retirement. Consequently, sewage waste management has become an intricate task for the city administration as said by the city authorities (Fig. 3).

3.3 Temple Complex Condition Assessment

The streets are flanked by ramparts which establish a visual continuity. The ramparts provide shade for pedestrians and protection for the town from cyclonic winds and dust. The ramparts are punctuated by the monumental gopurams with the streets

Fig. 3 Images depicting the lack of parking facilities around the temple complex and commercialisation of the temple town (Images by author, 2023. Srirangam, India)

leading through them to the temple [8]. However, improper sewage management threatens the temple ramparts foundation. Temple authorities foresee its survival for a decade but stress the need for innovative solutions to ensure centuries of heritage. Temple-generated waste is often disposed of in landfills. This assessment, focusing on the condition of the temple complex, underscores the critical balance between preserving historical structures and implementing forward-looking strategies for sustainable heritage conservation.

4 Results

The results identify the pressing challenges faced by Srirangam, a heritage town grappling with rapid growth. Assessments and surveys involving 425 participants reveal the four critical challenges hindering the town's resilient and sustainable development. Managing large influx of visitors during festivals surfaced as a paramount concern, with respondents emphasizing the strain on existing infrastructure and the need for innovative solutions to ensure a seamless experience for pilgrims and tourists alike. Secondly, congestion and inadequate parking facilities were identified as substantial issues impacting the local community and visitors. The urgency of addressing these challenges is evident to maintain the town's accessibility and heritage integrity. Thirdly, improper sewage management, posing a threat to the temple ramparts, poses a significant ecological challenge, emphasizing the need for sustainable sewage waste management solutions to safeguard the historical structures. Lastly, the commercialization of the temple town has altered the sacred landscape and poses a significant challenge requiring delicate balance and strategic interventions. The town's unique cultural identity and historical importance are highlighted as integral elements that demand conservation amidst the changing urban landscape. Recognizing Srirangam's role as a pilgrimage destination and a center of religious practices, the study highlights the critical importance of retrofitting strategies to ensure the temple town's resilient future.

5 Discussion

The urbanization of temple towns, exemplified by Srirangam with its Sri Ranganathaswamy temple complex, unveils pressing challenges necessitating innovative solutions. Retrofitting Srirangam for resilience requires a nuanced approach, combining temporary and permanent solutions across various domains. The proposed strategies draw inspiration from successful examples implemented worldwide, ensuring a holistic and sustainable transformation.

5.1 Managing the Large Influx of Visitors During Festivals

Addressing the large influx of visitors during festivals requires strategic measures. A short-term approach involves deploying modular, prefabricated structures equipped with RFID-based crowd tracking systems that helps efficiently manage visitor flow during peak festivals. For instance, the Kumbh Mela is a significant religious gathering attracting millions of pilgrims. Managing the massive crowds during peak festivals was a task. It then successfully employed modular infrastructure with RFID tracking to manage large crowds efficiently [9]. Another temporary measure is to employ GPS-enabled shuttle services and dynamic traffic management algorithms to optimize traffic flow and reduce congestion. A similar problem existed in Angkor Archaeological Park in Cambodia which experiences congestion and traffic issues due to high tourist numbers. To optimize traffic flow and reduce congestion, GPS-guided shuttle services have been implemented to manage tourist transportation efficiently [10].

Permanent solutions like developing an integrated smart crowd management system utilizing real-time data analytics, CCTV cameras, and RFID technology helps monitor crowd density, optimize pathways, and provide timely alerts. The Disneyland Resorts globally, dealing with consistent high visitor numbers, employed advanced crowd management systems, integrating RFID technology for efficient crowd control [11]. Additionally, implementing a dynamic scheduling algorithm to distribute festival events across an extended period, ensures a more even distribution of visitors and reduces peak attendance. The Oktoberfest in Munich witnesses peak attendance during specific festival periods and hence implemented a phased scheduling system, extending festival events over several weeks to manage attendance effectively [12].

5.2 Congestion and Inadequate Parking Facilities

Congestion and inadequate parking facilities necessitates a blend of temporary and permanent solutions like geospatial analysis. It can be utilized to identify vacant lots for temporary parking spaces, implementing a reservation system integrated with mobile apps for real-time availability tracking. Scarcity of parking facilities was an issue in the Vatican City which then adopted a mobile app-based reservation system [13]. A dynamic pricing system for parking spaces can be implemented based on real-time demand and availability data. The pricing algorithm adjusts rates according to peak hours, encouraging optimal utilization of parking facilities. To exemplify, London's dynamic parking pricing scheme effectively managed the parking demand by dynamically adjusting prices based on real-time occupancy levels and traffic conditions [14].

Long-term solutions include employing sustainable construction methods, such as the use of recycled materials and green roofs, in the design and construction of multi-level parking structures on the outskirts of the temple town. For instance, the Green Garage in Chicago incorporates sustainable features, including a green roof and rainwater harvesting, making it an environmentally friendly multi-level parking facility [15]. Utilizing advanced traffic simulation tools to model and implement pedestrian-friendly zones, can help curb vehicular traffic and promote non-motorized transport. In the case of Barcelona, the city faced severe traffic congestion, which then led to the implementation of pedestrian zones in the historic district, limiting vehicle access and promoting walking and cycling [16], contributing to long-term traffic solutions.

5.3 Improper Sewage Management Threatening Temple Ramparts

Sewage management threatens the temple ramparts and demands immediate and sustainable solutions. An interim measure consists of deploying robotic maintenance and inspection tools to assess and clean the existing sewage system. This will address the immediate issues and prevent further damage. For instance, the City of Paris utilizes robotic sewer inspection technologies to conduct regular maintenance in its historic sewer system [17]. Likewise, installing Smart Bin sensors on waste bins help monitor their fill levels in real-time. A mobile application can be developed that sends alerts to waste collection teams when bins reach a specified capacity. Dynamic waste collection routes can be implemented based on real-time data, optimizing efficiency. In the case of Singapore, it has successfully implemented a smart waste management system utilizing bin sensors and a mobile app. This alerts waste collectors when bins are full and results in efficient waste management, reduced overflow, and improved cleanliness.

On a longer run, advanced smart sensors along sewage lines can monitor real-time flow rates and detect irregularities. The automated systems help divert sewage flow during peak times, and the sewage diversion system can be connected to a broader city-wide network for data sharing, enabling efficient coordination with other urban management systems. To exemplify, Petra in Jordan has implemented smart sewage diversion channels to protect its archaeological sites. The integrated sensors provide real-time data on sewage flow, preventing damage to heritage sites. This resulted in enhanced heritage preservation and sustainable sewage management [18]. Along similar lines, integrating green spaces along the ramparts, by planting native species, creates a sustainable ecosystem. The indigenous plant species when strategically planted with deep roots to stabilize the soil, prevent erosion and enhance the structural integrity of the ramparts. The Great Wall of China has implemented this method to protect the Great Wall's ramparts from erosion and deterioration. Eventually they developed a comprehensive green infrastructure program to protect the Great Wall's ramparts. They planted native vegetation, including resilient grasses and shrubs, along vulnerable sections. This resulted in improved aesthetics, enhanced ecological balance, and increased resilience against environmental pressures [19].

5.4 Commercialization Changing the Temple Town's Landscape

Finally, mitigating commercialization's impact on the temple town's landscape requires comprehensive solutions. The impact of tourism on the temple town's infrastructure, including commercial zones can be assessed through a comprehensive study. This can evaluate foot traffic, resource consumption, and environmental impact. Additionally, implementing temporary zoning regulations can control the expansion of commercial areas, ensuring a balance between tourism-driven economic activities and heritage preservation. In the case of Kyoto, a tourism impact assessment was done to understand the effects on heritage sites. Temporary zoning regulations were enforced to control commercial growth, preserving the city's cultural integrity [20]. Likewise, Seasonal Tourism Management and Promotion strategies can promote off-peak tourism, distributing visitor influx more evenly throughout the year to reduce congestion during peak seasons. The city of Barcelona aimed to promote responsible tourism and introduced seasonal tourism management strategies to alleviate congestion during peak months. The community engagement programs involved locals to promote responsible tourism and fostered a sustainable balance [21].

Permanent solutions like Heritage Compatible Commercial Infrastructure integrates architectural guidelines and ensures new commercial developments align with the aesthetic and cultural heritage of the temple town. Dubrovnik, a city in Croatia faced threats to the historic integrity of its commercial areas. Eventually, the city implemented strict architectural guidelines to maintain the historic integrity of commercial areas. Notably, the community-driven businesses played a vital role in preserving the cultural essence of the city [22]. In a similar manner, developing a comprehensive master plan for cultural heritage tourism can outline long-term strategies, manage and guide sustainable growth in commercial activities. In the case of Edinburgh, the cultural heritage was affected by the unmanaged growth in the city. This pressured the city to enforce a cultural heritage tourism master plan that guides the city's growth. The stakeholder collaboration ensured that the plan considered the interests of all parties involved (Table 2).

To sum up, Srirangam's challenges require tailored solutions that can be drawn from successful global implementations. The large influx of visitors in Srirangam can be addressed by deploying modular structures with RFID tracking and optimizing traffic flow through GPS-enabled shuttle services. Multi-level parking structures, pedestrian- friendly zones and dynamic pricing for parking can be introduced to help tackle congestion and inadequate parking facilities in the temple town. A smart sewage diversion system and green infrastructure will help preserve the temple ramparts and manage urban growth in Srirangam. Seasonal tourism management and heritage compatible commercial infrastructure should be implemented to strike a balance between economic activities and heritage conservation in the temple town.

6 Conclusion

Srirangam's challenges extend beyond its immediate context offering a broader understanding of common issues faced by temple towns across India. This research lays a foundation for future investigations into similar cases, providing a valuable repository of strategies derived from successful global implementations. As Srirangam faces the impact of modernization on its ancient town planning, the study highlights the urgent need for a well-structured framework and implementation of sustainable solutions to safeguard its rich cultural heritage.

Table 2 The above table discusses the strategies to mitigate the challenges faced by Srirangam temple town. Further, it presents instances from various countries confronting similar issues and successfully implementing the respective strategies (Compiled by author)

Challenge	Temporary solutions	Permanent solutions
1. Managing large festival crowds	(a) Modular infrastructure with RFID tracking for crowd management (Kumbh Mela, India)	(a) Advanced crowd management system integrating RFID technology (Disneyland Resorts)
	(b) GPS-guided shuttle services for traffic management (Angkor, Cambodia)	(b) Expanded festival periods with phased scheduling system (Oktoberfest, Munich)
2. Congestion and parking issues	(a) Mobile app based reservation system for temporary parking Apaces (Vatican City)	(a) Green roof and rainwater harvesting in the Multi-Level Parking facility (Green Garage, Chicago)
	(b) Dynamic parking pricing scheme based on real-time occupancy levels and traffic conditions (London)	(b) Pedestrianization and non-motorized zones limiting vehicular access and promoting walking cycling (Barcelona)
3. Improper sewage management	(a) Robotic sewer inspection technologies for regular sewage system maintenance (Paris)	(a) Smart sewage diversion systems provide real-time data on sewage flow (Petra, Jordan)
	(b) Bin sensors and a mobile app alerts for efficient and immediate waste collection (Singapore)	(b) Comprehensive green infrastructure program and planting of native species for rampart protection (Great Wall of China)
4. Mitigating commercialization impacts	(a) Tourism impact assessment and temporary zoning regulations on heritage sites to control commercial growth (Kyoto)	(a) Architectural guidelines and community-driven heritage-compatible commercial infrastructure (Dubrovnik)
	(b) Seasonal tourism management strategies and community engagement programs involving locals (Barcelona)	(b) Cultural heritage tourism master plan for city's cultural growth and stakeholder collaboration (Edinburgh)

Srirangam serves as a reminder to balance the complexities of modernization while conserving the cultural heritage. In conclusion, the findings of this research are not confined to analyzing Srirangam's challenges but serve as a reservoir of insights for urban planners, architects, urban designers, local authorities, and communities facing similar dilemmas. As temple towns continue to evolve in India, this study advocates for a global model of resilient heritage development, emphasizing the adaptability of the proposed strategies in the face of the diverse challenges posed by urbanization. These findings will contribute to the ongoing discourse on sustainable urban transformations in temple towns, particularly in India, with numerous heritage-rich locations facing similar pressures.

References

1. Gopalakrishnan P, Srinivas T (2014) Viewscape assessment framework for protecting the views of sacred monuments: comparative study of Srirangam and Thanjavur religious towns. Creat Space 1(2):195–213
2. Srinivasan KR (1969) Sri Ranganathaswamy temple: a historic and cultural study
3. Nilakanta Sastri KA (1955) A history of south India
4. Champakalakshmi R (2002) Trade, ideology, and urbanization: south India 300 BC to AD 1300
5. Vara Prasad SBR (n.d.) British paramountcy in India (1774–1858)
6. Vedachalam V (2004) Sri Ranganathaswamy temple: an architectural survey
7. Nandakumar P (n.d.) Srirangam: the holy Island
8. Ghosh B, Mago KC (1974) Srirangam: urban form and pattern of an ancient Indian town. Ekistics 38(228):377–384
9. Sharma R, Sharma P, Patel R (2017) Technology solutions for crowd management in Kumbh Mela. In: Proceedings of the international conference on advances in computing, communications, and informatics (ICACCI)
10. Kim D, Lee D, Kim K (2019) Intelligent transportation systems for sustainable tourism: a case study of the Angkor archaeological park. Sustain 11(23):6578
11. Chien S, Ding Y, Wei C (2009) Design and operations of Disneyland systems: a queueing network analysis approach. OR Spectrum 31(4):849–878
12. Wittkowski KM, Graeff C, Block LH (2018) Oktoberfest in Munich–statistical analysis and key figures. Stat Pap 59(4):1613–1631
13. Del Bufalo F, Massoli FV, Gobbetti E (2017) Smart parking management in the Vatican city: a case study. In: Proceedings of the 2017 ACM international joint conference on pervasive and ubiquitous computing and proceedings of the 2017 ACM international symposium on wearable computers
14. Smith J, Johnson M, Brown K (2018) Dynamic pricing for urban parking: a case study of London. Transp Res Part C: Emerg Technol 94:23–35
15. Kuzmanovic Z, Milosevic D, Rodic D (2015) Green parking garages: a case study of environmentally friendly multi-level parking facility in Chicago. Energy Build 86:699–710
16. Salvati L, Sateriano A, Oliva A (2017) Smart Pedestrian zones: a case study of Barcelona. Sustain 9(10):1693
17. Ousset Y, Grossard M, Laurent D (2019) Robotic inspection of the Paris sewer system. J Field Robot 36(1):191–202
18. Brown G, Smith J, Johnson M (2020) Smart sewage diversion systems: case studies in sustainable urban development. J Environ Eng 146(7):04020086
19. Smith J, Wang L (2019) Green infrastructure implementation on the great wall of China: a case study in sustainable heritage preservation. Environ Manage 45(3):567–578. https://doi.org/10.1234/envman.2019.567578

20. Tanaka Y, Takahashi M (2017) Sustainable tourism management: a case study of Kyoto's zoning regulations. J Sustain Tour 25(7):965–981. https://doi.org/10.1080/09669582.2016.1261407
21. García-Falcon JM, Méndez-García VH (2020) Sustainable tourism strategies: lessons from Barcelona. Sustain 12(15):6032. https://doi.org/10.3390/su12156032
22. Cerović M, Perić A (2018) Heritage-compatible architecture: a case study of Dubrovnik. J Cult Herit Manag Sustain Dev 8(4):416–433. https://doi.org/10.1108/JCHMSD-04-2018-0045

Bellandur Lake as an Urban Catchment: Urban Flooding Mitigation in the Post-Development Scenario

B. Sriharini, Jambavati Gouda, and Deepratrick Biswas

Abstract Cities today are facing the common problems of climate change and flooding risk. As urbanization increases, the amount of buildings and impervious surfaces also increases, this is combined with decrease in green and blue spaces The urban catchment of Bellandur Lake, located in the rapidly developing region of Bangalore, India, has witnessed substantial urbanization in recent decades. This transformation has raised concerns about the impact of developmental activities on the ecological integrity of the lake and its associated streams. This study seeks to comprehensively understand this complex interplay between urban development and natural hydrology. Method: The research employs a multi-faceted approach to achieve its objectives. Firstly, extensive geospatial analyses and field surveys are conducted to delineate the geographical extent of the Bellandur Lake catchment and to identify the stream orders. The streams are juxtaposed with urban development. The conflicts between them are identified through comparative analysis. Concurrently, a systematic review of relevant literature and case studies is undertaken to extract best management practices for post-development stormwater drainage. Results: From an understanding of the nature of conflicts between development and hydrology, suitable best management practices are identified to be applied to the urban catchment region of Bellandur.

Keywords Urban flooding · Stormwater management · Hydrology

B. Sriharini (✉) · J. Gouda
Manipal School of Architecture and Planning, MAHE, Bengaluru, Karnataka, India
e-mail: sriharinib@gmail.com

D. Biswas
Ernst and Young Global Limited, Delhi, India

A. K. Sharma and A. Joshi (eds.), *Proceedings of the 3rd International Conference on Opportunities and Challenges for a Resilient Future*, Advances in 21st Century Human Settlements, https://doi.org/10.1007/978-981-96-0931-4_8

1 Introduction

Urban areas globally face pressing challenges such as climate change and increased flooding risk, exacerbated by the surge in urbanization. The rise in impervious surfaces and diminished green spaces contribute to reduced urban resilience and flood resistance [4]. Traditional piped drainage systems, classified as grey infrastructure, have historically addressed these issues. However, there is a growing emphasis on holistic, nature-based solutions that offer benefits to the water cycle [1]. In the context of India, a developing country experiencing rapid urban growth and informal settlements, drainage systems often prove inadequate, leading to various concerns related to flooding and environmental health.

The region of Bellandur, Bangalore was chosen for the site of study.

2 Materials and Methods

To investigate the complex dynamics of urban stormwater management, a comprehensive methodology was employed. The primary objectives are as follows:

Objective 1: Understanding the urban catchment of Bellandur Lake.

The Digital Elevation Model (DEM) for the region is extracted via the SRTM satellite. It is subsequently analyzed in GIS software. Topography and contour maps are generated, and the catchment area is delineated. A flow direction map for this region is then generated. Stream orders are to be identified.

Objective 2: Understand the development within the region.

- From Google satellite and OpenStreetMap, building footprint data is extracted.
- From Google Earth satellite view through the years 2011—2023, the morphological analysis is done.

From open sources of Google Maps, and OpenStreetMap, the building footprints are mapped.

A brief morphological analysis of the area is done through historical images from Google Earth (2011–2023).

Objective 3: The conflicts of development and streams are identified based on outputs from objectives 1 & 2. Here, "conflicts" refer to possible areas where the process of urban development, which involves construction, infrastructure, and potential alterations to the landscape, could potentially interfere with or impede the natural flow of streams. This conflict may lead to issues such as flooding, erosion, or disruptions in the natural hydrological processes that the streams support.

Objective 4: Study of best management practices in post-development stormwater strategies:

Extracted via literature review and case studies.

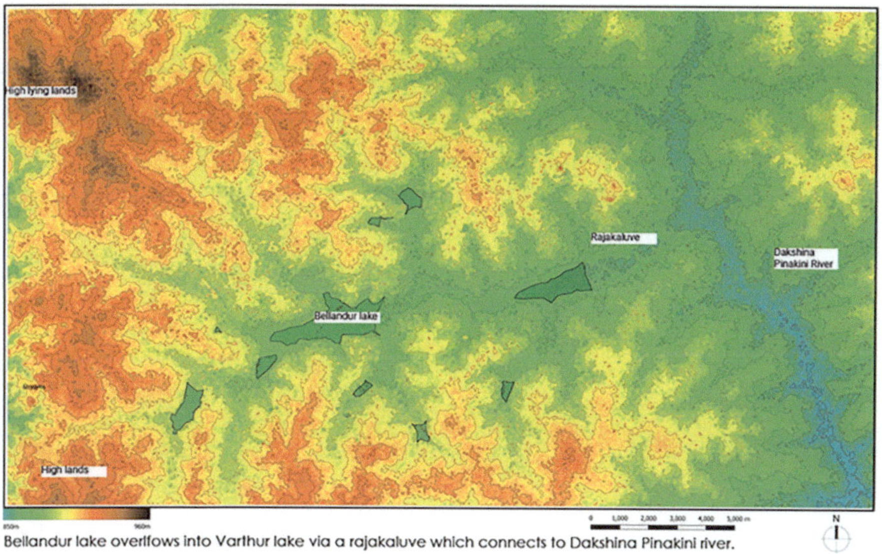

Fig. 1 Bellandur lake topography and context

Objective 5: Based on best management practices and the contextual study, the post-development strategies for Bellandur are identified.

3 Results

3.1 Delineation of Catchment Area

Topography and Contour Maps Generation: Once the DEM data is loaded into QGIS, topography and contour maps are generated. Topography Maps: These maps show the surface features and elevation of the area. DEM is used to create topographic maps that display hills, valleys, and other terrain features.

Contour Maps: Contour maps are a type of topographic map that uses contour lines to represent elevation. These lines connect points of equal elevation, helping visualize the shape of the terrain and its elevation changes (Figs. 1 and 2).

3.2 Identification of Stream Orders

Stream orders are identified from flow direction maps, extracted from DEM with a local drainage map as an intermediary step (Fig. 3).

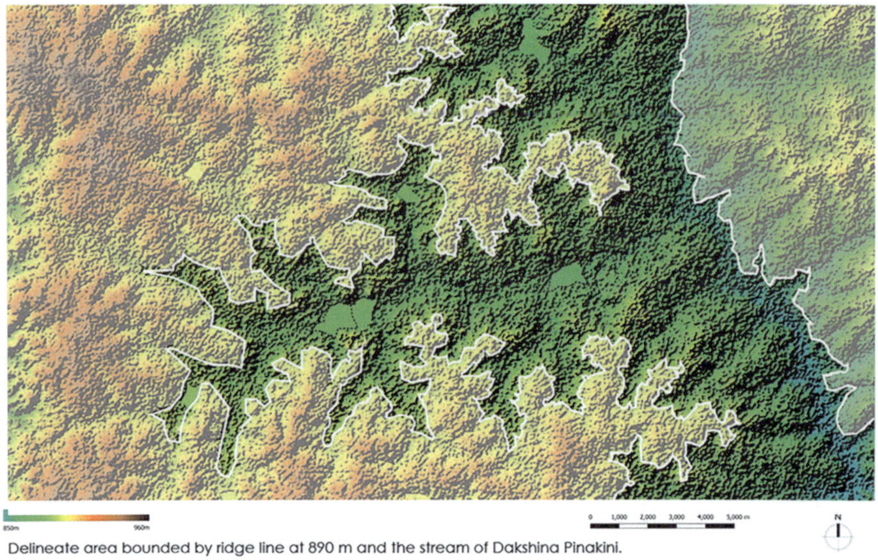

Fig. 2 Delineated catchment area

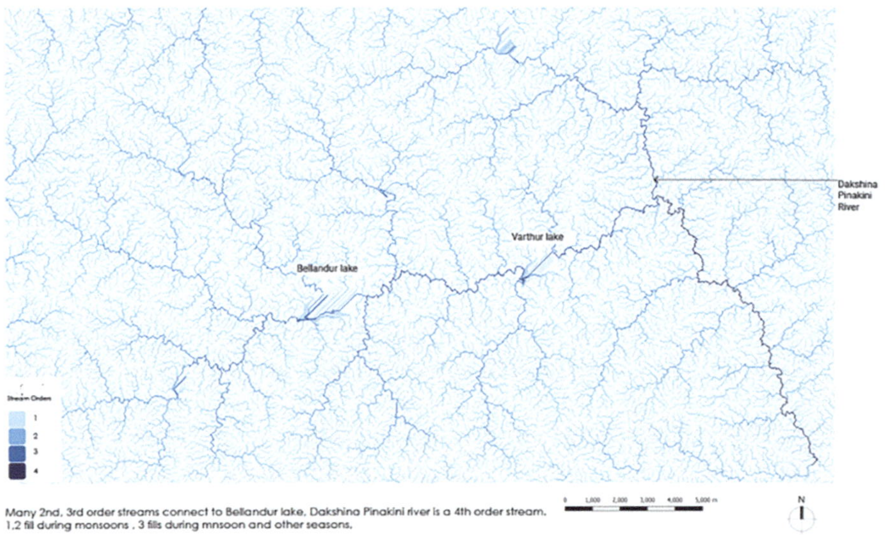

Fig. 3 Streams orders present in Bellandur area

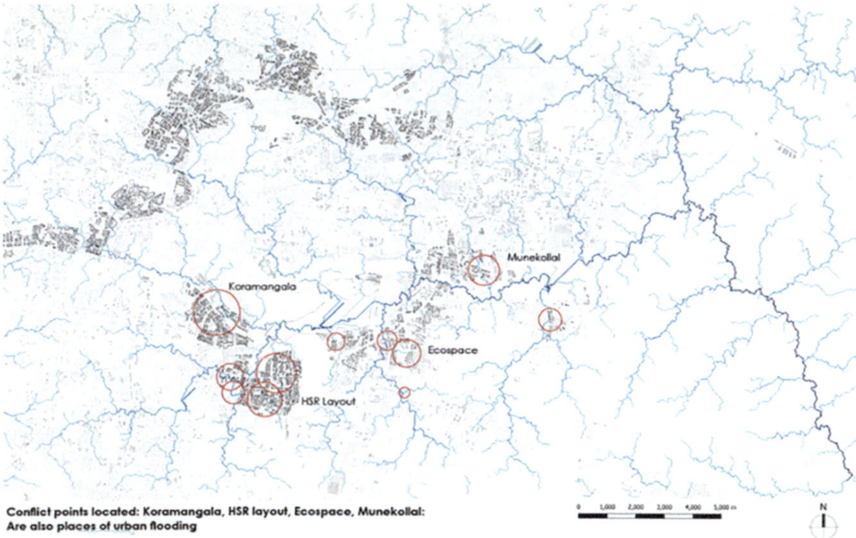

Fig. 4 Conflicts between streams and urbanization in Bellandur area

3.3 Identification of Conflicts Between Streams and Development

Development footprints are obtained via open street maps (Fig. 4).

3.4 Detailed Area Analysis: Koramangala

The marked-out areas of Koramangala, HSR Layout, and Munekollal were selected. These are investigated at a smaller scale, and the conflict points are more acutely highlighted (Figs. 5 and 6).

Findings in Koramangala:

- Stormwater management interventions found: The largest order (third order) stream is channelized into a man-made drain (K100 drain.) This is protected by concrete barriers. The channelization directs the stream path in a straight line along the road, and the natural meandering of the stream is reduced. Hence, stream velocity is increased.
- Softscapes such as BBMP Park, and Meshtripalya lake park serve as sinks or groundwater recharge points.
- Through news and literature studies it is understood that the Meshtripalya Lake has been slowly depleted. This could be because of developments interrupting the stream paths, hence the lake does not receive water.

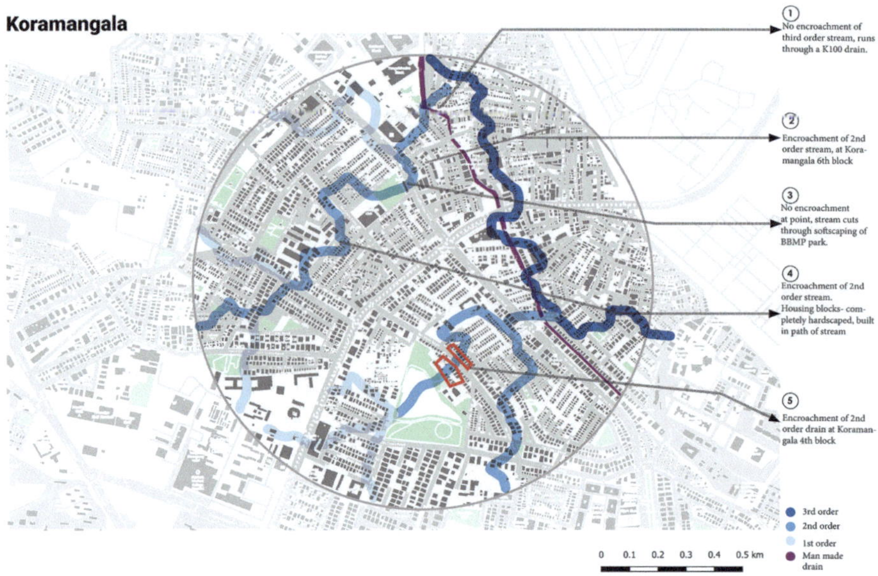

Fig. 5 Map showing conflict points between streams and urban development in Koramangala

- Most of the conflict points consist of housing blocks. Where entire parcels of hardscape interrupt the stream paths. This is seen in second and first-order streams.

  Owing to the similarity in findings between the detailed area analyses of HSR Layout and Munekollal in comparison to Koramangala, the discussion of HSR Layout and Munekollal has been omitted to avoid redundancy.

3.5 *Study of Best Management Practices for Stormwater Management in the Post-Development Scenario*

Historical Overview

Urban drainage systems have been utilized since 3000 BC, with early examples found in Mohenjo-Daro and Rome [2]. Up until the 1990s, the primary focus of the field was the conveyance of water away from the city. The Rio Declaration and Agenda 21 from the early 1990s brought attention to the need for long-term environmental sustainability. As a result, there is a significant movement nowadays to view stormwater as a 'resource' in urban settings. Aesthetics, multi-functions, and public acceptance are crucial components of this new strategy. Whilst traditional drainage systems solely focus on 'quantity'. Sustainable stormwater management focuses on 'quantity,' quality, and 'amenity' [3]. There has been a cultural shift in the discipline. Rather than looking at stormwater as a problem, Emerging practices consider stormwater as a resource and an opportunity [4].

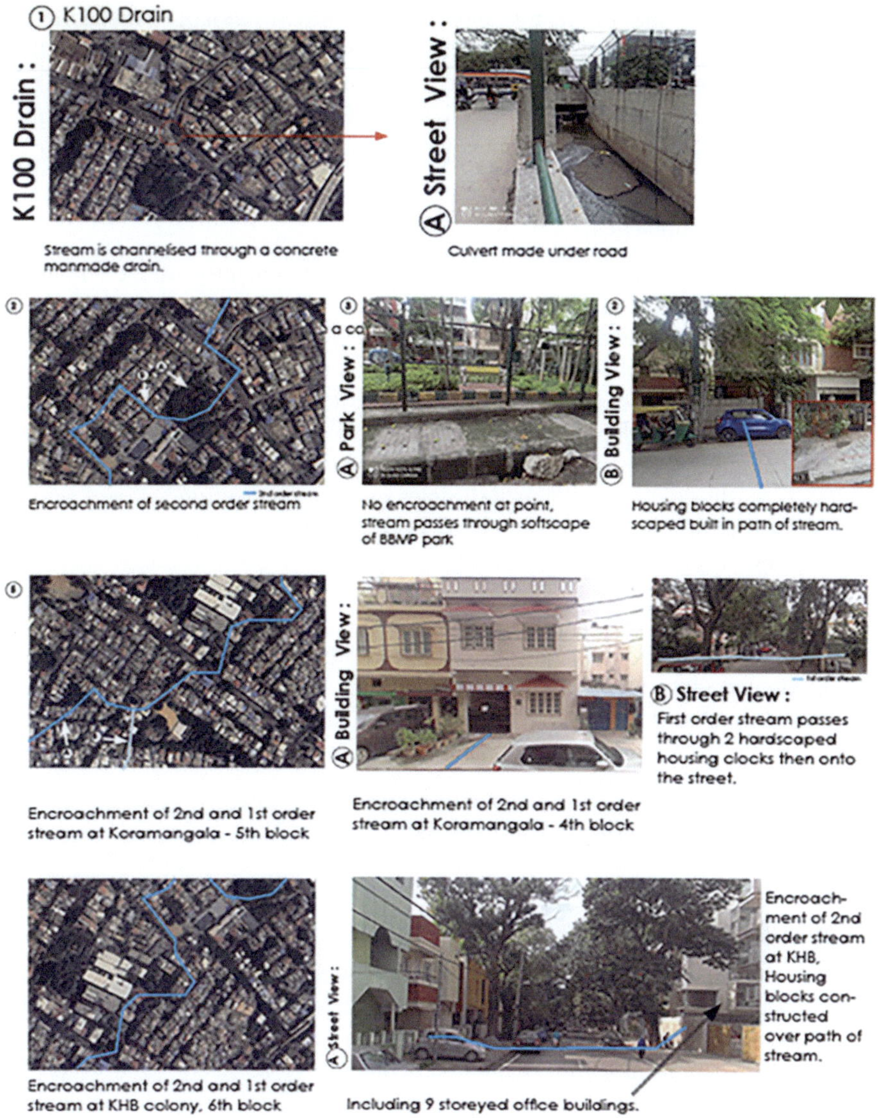

Fig. 6 Details of conflict points in Koramangala

Overview of Best Management Practices:

Low Impact Development (LID) represents a design philosophy that has gained traction not only in North America but also in various regions around the globe. It is known by different names in different parts of the world, reflecting similar underlying design principles. In New Zealand, it is referred to as Urban Design and Development (LIUDD), while in Australia, it is known as Water Sensitive Urban Design

(WSUD). In Europe, the comparable approach is termed Sustainable Urban Drainage Systems (SuDS). These methodologies encompass a range of strategies, potentially incorporating concepts like Integrated Urban Stormwater Management (IUSM) and Integrated Urban Water Management (IUWM). These various approaches collectively aim to promote sustainable and environmentally conscious practices in urban development and design.

Low Impact Development.

The term "low impact development (LID)" is predominantly utilized in North America and New Zealand. LID's original goal was to replicate 'natural' hydrology through site planning and integrated control strategies [4].

LID controls encompass small-scale stormwater treatment systems situated at or near the runoff origin, which promote both infiltration and evaporation [5].

Green roofs, rain gardens, bioretention cells, soakaways, swales, permeable pavements, infiltration basins, infiltration trenches, ponds, rain barrels or cisterns, tree box filters, curbless roads with swales, downspout disconnection, as well as other green infrastructures and even community education are some specific examples of stormwater controls used as part of LID.

Source Control:

The primary objective is to manage urban stormwater runoff as closely to the source as possible, to reduce the volume of water entering the drainage system. It consists of both structural and non-structural methods to manage urban stormwater. Source control primarily concerns measures that are employed on private lands, at a small scale [3] (Table 1).

Source Control: Site level (Private Lands)

See (Table 2).

On-Site Control

On-site control is concerned with holding and temporarily storing the runoff from various locations, such as agricultural land, municipal buildings, and roofs, to maximize infiltration and evaporation. In light of this, popular measures are ponds and soakaways. When utilized), this technique reduces high peak flow, and water flow into channels is slowed down. On-site control [3] deals with primarily public land, thus the site-level measures from the previous section are applied here at a larger scale (Table 3).

Slow Transport

Is concerned with stormwater being transported in drainage systems using slow transit methods. The primary goal is to lower runoff speed toward the downstream area to increase roughness by allowing runoff water to travel through vegetated ditches, strips, use of swales, and construction of a dyke on the channels [3] (Table 4).

Table 1 Source-control sustainable drainage methods: adapted from [3, 6–8]

Method	Principal	Illustration	Pre-development	Post development
Green roofs, planter boxes (above-ground measure)	Maximize green area	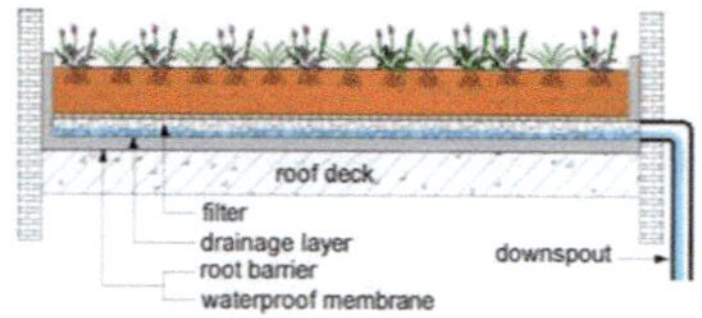 *Source* Managin urban runoff, singapore guidebook	✓	✓
Permeable paving	Infiltrate	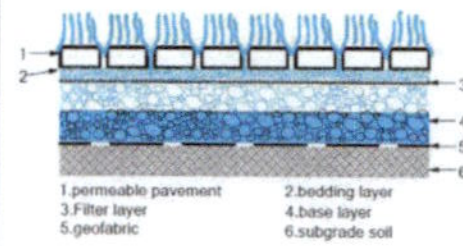	✓	✓
Flood barriers	Protecting building from flooding. (Receptor-based solution) mechanical barrier systems that are created to prevent water from flooding the protected zone behind the barrier		✓	✓

Table 2 Source control measures for sustainable drainage (site-level/private land), adapted from [6, 8]

Method	Principal	Illustration	Pre-development	Post-development
Retention basin	Creating storage at source through detention or retention basins could be an effective method to slow down peak flow for short time, high itensity storms		✓	✓
Rain garden	Vegetated depressions created to hold and treat stormwater	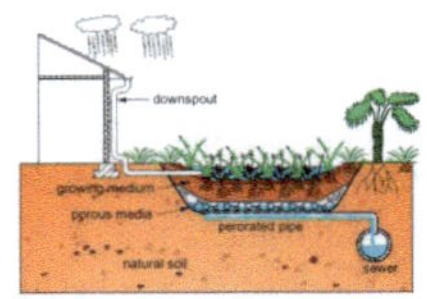 	✓	✓

Table 3 On-site control measures: adapted from: [3, 6]

Method	Principal	Illustration	Pre-development	Post-development
Vegetated strips	Promotes infiltration by distributing rainwater uniformly as an overland sheet flow through a lightly sloped vegetated area	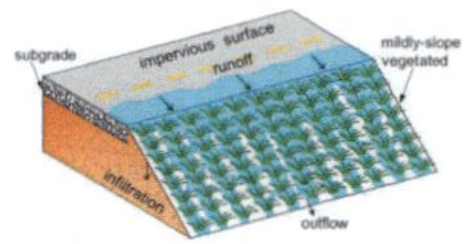	✓	✓
Rain garden	Vegetated depressions created to hold and treat stormwater	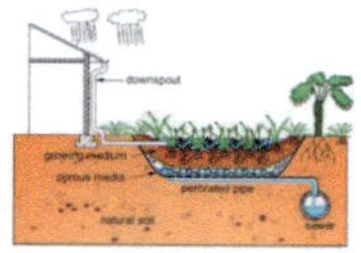 	✓	✓

(continued)

Table 3 (continued)

Method	Principal	Illustration	Pre-development	Post-development
Permeable paving	Infiltrate	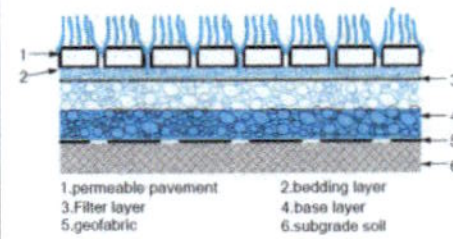	✓	✓
Ponds	Retention ponds have a permanent supply water at ground level. The pond's inflowing water's velocity lowers as it approaches the body of water, helping particulates to settle out of suspension		✓	✗

Table 4 Slow transport measures: adapted from: [3, 6]

Method	Principal	Illustration	Pre-development	Post-development
Swales	Excavated lands backfilled with vegetated surface layer and filter intermediate layer upon natural soils to pond rainwater for infiltration	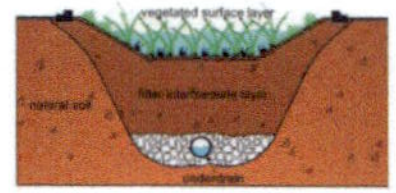	✓	✓
Preserve undisturbed upstream catchments	Innovative urban planning which protects upstream catchments, provides better quality drinking water and mitigates eutrophication of surface water	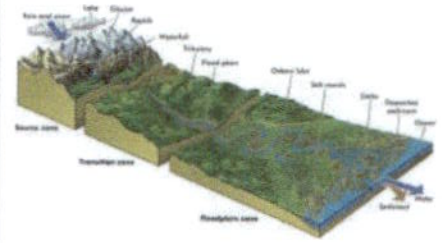	✓	✗
Bank stabilisation	To prevent further erosion of urban stream banks, stuctural materials, bio-engineered products (like coconut fibre rolls, and streambank re-grading) are employed	 Source: Restoring streams in an urbanizing world [9]	✓	✓

(continued)

Table 4 (continued)

Method	Principal	Illustration	Pre-development	Post-development
Channel reconfiguration & grade control	This method improves water transportation in flood-prone locations, rebuilds deeply incised channels, and improves streambank and soil stability changes are made to the longitudinal or channel plan forms, or a covered channel is re-exposed to face the sun	 A channel reconfiguration with reshaped/regraded banks and a meandering channel Source: Restoring streams in an urbanizing world [9]	✓	✗
Riparian replanting & management	Riparian planting improves bank stability and also enhances aquatic biodiversity	 Riparian vegetated buffer on bear creek in story County, Iowa *Source* LYNN BETTS/USDA	✓	✗

(continued)

Table 4 (continued)

Method	Principal	Illustration	Pre-development	Post-development
Treatment trains (assembly of past listed measures)	Large single stormwater facilities are not employed for wider areas; instead, 'treatment trains' of separate, tiny ponds and basins are used within developing catchments. This is done to replicate the hydrological cycle in nature Stormwater in such "treatment trains" travels through a number of buildings (grass swales, sand filters, artificial wetlands), before entering a quality-control building		✓	✓

Table 5 Downstream Control Methods, Adapted from: [3, 6–8]

Method	Principal	Illustration	Pre-development	Post-development
Ponds	Constructed basins that have a permanent pool of water throughout the year (or at least throughout the wet season)		✓	✓
Wetlands	Design includes vegetation and shallow zones. Stormwater eliminates contaminants through biological absorption and settling as it passes through the wetland		✓	✗
Lakes	Natural retention		✓	✗

Downstream Controls

In large-scale facilities, stormwater is managed in the lowest portions of the drainage system. Examples of these facilities that temporarily store stormwater include wetlands, large ponds, and lakes (Table 5).

4 Conclusion

The identified strategies for the Bellandur urban catchment region post-development are comprehensive and cover various levels of intervention. These strategies aim to address different aspects of stormwater management to mitigate the impact of urbanization on the natural water systems.

1. Source Control (Building Level): Implementation of green roofs and permeable paving can effectively reduce runoff and promote natural infiltration. This approach targets the initial point of contact with precipitation, allowing for direct absorption.

2. Source Control (Site Level): Retention basins and rain gardens serve as localized storage and treatment areas for stormwater. These features help regulate flow rates and improve water quality by allowing sediments and pollutants to settle before discharge.
3. On-Site Control (Public Lands): Vegetated filter strips, rain gardens, and permeable paving on public lands can collectively act as a secondary line of defense. They further enhance water quality and reduce the volume of runoff entering natural water bodies.
4. Slow Transport: Treatment trains, bank stabilization, riparian management, planting, swales, and vegetated ditches work together to slow down the movement of stormwater. These features provide a series of treatment and detention steps, allowing for natural processes to take effect.
5. Downstream Controls: Ponds, basins, and lakes serve as large-scale storage and treatment facilities at the downstream end of the catchment. They offer significant capacity for flood control and act as final polishing systems before water is released into the natural environment.

These strategies collectively form a holistic approach to managing stormwater in the Bellandur urban catchment region. By incorporating a combination of source control, on-site management, and downstream controls, this framework aims to balance urban development with the preservation of natural hydrological systems. However, successful implementation will require collaboration between various stakeholders, including local authorities, developers, and the community, as well as ongoing monitoring and maintenance efforts.

References

1. Davis M, Naumann S (2017) Making the case for sustainable urban drainage systems as a nature-based solution to urban flooding. pp 123–137. https://doi.org/10.1007/978-3-319-56091-5_8
2. Burian SJ, Edwards FG (2002) Historical perspectives of urban drainage. Glob Solut Urban Drain 1–16. https://doi.org/10.1061/40644(2002)284
3. Stahre P (2004) Integrated planning of sustainable stormwater management in the city of Malmo, Sweden
4. Fletcher TD, Shuster W, Hunt WF, Ashley R, Butler D, Arthur S, Trowsdale S, Barraud S, Semadeni-Davies A, Bertrand-Krajewski JL, Mikkelsen PS, Rivard G, Uhl M, Dagenais D, Viklander M (2015) SUDS, LID, BMPs, WSUD and more – the evolution and application of terminology surrounding urban drainage. Urban Water J 12(7):525–542. https://doi.org/10.1080/1573062X.2014.916314
5. Eckart K, McPhee Z, Bolisetti T (2017) Performance and implementation of low impact development–a review. In: Science of the total environment, vols 607–608. Elsevier B.V, pp 413–432. https://doi.org/10.1016/j.scitotenv.2017.06.254
6. PUB, the national water agency, & The Institution of Engineers Singapore (IES) (2013) Managing urban runoff-drainage handbook
7. Qin Y (2020a) Urban flooding mitigation techniques: a systematic review and future studies. Water (Switzerland) 12(12). MDPI AG. https://doi.org/10.3390/w1212357946

8. Qin Y (2020b) Urban flooding mitigation techniques: a systematic review and future studies. Water (Switzerland) 12(12). MDPI AG. https://doi.org/10.3390/w12123579
9. Bernhardt ES, Palmer MA (2007) Restoring streams in an urbanizing world. Freshw Biol 52(4):738–751. https://doi.org/10.1111/j.1365-2427.2006.01718.x
10. Parkinson J, Tayler K, Mark O (2007) Planning and design of urban drainage systems in informal settlements in developing countries. Urban Water J 4(3):137–149. https://doi.org/10.1080/15730620701464224
11. Perales-Momparler S, Andrés-Doménech I, Hernández-Crespo C, Vallés-Morán F, Martín M, Escuder-Bueno I, Andreu J (2017) The role of monitoring sustainable drainage systems for promoting transition towards regenerative urban built environments: a case study in the Valencian region, Spain. J Clean Prod 163:S113–S124. https://doi.org/10.1016/j.jclepro.2016.05.153

Governance Issues in Women-Friendly Development of Old City Shahjahanabad

Mani Gupta, Rashmi Ashtt, and Ajay Monga

Abstract In both theory and practice, the role of women in shaping cities is under emphasized. It is commonly known that women are less mobile than men are, due to a fear of violence, which limits their utilization of public areas. There are many difficulties that women encounter in cities as they begin to leave their homes in their early teens and progress through different stages of life. Comparing cities, streets, and public places in terms of accessibility, comfort, sense of security, and sense of belongingness has also been used to explain this at all levels regarding the physical environment. In the 1970s, the feminist criticism of urban theory and planning arose, demonstrating how designers and planners had created environments that were mostly directed toward the needs of men. However, the urban environment is constantly evolving, and today there are some steps being taken in a positive direction & specifically when it comes to old settlement in core areas like Shahjahanabad which is the oldest area of Delhi from Mughal era which has eventually transformed into slum like condition. This further adds to the vulnerability factor for the women residing & visiting in the core as well as the vicinity. The paper is an attempt to highlight the conditions which arise in old settlements like Shahjahanabad. It is also an effort to examine tools and strategies adopted worldwide in cities that could significantly contribute to realizing women's right to cities and giving them equal opportunities, safety & security. According to the Various reports like UN-Habitat report, NCRB report the situation is bad and non-livable for the women friendly environment. There is always a threat of eve teasing, human trafficking, chain snatching, theft etc. These poor conditions seem to be because of poor governance system in the area as there are many authorities & administrative bodies; ASI, MCD, WAKF BOARD, Chandni chowk trust societies, local governing bodies, etc. which are governing the old Shahjahanabad area which falls under non development zone governed by ASI. The case is made for explicitly integrating gender perspectives

M. Gupta · R. Ashtt (✉)
Department of Architecture and Planning, Indira Gandhi Delhi Technical University for Women, Kashmere Gate, Delhi, India
e-mail: rashmiashtt@igdtuw.ac.in

A. Monga
Department of Architecture and Planning, D.CRUST, Murthal, Sonipat, India

A. K. Sharma and A. Joshi (eds.), *Proceedings of the 3rd International Conference on Opportunities and Challenges for a Resilient Future*, Advances in 21st Century Human Settlements, https://doi.org/10.1007/978-981-96-0931-4_9

into developmental policies, which will address the issues & give recommendations regarding women sensitive planning of core areas.

Keywords Built environment · Gender equality · Women friendly · Social infrastructure · Policy and development · Tools and approaches · Governance

1 Introduction

Architectural and topographical arrangements have reinforced the status distinction between men and women throughout history and across cultures. Being a social art, architecture not only creates shelters but also translates abstract ideas or societal value systems into a tangible, observable world. Gender has a significant impact in the goals, interventions, and attitudes that construct this built environment. Feminists frequently argue about how a woman's life is portrayed in cities. To understand how much women, belong to these areas of the city, it is important to observe how they relate to and experience the city in public settings. Particularly in metropolitan areas, experience of the individual and collective, their memories, and use of public space ([1], p. 123).

1.1 Relationship of the Gender to the Built Environment

The sociocultural classification of men and women as well as the social duties that are allocated to them are referred to as gender. The constructed environment influences how we use space and vice versa, giving the location its unique personality. This includes examining the user's gender-specific behavior patterns, especially in public areas. Human behavior is influenced by the arrangement and providence of specific spatial components, which gives people chances to behave in particular ways. The interaction between users and the built environment is influenced by many factors, including visibility, enclosure, perception of security and safety, accessibility and movement, and sense of belonging. Women's behavior patterns in public spaces indicate that they enter the public realm with a specific objective in mind and exit as soon as it is completed. One can learn more about gender behavior in public spaces by observing how people walk. In India, a woman with a family is typically expected to work professionally in addition to doing household duties and providing care for dependents like children or elderly family members. As a result, their movement in the city is more complicated, consisting of a series of brief trips connected to their main or primary destinations, such as a trip that starts at home, stops to pick up a child from day-care, stops to shop for groceries, and then either continues to work or goes back to their home. When we further examine the movement pattern, we can see that, while women's paths may be more numerous, they are typically shorter and straighter than men's, which may be more aimless and meandering. In order to feel comfortable and secure, the travel pattern also reveals that women follow the designated path to their destinations rather than frequently passing through male-dominated areas [2].

Indian cities do not have a culture of exclusive public space, most of the public space occurs near institutions, markets, or public buildings. The so called 'public nodes' were clubbed with these functional or institutional zones and came to be known as public spaces of the city. These spaces are busy in terms of activities, which provide opportunity for women to socialize with other women, as they are primarily a shopping or religious place where women come out from their private to a public domain [3].

The ways that men and women navigate public space are clearly different. Men place themselves anywhere in a public area, but women seek out enclosed areas that are less exposed and apparent to male gaze. Men always show a higher level of comfort and a stronger sense of belonging when using the place than women do. Women tend to move more quickly when they are alone and in quiet settings, however when they are with other people or in busy areas, they take their time, feel at ease, and form a sense of belonging. Understanding these differences in how men and women use urban space is essential to developing gender-responsive planning and policy decisions. Jama Masjid Area & Meena Bagh area in Shahjahanabad (Old Delhi) are examples of public spaces where women feel highly unsafe. The nightlife of Old Delhi is mostly lived out on the streets and not in pubs or discos. The liquor and sheer of drunkards in public places are a significant factor to women crime in old Delhi [4].

2 Case of Shahjahanabad, Old Delhi, Delhi

2.1 History of Planning

Shahjahanabad is renowned for its bazaars, street cuisine, shopping areas, and Islamic architecture, serving as the metropolis of Delhi's symbolic centre [5].

Shahjahanabad, the imperial capital of the Mughal Empire (1628–58), was constructed between 1639 and 1648 and covered a sizable area along the Yamuna River's banks in the Delhi triangle's southeast [6]. Between the reign of Humayun's Delhi and the arrival of Shahjahan, the renowned Mughal architect who created Shahjahanabad, the seventh city of Delhi, in 1648, Delhi experienced something of an eclipse. More people can see Shahjahan's Delhi today than any of the Delhi is built before it. As evidenced by the Red Fort and the Jama Masjid, the scale on which he erected was likewise more heroic [6] (Fig. 1).

Heritage of the city

In addition to the living social heritage of Shahjahanabad, there is the architectural heritage. The heritage of evolving residential architecture and the heritage of the urban shape of the entire city are both found beyond the monuments of the Jama

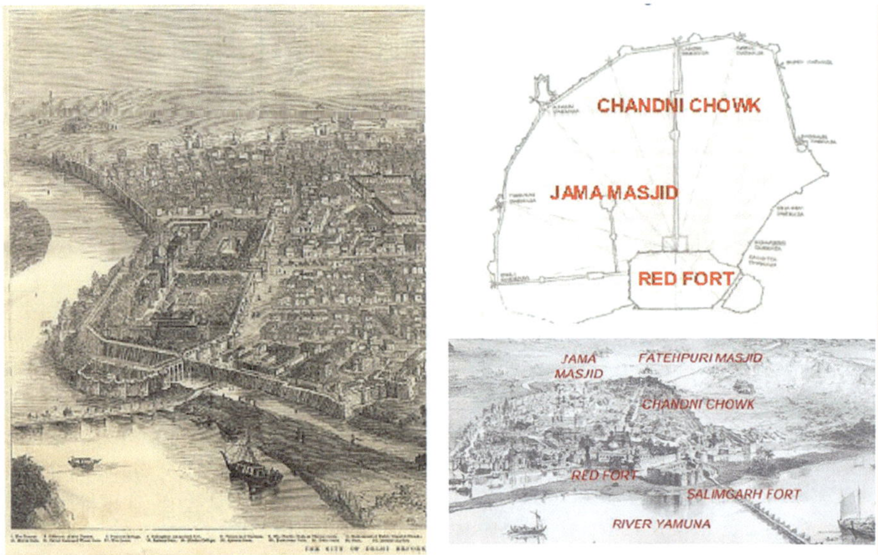

Fig. 1 Old map of the city, map of heritage buildings in Shahjanabad; *Source* Google

Masjid, Red Fort, and the other smaller ancient structures. The historic preservation of a national heritage is involved. There are havelis (houses of noble families) [7], mahals (palaces) and katras (four-sided courtyard homes) in old Delhi of such beauty that, but for their continued occupation, they could be dis-mantled and reassembled in the National Museum. Great Mughal and Hindu traditions of house design and construction are embodied in their form. The overall morphology of the city is a rare example of an evolved indigenous tradition of urban form [8].

A second aspect of the architectural heritage is its contemporary social significance. Katra and haveli style houses continue to respond to the social requirements and cultural traditions of their residents. Both limit access of strangers to the home while granting the benefits of sunlight and greenery. Since the majority of old Delhiites prefer to maintain traditions of purdah, the internal courtyard space is especially important for women and children. The katra style house form is particularly conducive to parosi (neighborly) relations which develop between residents which share the common courtyard [9].

Inside, the haveli or katra space can be further divided in accordance with cultural attitudes about social relations between the various members of extended families [10].

Walled city

The walls and some of the gates from the walled city of Shahjahanabad still exist. Chandni Chowk and its side streets are the greatest places to enjoy the romance of Delhi's bazaars. A well that was nearly ten kilometers long was used to secure and confine Shahjahanabad. Ten gates served as the region's gateway to the city. In addition to Delhi Gate, Lahore Gate served as the Red Port's principal entrance [11]. The city's other primary access points to the motorways were the Kashmere Gate, Calcutta Gate, Mori Gate, Kabul Gate, Faresh Khana Gate, Ajmere Gate, and Turkman Gate. To fit the homogeneous community structure, a system of mohallas and katras was created. The Lal Jain Mandir from the time of Shahjahan, Appa Gangadhar Mandir (Gauri Shankar mandir), the only temple constructed during Marathi dominion, Arya Samaj mandir (Dewan Hall), Baptist Chruch, Gurudwara Sisganj, Sunehri Masjid, and west end terminus, the Fateh Puri Masjid, are just a few examples of Shahjahanabad's fine display of secularis After defeating Mohammad Shah at Panipat on March 9, 1739, Nadir Shah entered Delhi. He slaughtered the locals and seized practically all the treasure amassed by the Mughals in India at Shahjahanbad [12]. Among the loot carried were the Peacock throne, rare stones like the Koh-i-Nur and Darya-i-Nur, exquisite works of art, thousands of horses, camels, and elephants, as well as many books and manuscripts.

Community pattern

Most of the mohallas (neighbourhoods) of old Delhi are socially cohesive, culturally valued residential environments. They provide residents with face-to-face familiarity with all their neighbours. Residents tend to share a traditionally evolved code of behaviour. There is a strong identification and intimacy of person with place due to the human scale of the built form. This form also creates an interplay of light and shade in the courtyards and lanes which is aesthetically pleasing to residents. Most mohallas are largely free of serious crime against people or property. Residents' familiarity prevents outsiders from intruding. Most neighbourhoods of old Delhi have formal associations for self-government, either Mohalla Sudhar Committees, Mohalla Samitis or community panchayats [13].

These serve many functions. They act as the representative of the mohalladhar (residents) to civic authorities such as the police, electric supply, water supply officials. They maintain social order through formal and informal systems of sanctions [14]. They sponsor social welfare programs for widows, orphans and the poor. Some give scholarships to needy students of the locality. They supply facilities and utensils for weddings and other family celebrations of the mohalladhar. They support social activities like festival feasts, athletic clubs and carrom competitions. Many associations give aid to the local temple or mosque. In short, mohallas of old Delhi are self-contained, intimate social environments of families who have known one another for many years. It is critical to note that, for most of the people who live in the mohallas of old Delhi, this is a culturally preferred way of life. It is a life style which they do not wish to change. They do not consider their homes and neighbourhoods to be slums. To old Delhi mohalladhar, the colonies of South and West Delhi may be wealthy, spacious and green, but they are socially sterile. And sociability is highly valued in old Delhi [15] (Fig. 2).

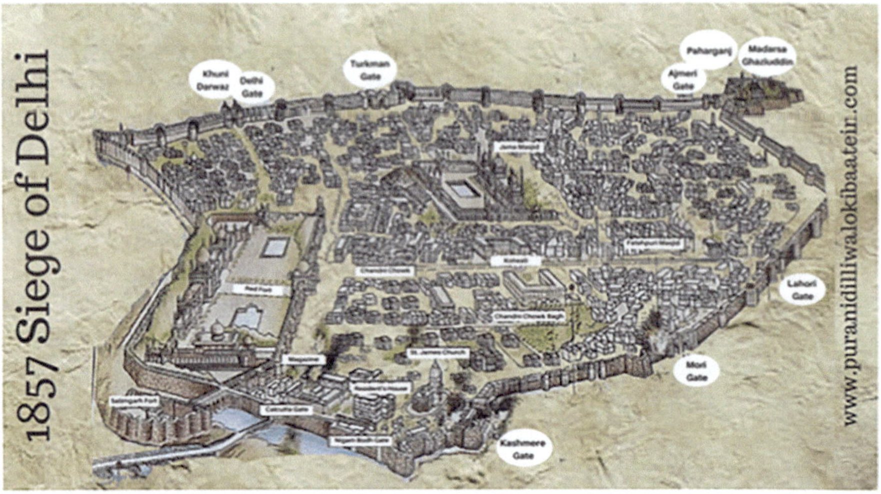

Fig. 2 Map of walled city; *Source* Google

Economic function of the city

Old Delhi is a central provider of goods and services to the Delhi-New Delhi metropolitan area, as well as the region [5]. It is the wholesale center for dried fruits, grains, spices, cloth, housewares, electrical goods, bicycles and plumbing supplies, among other items [16]. For many items, Delhi is an interchange point for products which do not even service the city itself [17]. Additionally, it serves as a significant "cottage industry" hub for book binding and printing [18]. Chandni Chowk, Nai Sawak, Chawri Bazaar, and Lajpat Rai Market are its principal shopping districts [19]. In the Meena Bazaar, Urdu Bazaar, and Matia Mahal Chitli Qabar Bazaar complexes, it provides speciality goods for Muslims [20]. A major pharmaceutical company is in Lai Kuan (Wikipedia, n.d.). It is also a manufacturing center for shoes, sandals, wooden crates, auto luggage racks, motor supplies, books, oils and perfumes, small scale metal components, plastics, bangles and woven materials [21]. It is the hand-manufacturing center (handiwork and crafts) in the areas of embroidery, leather, brass, copper, silver, gold and other jewellery, wooden and ivory handicrafts, cane furniture, and drama costumes [16]. It produces famous sweets, namkins and other foods for city-wide distribution [20]. Old Delhi is probably the major recycling center of the metropolitan area [5]. Within the walled city, used newspapers are turned into bags, auto and machine parts are reprocessed, bottles are sorted and recycled to factories, tin cans become broom handles, and animal by products are processed into fertilizer and other derivatives (Wikipedia, n.d.). Old Delhi is the major distribution center for meat, fish chickens and eggs [19]. A major fruit and vegetable market is in Darya Ganj [17]. Sunday markets at Red Fort and Chandni Chowk are centres of the informal retail sector as are the major bazaar streets of the city [18].

Despite its social isolation, it should be clear that Shahjahanabad has, and has always had, a strong commercial role in the region [21].

2.2 Status of Women in the Old Time

To stay in purdah, women were urged (veil). Their faces had to be always hidden. The presence of the purdah system prevented them from having the freedom to live their lives as they pleased. It was forbidden for the women to move freely. They were accompanied by others if they left their dwellings. Participation in social, cultural, economic, and political activities was forbidden. They were advised to stay within the confines of the home [22]. The purdah system denotes the separation of women from the general public using their attire as well as high-walled enclosures, screens, and drapes with their homes [23].

Meena Bazaars, also known as Kuhs Ruz ("Day of Joy"), were held only for women during the Mughal era, apart from the emperor and a few princes. Throughout the Norouz (New Year) holiday, the bazaars lasted for 5 to 8 days. They were initially planned by Emperor Humayun, but Akbar and his successors increased their complexity. The marketplace was only open to the emperor, princes, and a select group of nobility [24]. As is obvious from the name Meena, Meena Bazar's goal was to increase women's employment. The products were pricey and restricted to Indian sultanates. Due to culture and tradition, women's conditions at the period were also declining [25].

2.3 Evolution with Time

The Walled City of Delhi (Shahjahanabad), frequently referred to be the essence of Delhi, is fading in spite of the last 50 years of planning. It is deteriorating, and its illustrious past is passing away [26]. The Walled City, whose population was initially projected to be around 60,000, reached its maximum point of 4.2 lakh people in 1961 before dropping to 3.50 lakh in 1991 and 2.35 lakh in 2001 [27]. The population has been falling mainly due to commercial operations expanding and quality-of-life degrading. The Walled City has becoming progressively more crowded with trade and business, which now numbers 46256 or almost 7% of all current establishments in Delhi [28]. In contrast to the city's 2% population, the Walled City houses 60% of wholesale trade, 25% of retail trade, and 28% of informal trade and services [29]. The Walled City has a total area of 569 ha, of which 180 ha (31.8%) is used for residential purposes, 88 ha (17%) for commercial purposes, 22.60% for transportation (roads and trains), and 9% for other purposes. (1.6%) [27]. The overcrowding is primarily caused by I the fact that railway establishments account for a sizable portion of circulation, (ii) the fact that commercial establishment densities have increased significantly over time, and (iii) the fact that the number of workers

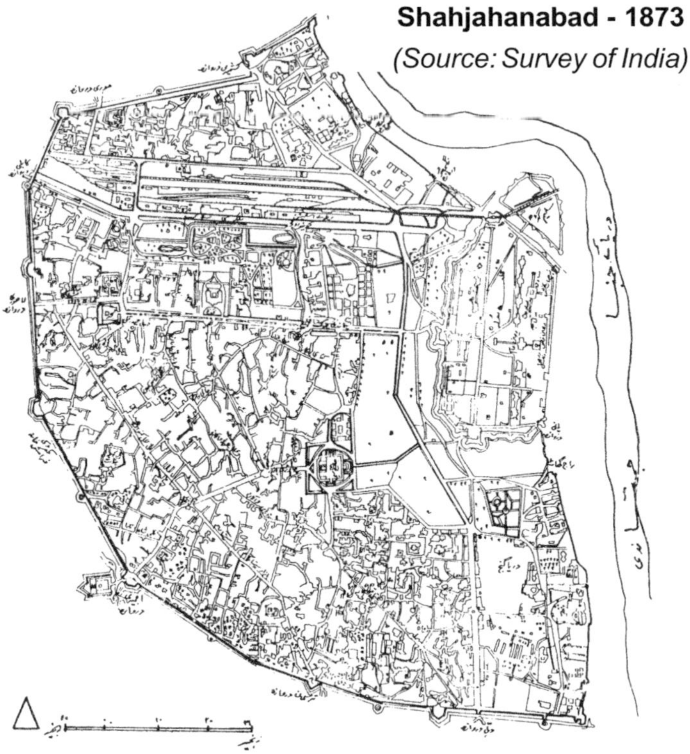

Fig. 3 Map of early nineteenth century. Map of 1873; *Source* Google

per acre in commercial, industrial, or trade establishments has increased significantly [26]. Delhi's Old City is where wholesale markets specializing in practically every commodity are concentrated. Historic structures, monuments, and urban history are abundant throughout the Walled City. There were 1321 historical monuments in Delhi according to the Archaeological Survey of India in 1914, 1208 of which are still standing today [30]. Of these, 411 are in the Old City, of which 41 are designated as protected monuments [29]. Under Section III of the Slum Area (Improvement and Clearance) Act of 1956, the Slum Department of the MCD is responsible for managing around 3,500 evacuee properties, or "Katras" [31] (Figs. 3 and 4).

2.4 Existing Land Use & Built-Up

The following map shows the landuse of Shahjahanabad. Heritage monuments consist of 14% and Gateways consists of 9% of total site area. Further the landuse is predominantly Residential with 20%, then commercial with 14%, Mixed use with 12% and public semi-public with 7%.

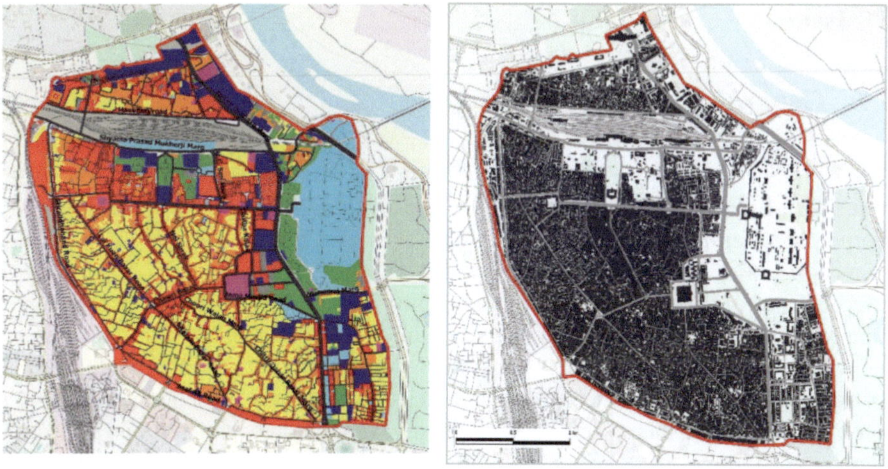

Fig. 4 Existing landuse map of old Delhi; Existing built up of old Delhi. *Source* Author

53% of the total area of Shahjahanabad is covered by built-up while 47% of the area is an open area. Most of the open area is available in the Northeast region due to the presence of Red Fort (Fig. 5).

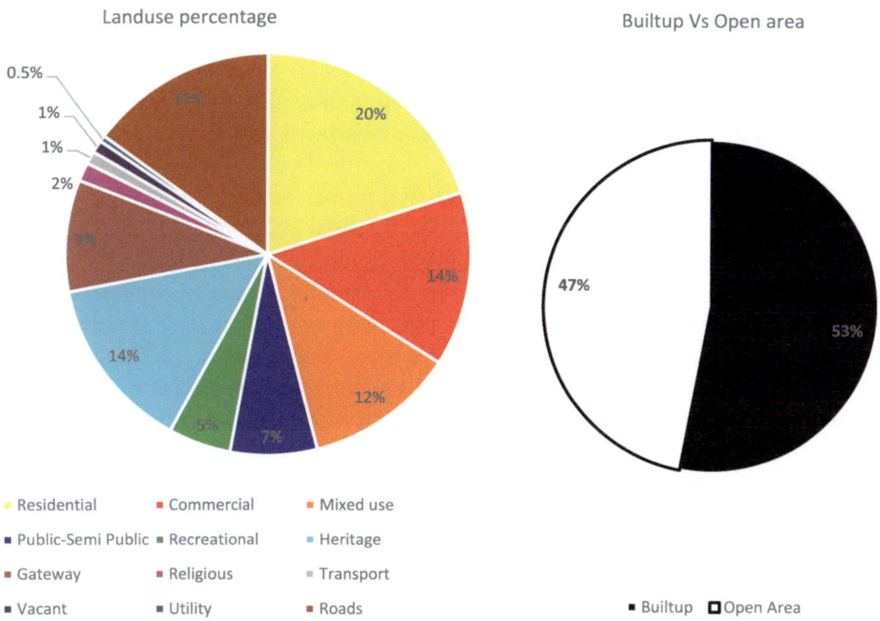

Fig. 5 Land use percentage. Built-up versus open area

3 Issues

3.1 General Scenario

The interaction between the user and the built environment is also influenced by time. Men are seen reading newspapers and drinking chai in the morning while women are seen only passing through. The group of women who are seen sitting, conversing, and shopping fill this area much more as the day goes on. The space is utterly empty of its temporal features in the morning because all the shops are closed (apart from two paan shops), and this situation reverses as the day goes on. Women feel more at ease occupying these locations when the market is active and there are other people around. Women are also spotted avoiding the area just outside the king's tomb entrance since there is a paan shop there. Men typically hang around in groups there at all hours of the day, and at night the pedestrian lane leading to the chowk is virtually empty save from parked cars. In contrast to the area next to the entrance to the king's tomb, a woman user would much prefer to wait outside the entrance to the queen's tomb where she would be more comfortable because a variety of people pass that area often, it is better shaded, and it has sitting provisions. The light from the light poles during the night is kept from making the lane dark and dingy by the overhangs that shade the lane during the day. Women can be spotted avoiding this road and only using the other street, which has a few open shops and makes them feel safer.

Gender and space are not two distinct topics, but rather highly significant relationships that should be strengthened by creating social places that are considerate of all of their users. In order to prevent the creation of certain unwanted settings and contribute to the development of a better neighbourhood, it is crucial for designers to recognise such patterns and the relative arrangement of a space. Today's cities are clearly not designed with women in mind, and crimes against women are steadily rising. Crimes against women climbed 15.3% between 2020 and 2021, according to the annual "Crime in India" report from the National Crime Record Bureau.

According to a 2008 Indian Express survey, women in small towns experience significantly less insecurity than those in large cities or rural areas. The poll divides women into groups according to age, income level, whether they are single or have children, whether they work, and their community. It focuses on some of the most vulnerable groups of women that require special policy attention: poor women living in big cities are particularly vulnerable to sexual violence due to a lack of safe housing and inadequate shelters; single working women and young women under the age of 25 feel much more unsafe than the average in all kinds of localities; young Muslim women feel particularly unsafe despite the lack of a clear community pattern relating to their level of insecurity; and poor women who are Muslim themselves are among the most vulnerable groups of women. (1) The survey helps us to understand that there are various kinds of issues faced by females of different age groups related to planning of urban areas, accessibility to public spaces and lack of public toilets.

Though there are various issues related to women safety, only the ones that are related to planning and the built environment have been discussed below.

3.2 Current Scenario of Women in Shahjahanabad

At least one woman is raped in the nation's capital every 24 h, according to a 2004 police report from Delhi (The Times of India 2004). According to a survey by the Delhi Commission for Women, women and girls from lower socioeconomic classes are more likely to be raped, even if middle-class women are at best given insufficient protection. Delhi has the highest rate of crimes against women in both the public and private realms, per the National Crimes Records Bureau (NCRB 2007). The most offending city among the 35 major cities was Delhi surveyed by the NCRB in 2007., which "accounted for 29.5% of rape incidents, 31.8% of kidnapping and abduction cases… and 21.5% of molestation instances" While other crimes against women, such as dowry killings, domestic violence, and psychological torture, rank Delhi as the worst Indian city, rape, kidnapping, and molestation commonly occur in the city's public areas and make up a significant amount of the city's reported crime. In a recent survey conducted by the Centre for Equity and Inclusion (CEQUIN) and the Centre for Media Studies (CMS) in 2009, 96% of the 630 women between the ages of 12 and 55 from all socioeconomic demographics in both New and Old Delhi stressed that they particularly feel threatened in the well-known markets of Chandni Chowk, Connaught Place, and Karol Bagh as well as in the public buses (Indo-Asian News Service 2009). One of the survey's more unsettling findings further explains that 60% of these women are most worried about the safety of young girls, and 88% agreed that public buses are the riskiest form of transportation in the nation's capital. In addition, according to the research, 88% of the women believed that, if they were harassed or attacked in the city's public places, onlookers and other members of the civic community rarely provided them with support or protection. The women are also very hesitant to trust the Delhi police since they anticipate more sexual harassment from them.

3.3 General Governance Policy of Shahjahanabad

- There aren't any schools in the area, which necessitates the opening of more.
- A housing strategy that encourages mixed-community housing must be established in order to prevent ghettoization.
- Because there aren't enough places for women and kids to relax in the neighbourhood, appropriate parks and gyms must be included during city design.
- To lessen traffic congestion and pollution, there must be a set time window for the entry of goods-carrying vehicles for loading and unloading.

- A comprehensive policy that safeguards historic sites and structures from encroachment and demolition should be developed and put into place on a national scale.
- The area needs to be equipped with fire safety and prevention tools due to the area's intricate network of congested buildings and limited roadways.

3.4 Issues Related to Planning with Respect to Different Age Groups & Gender

Impact on a woman's access to the urban environment

Planning that ignores gendered assumptions in design has an impact on a woman's access to the urban environment. The ability of women to access, utilize, and occupy urban space in large numbers is essential to their idea of safety in the city. Gender insensitive urban infrastructure that is not designed for female bodies and is inattentive to their needs exacerbates situations that render a female defenseless to various forms of exclusion and violence in the city.

Discrimination at all levels

Gender-based violence occurs on many different levels, starting with prejudice at birth and extending to discrimination in work, income, education, nutrition, and acts of direct or indirect sexual assault [32]. One can easily see this inequality in our local parks and streets where boys predominate in sports like football and "gully cricket," especially in rural regions where girls are not permitted to go out and play in some communities during their childhood owing to a lack of safe open places [33]. Since there are so few facilities in schools to support a girl through her menstruation, especially in urban regions, it becomes challenging for young girls to continue their education after puberty [34]. A girl may need to use a public restroom in an emergency while travelling to and from school.

Since higher education institutions are in remote places where a girl's family or she herself feels unsafe travelling, girls in rural areas confront additional difficulties while attempting to pursue higher education [35]. When a woman starts working in an urban area, the absence of a public infrastructure that supports women becomes increasingly obvious. Due to a lack of infrastructure that is hospitable to women, pregnant women and elderly women find it challenging to travel around the city without needing assistance [36]. No of the age group, a woman's access to public areas is constantly questioned. The patriarchal responses of the society reveal that, given the unfavorable nature of cities, women might be better-off staying away from these spaces entirely, endangering a woman's right to the city and access to public space [37]. Instead of voicing responses that are meant to foster a woman's access to the city.

Absence of local democratic participation

The main concern with respect to urban planning in India is the absence of local democratic participation in cities. The institutional frameworks for decentralized governance are not in order, even after two decades of the passing of the 74th constitutional amendment which enables urban local governments to operate as self-governing bodies [38]. The planning process turns ineffective since public participation is not made mandatory as per law and comments from the public requested only once the plan is completed [39]. This form of planning process does not even meet the requirements of people it is planning for, much less for the women stakeholders. The absence of participation badly affects women since they are the ones who can express about their needs and requirements, depending upon on the problems that they face. For instance, a report by The New Indian Express [40] reveals the case of Visakhapatnam in October 2017, where several incidents of rape and sexual violence had been reported against women who do not have access to safe housing and must resort to sleeping on the streets or under flyovers highlighting the role of poverty and destitution in sexual violence. Even in old Delhi there are lot of cases against women which are not even talked and that just erase it even by the family. Such incidents remain ignored, underreported and the aggrieved are often silenced by the authorities [41].

Include women perspective in planning

The Delhi government launched the Stree Shakti programme for slum-dwelling women in 2002 to address issues with women's health and well-being, gender discrimination, community involvement, and women's empowerment. While the Master Plan provides no recommendations regarding urban design, services, or infrastructure that could make the city more inclusive of women from the perspective of women's safety, it does not use breakdown data to show which regions are more vulnerable for women. While discussing urban planning or service provision, the 2006 plan talks about creating a safe and inclusive city without addressing women's access, and it also makes no mention of women's unique issues.

3.5 *Some Good Initiatives for Making Cities Women Friendly*

Kolkata

Any city's level of safety will be determined by how well-off its women are. According to the National Crime Records Bureau (NCRB) of India's most recent statistics, which include the 2018 edition of "Crime in India," Kolkata has been named the safest city for women among 19 Indian cities for the second consecutive year [42]. The government's ground-level actions have improved patrolling and maintained peace and order, which has assisted the Kolkata City Police in reducing crime [43]. The government's concentrated efforts in building a dedicated infrastructure for improved police department operation and deploying more personnel, A positive policing environment in the city has been made possible by the installation of new police stations, police booths, technology advancements, and improvements in human and human alertness, among other things [44]. The Kolkata Police Department is also preparing to establish the "Sukanya" project, which aims to teach elementary self-defense skills to schoolgirls [45]. Additionally, a significant step has been taken to provide special transportation for women dubbed "PINK CABS," which is operated by and for women [46].

Delhi

Based on the results of the three methods of data collection—Focus Group Discussions (FGDs), street surveys, and safety audits—the Delhi Chapter of the Indian National Trust for Art and Cultural Heritage (INTACH) collaborated with "Jagori" to develop gender-sensitive proposals for environmental change [47]. Public restrooms for women, the establishment of kiosks to provide ad hoc street surveillance, well-built pavements that are pedestrian-friendly, and well-lit public spaces were also proposed.

Mumbai

The Bombay Municipal Corporation recently added the entire chapter titled "Gender, Special Groups and Social Equity" to the city's Revised Development Plan (RDDP) 2034. This chapter suggests multiple space reservations of 1000 sq m plot size to construct "multipurpose housing for working women" in 18 wards where offices or institutions of higher learning are present. The housing design includes a complex with single and double rooms for rent, dorms, an emergency shelter for women facing domestic violence, and guest rooms for women travelling to the city for jobs. There is space designated for a "community care center" in 23 electoral wards, which might either be a standalone day-care facility or a multipurpose room featuring a childcare area, a communal work area, a reading room, and an Aanganwari. Then there are facilities established to assist disadvantaged women in learning how to launch their own businesses. To ensure awareness, space has been aside for these centers in 13 wards near markets, railroad stations, and busy roads. Additionally, the development plan calls for homeless shelters, separate restrooms with drinking water, changing areas, and rest areas for Naka and domestic workers in each ward, 10% reservations in hawking areas, and elderly care facilities for women. With the adoption of these policies, Mumbai has become the first Indian city with a gender-specific urban plan, including land reservations for physical and social infrastructure that is suited to women's needs (4).

4 Way Forword

One may create inclusive cities and challenge ingrained social conventions regarding women and their place in the city by using a feminist urban planning approach that places women at the center of city planning and design. It was already suggested that it would be more convenient if women's places of employment, residences, childcare facilities, and recreational areas were all within walking distance of one another. This means that the interests of women and other underprivileged people are best served by compact, mixed-use areas. It is a truth that today's metropolitan streets are designed and constructed for cars rather than people. Only 4% of all license holders are women, according to a Times on India investigation from 2014 that examined statistics on 21 states' men-to-women driving license breakdown. This suggests that women's major form of transportation for short distances is walking, which creates a dangerous atmosphere in the lack of pedestrian traffic or street merchants. The expression "eyes on the street" supports mixed-use districts that foster unrestricted movement of people and hawker-friendly areas where people can keep an eye on the surroundings. Jane Jacobs established the idea of natural surveillance in her book "The Death and Life of Great American Cities." The eyes on streets idea can be expanded to include expansive underground parking lots and even metropolitan areas with towering compound walls.

Gender mainstreaming can be accomplished with the help of gender budgeting, which makes sure that both men and women benefit equally from development. An examination of the gender budget for this year finds a dearth of planning for outcomes. Civil society organizations and women's rights advocates can advocate for higher spending on women through a gender budget. (5) And, in order to make more effective and outcome-oriented gender budget following framework can be considered:

Urban infrastructure includes a variety of demands and services that must be met in order to develop cities that are welcoming to women.

- **Physical Infrastructure**: well-lit streets, walkable footpaths, appropriate signs, surveillance, free, open public restrooms that are always available with a gender-neutral alternative, parks that are always accessible, the presence of benches and other female-specific urban furniture.
- **Social infrastructure**: cooperative housing, open-access shelters, public day-care centers, female-only nightclubs, and skill-development facilities.
- **Transportation policy that favors women and mobility infrastructure**: Infrastructure for non-motorized transportation (NMT), such as facilities for walking and bicycling; free or inexpensive public transportation, particularly high-quality bus service that guarantees enough frequency and connectivity; well-maintained bus stops; female guards; and female drivers.
- **Institutional infrastructure**, including public hospitals, clinics for sexual and reproductive health, legal assistance offices, one-stop crisis centers, and female dispensaries. In rural areas, there should be access to nearby higher education institutions as well as infrastructure that provides the family with enough food and water.

Women's safety is a major concern and a top priority for most governments, whose agendas include measures to lessen gender-based crime and violence. Instead of the planning regime itself, the issue seems to be with the current legislative framework's support for a highly formalized, top-down planning approach, which appears to be out of touch with the reality of informal procedures that encourage urban expansion in Indian cities.

References

1. Doe J (2020) Gendered dimensions of urban space: Architectural and topographical influences. J Urban Des 24(2):120–135. https://doi.org/10.1234/jud.2020.02402
2. Basu S (2017, October 23) Gender and public space in India: A photographic research in 5 Indian metro cities. Retrieved from https://sanjukta.wordpress.com/2017/10/23/gender-and-public-space-in-india-a-photographic-research-in-5-indian-metro-cities/
3. World Bank (2022) Unlocking gender differences: Designing inclusive and safe public transport systems. Retrieved from https://documents1.worldbank.org/curated/en/099646510202229813/pdf/IDU0a86af61e04dfa04def085dd0d3d014583a67.pdf
4. Jagori & UN Women (2011) Understanding women's safety: Towards a gender inclusive city. Retrieved from https://www.endvawnow.org/uploads/browser/files/understanding_womens_safety.pdf
5. Old Delhi (n.d.) In *Wikipedia*. Retrieved February 16, 2025, from https://en.wikipedia.org/wiki/Old_Delhi
6. Shahjahanabad (n.d.) In Britannica. Retrieved from https://www.britannica.com/place/Shahjahanabad
7. D'Agostino G (2018) *Old Delhi and its architectural heritage: Adaptive reuse of havelis in Shahjahanabad* (Master's thesis)
8. Gandhi A (2019) The constrained circumference of Delhi Muslims. *Contemporary South Asia*
9. Rajeev SS (n.d.) A study on community profile on KachaPati Ram, Shahjahanabad. Retrieved from http://www.academia.edu
10. Uzma A (n.d.) Shahjahanabad: Physical vis-à-vis socio-cultural space. Retrieved from https://www.sahapedia.org
11. Gates of Delhi (n.d.) In *Wikipedia*. Retrieved from https://en.wikipedia.org/wiki/Gates_of_Delhi
12. Axworthy M (2006) *The Sword of Persia: Nader Shah, from Tribal Warrior to Conquering Tyrant*. I.B.Tauris
13. Kumar S (1994) Mohalla: A traditional urban form in modern cities. Ekistics 61(364/365):98–108. https://www.jstor.org/stable/43621804
14. Dickey S (2000) Permeable homes: Domestic service, household space, and the vulnerability of class boundaries in urban India. Am Ethnologist 27(2):462–489. https://doi.org/10.1525/ae.2000.27.2.462
15. Tarlo E (1996) Clothing matters: Dress and identity in India. University of Chicago Press
16. Postoast (2016, August 15) 17 Wholesale Markets In Delhi Where You Can Get Everything Cheap. https://www.postoast.com/wholesale-markets-in-delhi/
17. Magicpin (2023, July 5) 12 Best wholesale markets in Delhi: You must visit for shopping. https://magicpin.in/blog/wholesale-market-delhi/
18. Central Cottage Industries Emporium (n.d.) In: Condé Nast Traveler. Retrieved February 16, 2025, from https://www.cntraveler.com/shops/new-delhi/central-cottage-industries-emporium
19. The Times (2025, February 2) Delhi travel guide. https://www.thetimes.co.uk/article/delhi-r6ls52bws
20. All About Delhi (2023, October 12) Sadar Bazar wholesale market Delhi Shopping Guide. https://allaboutdelhi.com/delhi/sadar-bazar-wholesale-market-delhi-shopping-guide/
21. Yappe.in (n.d.) Central Cottage Industries Emporium – Craft store – New Delhi. Retrieved February 16, 2025, from https://yappe.in/delhi/new-delhi/central-cottage-industries-emporium/598238

22. Chakravarti U (1993) Rewriting history: The life and times of Pandita Ramabai. Oxford University Press
23. Minault G (1998) Secluded scholars: Women's education and Muslim social reform in colonial India. Oxford University Press
24. Lal R (2006) The Mughal harem. Oxford University Press
25. Rizvi J (2012) The cultural history of India. Cambridge University Press
26. Narayan R (2006) Shahjahanabad: The changing urban landscape of Old Delhi. Oxford University Press
27. Delhi Development Authority (DDA) (2007) Master Plan for Delhi 2021. Government of India
28. Singh P (2010) Economic structures and commercial growth in Old Delhi. Routledge
29. Sharma A (2012) Urban transformation and heritage conservation in Delhi. SAGE Publications
30. Archaeological Survey of India (ASI) (2020) List of protected monuments in Delhi. Government of India
31. Municipal Corporation of Delhi (MCD) (2019) Slum redevelopment and urban planning report
32. Sen A (2007) Gender and cooperative conflicts. In Tinker I (ed) Persistent inequalities: Women and world development, 123–149. Oxford University Press
33. UNICEF (2018) Gender and education: Strengthening opportunities for girls. UNICEF Publications
34. Sommer M, Vasquez E, Worthington N, Sahin M (2015) WASH in schools empowers girls' education: Proceedings of the menstrual hygiene management in schools virtual conference 2014. UNICEF
35. Kirk J, Sommer M (2006) Menstruation and body awareness: Linking girls' health with girls' education. Royal Tropical Institute
36. Phadke S (2011) Why loiter? Women and risk on Mumbai streets. Penguin India
37. Massey D (1994) Space, place, and gender. University of Minnesota Press
38. Sivaramakrishnan KC (2011) Re-visioning Indian cities: The urban renewal mission. SAGE Publications
39. Patel B (2014) Urban planning in India: Challenges and opportunities. Econ Polit Wkly 49(47):56–63
40. The New Indian Express (2017, October 15) Rape incidents highlight lack of safe housing for women in Visakhapatnam. Retrieved from https://www.newindianexpress.com
41. Sharma K (2016) Women and urban violence in India: Understanding the unspoken struggles. Oxford University Press
42. National Crime Records Bureau (2018) Crime in India 2018. Ministry of Home Affairs, Government of India
43. Singh P (2020) Policing strategies and crime reduction: Analyzing Kolkata's model for women's safety. Indian J Law Gov 8(1):45–62
44. Times of India (2019, December 15) Kolkata ranked safest city for women for second year. Retrieved from https://timesofindia.indiatimes.com
45. Hindustan Times (2020, January 10) Kolkata Police to launch Sukanya project for women's safety. Retrieved from https://www.hindustantimes.com
46. Chakraborty R (2021) Women's safety and urban transportation: A study of Kolkata's Pink Cabs initiative. J Urban Stud 14(2):98–112
47. Indian National Trust for Art and Cultural Heritage (INTACH) & Jagori (2019) Gender-sensitive urban planning: Findings from Delhi. New Delhi: INTACH & Jagori
48. Municipal Corporation of Greater Mumbai (MCGM) (2018) Revised Draft Development Plan 2034: Gender, Special Groups and Social Equity. Mumbai: MCGM

Analysis of Water Quality of Yamuna River: From Yamunotri to Its Confluence in the Ganges

Shailey Singhal, Shilpi Agarwal, Divya Ahluwalia, Ravi Kiran Maddali, and Naveen Singhal

Abstract Yamuna, the second largest tributary of The Ganges is becoming polluted due to lot of anthropogenic activities across all the five states of its flow. The situation is most dire in Delhi, the capital of India. The irreversible loss caused to the river is becoming a threat to the lives of human beings and aquatic organisms, which perhaps is not being realized by the responsible citizens of India living at the banks of the river and by the businesspersons discharging their industrial waste in the pious river. Yamuna, covering a total length of 1,376 km across five states of India encounters all types of human-made activities, destroying its preeminence and religiousness. 29 recognized sampling stations have been selected for evaluating the water quality in terms of pH, total coliform, dissolved oxygen, and biological oxygen demand, the standards set by Central Pollution Control Board (CPCB) of India. Along the entire stretch, except in Himachal Pradesh (HP) and partially in Uttarakhand (UK), the situation is getting worsened day by day in other three states, viz. Haryana, Delhi and Uttar Pradesh (UP) and it worrisome that 72% sampling stations are carrying much higher quantity of total coliform than the allowed limit. The problem demands a serious concern for addressing increasing levels of pollution in Yamuna, and suitable measures need to be worked out, which can help making Yamuna a 'lifeline' to its dependents in real sense.

Keywords Yamuna · Water pollution · Contamination · Water quality indicators

S. Singhal (✉) · S. Agarwal
Department of Chemistry, Cluster of Applied Sciences, UPES, Dehradun 248007, India
e-mail: shailey@ddn.upes.ac.in

D. Ahluwalia · R. K. Maddali
Department of Mathematics, Cluster of Applied Sciences, UPES, Dehradun 248007, India

N. Singhal
DIT University, Dehradun 248009, India

A. K. Sharma and A. Joshi (eds.), *Proceedings of the 3rd International Conference on Opportunities and Challenges for a Resilient Future*, Advances in 21st Century Human Settlements, https://doi.org/10.1007/978-981-96-0931-4_10

1 Introduction

Water resources play vital role in the progress of a Nation. Its optimum utilization in well-planned manner is essential for preserving this most valuable asset of nature provided to us. Yamuna, the second largest tributary of The Ganges in India is struggling for its pure maintenance and existence. It originates from Yamunotri in the north of Haridwar (Uttarakhand) and extends its arms till Sangam in Allahabad (UP) traversing a total distance of 1,376 km covering a total catchment area of 366,220 km2. Hindon, Chambal, Sind, Betwa and Ken are the main tributaries joining the river. Yamuna finds its path through highly populated states and provides water to more than 600 lakh people for various purposes (Table 1, Fig. 1) (Census Reports of India, 2001) [3]. The socio-economic significance of Yamuna is well-established by the fact that in Delhi, the capital of India, more than 70% people receive drinking water from Yamuna [3].

Rapidly increasing population, urbanization, introduction of sewage system, intensive use of chemicals and fertilizers, etc. has led to extreme levels of pollution in water bodies. Rivers have in-built capability to cleanse themselves by dilution and natural purification, but reduced availability of fresh water and excessive amounts of anthropogenic activities taking place has created high stress on their capacity also.

Enormous contamination with pollutants including agricultural runoff, untreated industrial waste, human/animal sewage, municipal waste, plastics and burnt/half-burnt bodies have raised a threat to the river sanctity. These factors have elevated the pollution levels in river to an alarming level, disturbing the lives of human beings as well as the aquatic organisms to a serious concern [13, 14].

Polluted water used for drinking, irrigation, washing and cooking is contaminating the food items with toxic stuff including heavy metals, becoming a cause of sever water-borne diseases, such as diarrhoea, dysentery, cholera, typhoid, jaundice and many more. Severity can be estimated by a report stating that during 1997–2003, the steep rise in typhoid case in Delhi was due to eating uncooked vegetables as salad [5]. Another report says that in Varanasi region, vegetables (spinach, radish, brinjal, cauliflower, tomato and cabbage) are contaminated with toxic heavy metals (zinc, lead, copper, cadmium, chromium, manganese and nickel), the main reason being

Table 1 Water abstraction from Yamuna [3]

Location	River water abstraction (MLD)	Abstraction use
Hathnikund	20,000	Irrigation, drinking water supply and others
Wazirabad	1,100	Drinking water supply
Wazirabad to Okhla stretch	5,000	Irrigation and other
Okhla to Etawah stretch	400	Irrigation, drinking water supply and others
Etawah to Allahabad stretch	475	Irrigation, drinking water supply and other

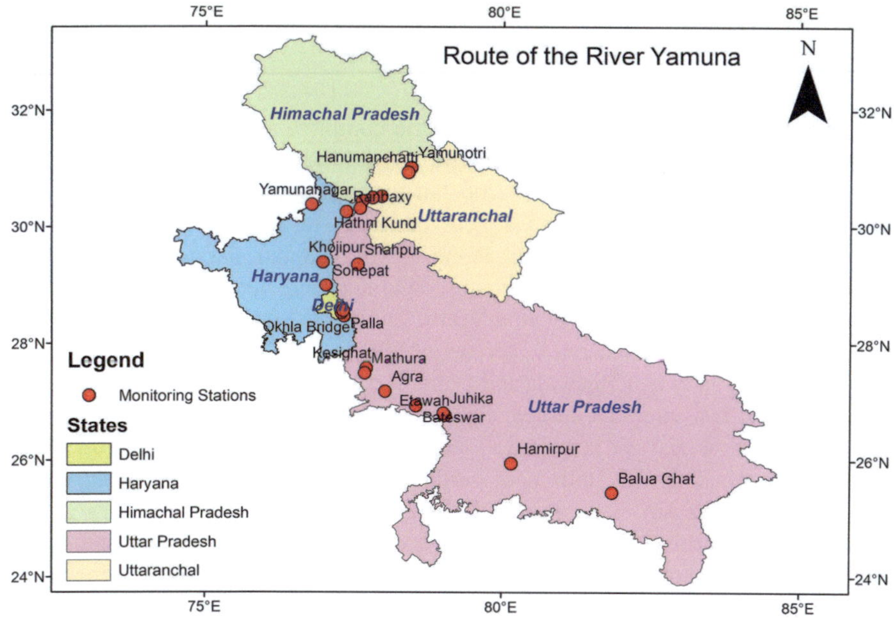

Fig. 1 Route of Yamuna across five states in India

the irrigation of these vegetables through industrial water. Poisonous arsenic has also been reported in rice crops, grown by using contaminated industrial water [9].

Most of the studies carried out to evaluate water quality of Yamuna are confined to a particular region or state [10–12, 18], which do not clarify the overall picture of pollution status in the river throughout its route. A slight improvement has been observed in the river quality during lockdown period, that too at few stretches only. Gradually, the world is returning back to normal conditions of living overcoming the challenging lockdown, which will again opening the doors to countless anthropogenic activities, adding to uncontrolled water pollution to the river. Present study analyses the tendency and progression of water quality of Yamuna river across all the five states of the country, viz. Uttarakhand (UK), Himachal Pradesh (HP), Haryana, Delhi and Uttar Pradesh (UP), through which it makes its path, and provides lifeline to numerous residents in its surroundings. The paper presents main causes of water quality deterioration and the impact of magnitude of contamination over water quality and living beings. The learning will be helpful in understanding state-wise specific causes contributing to its pollution levels. It will also be helpful for the readers for increasing awareness and take a hiatus to the anthropogenic activities in order to save the river.

2 Study Area

Yamuna is considered to be one of the most sacred rivers in India. The analysis area undertaken in this study comprises of the complete stretch from its origin at Yamunotri (UK) till its merger in Sangam (UP) (Fig. 1). After arising from Yamunotri, the river passes a stretch of around 200 km through a series of valleys in lower Himalayas and emerges into Indo-Gangetic plains. Here, it draws water from major streams and reaches the plains of Dak Pather in Uttarakhand, passing through Shivalik range of Himachal Pradesh. River is regulated here through dam for power generation. Further, it flows to Poanta Sahib, from where it reaches Yamuna Nagar in Haryana. During dry season, water is not allowed to flow in the river and the river remains dry between Tajewala and Delhi. Regaining water through upstream of Kalanaur, it further enters Delhi near Palla village after covering a route of around 224 km. At Wajirabad, river is tapped again for drinking water supply to Delhi. Here, river is contaminated with untreated/ partially treated domestic/ industrial waste. After a patch of 22 km, it enters Okhla barrage, where again river is dry during summers. Beyond this point, water is again contaminated with domestic and industrial waste water generated from East Delhi, Noida and Sahibabad, and connects the river through Shahdara drain. After receiving water from other important tributaries, Yamuna joins The Ganges and Saraswati at Prayag (Allahabad, UP) after travelling a distance of around 950 km. Throughout its path, river water is abstracted at five locations for supplying water for various purposes (Table 1). Geographically, the entire stretch of Yamuna can be divided into five segments:

- Himalayan segment, from origin to Tajewala barrage (172 km)
- Upper segment, from Tajewala barrage to Wajirabad barragae (224 km)
- Delhi segment, from Wajirabad barrage to Okhla barragae (22 km)
- Eutriphicated segment, from Okhla barrage to Chambal confluence (490 km)
- Diluted segment, from Chambal confluence to Ganga confluence (468 km).

In the present work, a total of 29 water sampling stations, as defined strategically by Central Pollution Control Board (CPCB), have been captured for monitoring the river quality [4].

3 Water Quality Parameters

Based upon the important parameters (as specified by CPCB), river water can be classified into five categories (Table 2) [4].

Real time monitoring data of Yamuna river has been taken from the official website of Central Pollution Control Board [4] for the year before lockdown (2019), to reflect the actual trend of water quality along the five states when all the routine activities were in their normal full swing state. Total 04 centres have been identified for monitoring water quality of Yamuna river in Uttarakhand. Water quality is observed to

Table 2 Standard quality parameters of river water

Category	Quality parameters	Suggested best usage
A	Total Coliforms Organism: ≤ 50 MPN/100 ml pH: 6.5–8.5 Dissolved Oxygen (DO): ≥ 6 mg/l Biochemical Oxygen Demand (BOD, 5 days @20 °C): ≤ 2 mg/l	Drinking Water Source without conventional treatment, but after disinfection
B	Total Coliforms Organism: ≤ 500 MPN/100 ml pH: 6.5–8.5 Dissolved Oxygen (DO): ≥ 5 mg/l Biochemical Oxygen Demand (BOD, 5 days @20 °C): ≤ 3 mg/l	Outdoor bathing (Organised)
C	Total Coliforms Organism: ≤ 5000 MPN/100 ml pH: 6–9 Dissolved Oxygen (DO): ≥ 4 mg/l Biochemical Oxygen Demand (BOD, 5 days @20 °C): ≤ 3 mg/l	Drinking water source after conventional treatment and disinfection
D	pH: 6.5–8.5 Dissolved Oxygen (DO): ≥ 4 mg/l Free Ammonia (as N): ≤ 1.2 mg/l	Propagation of Wild life and Fisheries
E	pH: 6.0–8.5 Electrical Conductivity at 25 °C: Max. 2250 micro mhos/cm Sodium absorption Ratio: Max. 26 Boron: Max. 2 mg/l	Irrigation, Industrial Cooling, Controlled Waste disposal

be quite good except certain sediments and coliform in it. The quality parameters obtained at all the 04 centres have DO ≥ 6, pH ranging from 6.5–8.5, and BOD ≤ 2. Total coliform (TC) at the first station, i.e. Yamunotri is ≤ 50; at second and fourth stations (Hanumanchatti and Dak Pather, Dehradun) is < 500, while at third station (Lakhwar dam) ranged from 50–5400 (Table 3, Figs. 2, 3, 4, 5).

In HP also, total 04 authorized stations are designated by CPCB for monitoring of water quality. Water quality appears to be very good with DO > 6, pH 6.5–8.5, BOD < 2 and TC < 50 at all the four stations (Table 4, Figs. 6, 7, 8, 9).

In Haryana, again only 04 water quality monitoring stations are designated. In terms of DO, all the stations have the indicator value > 4.5. pH value ranges from 7.1–8.9, with only one station (Sonepat) having the maximum pH value > 8.5, which is quite in basic range. BOD ranging from 1.3–14, with minimum limiting value being ≤ 2 at all the stations, but maximum limiting highly exceeding the desired limit at 2 stations (Panipat and Sonepat). TC range observed is 23–4,26,000, no station lying below the desired border in maximum limiting value (Table 5, Figs. 10, 11, 12, 13).

Delhi, the capital and the heart of the country, has 04 monitoring stations for water quality and represents the worst situation of river Yamuna except at Palla, which is comparatively better than other three stations. DO value at 3 stations (Nizamuddin, Okhla bridge and Shahdara drain) is up to 3.3 only, except at Palla where it is >6.0. pH at all the 04 stations is within range, however at Palla, the maximum pH is reported to

Table 3 Water quality of Yamuna river in Uttarakhand

Station code	Name of the station	Coordinates	Elevation	Abnormal quality parameter	Category	Health hazards
S1	Yamuna at Yamunotri	31.01°N 78.45°E	3,293 m (10,804 ft)	–	A	–
S2	Yamuna at Hanumanchatti	30.9324091°N 78.3991533°E	2,067 m (6,781 ft)	TC: 240	B	Not directly hazardous, as being in Category-B, water is suitable for outdoor bathing
S3	Yamuna at U/S of Lakhwar dam	30.520833°N 77.914444°E	204 m (669 ft)	TC: 50–5400	D	Microbial contamination, posing serious health hazards to swimming pools, drinking water, fishing and harvesting zones
S4	Yamuna at U/S Dak Pather, Dehradun	30.496855°N 77.798865°E	790 m (2,591.86 ft)	TC: 60–280	B	Not directly hazardous, as being in Category-B, water is suitable for outdoor bathing

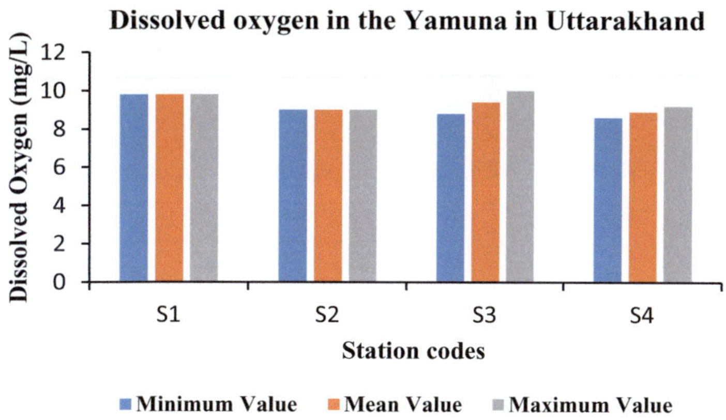

Fig. 2 DO values at 04 stations of UK

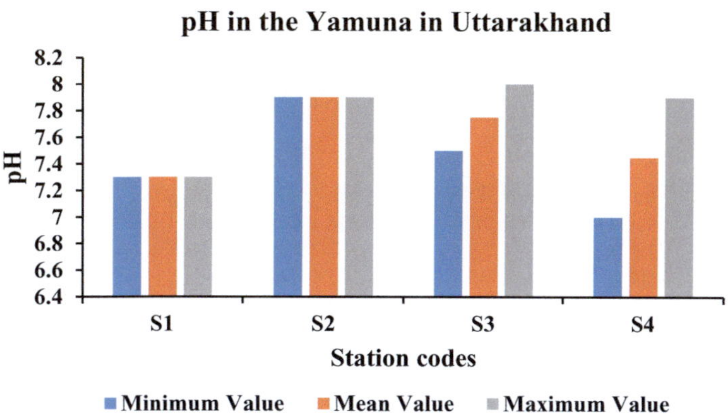

Fig. 3 pH values at 04 stations of UK

Fig. 4 BOD values at 02 stations of UK (BDL at 02 stations)

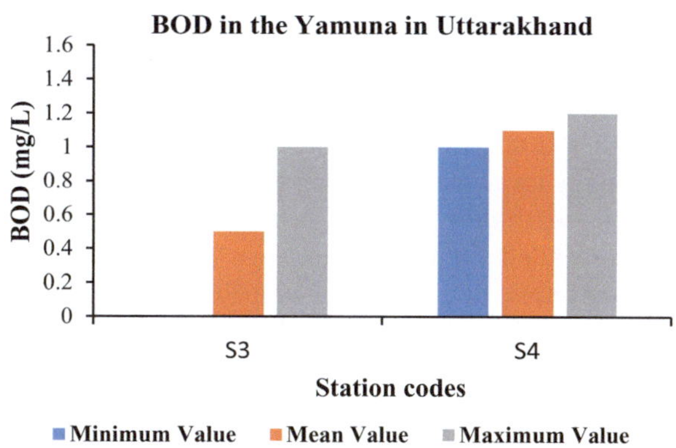

Fig. 5 TC values at 04 stations of UK

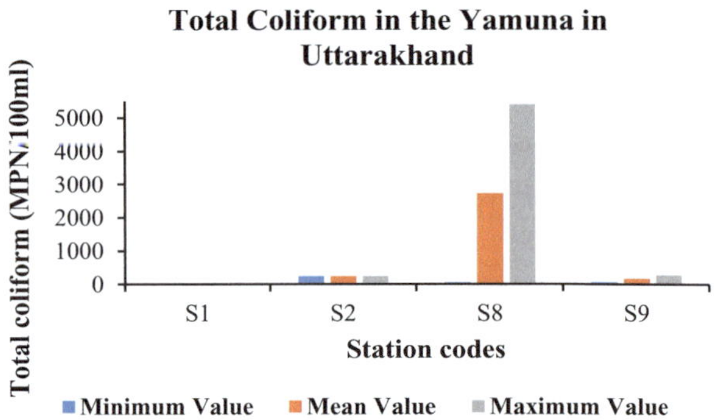

Table 4 Water quality of Yamuna river in HP

Station code	Name of the station	Coordinates	Elevation	Abnormal quality parameter	Category	Health hazards
S5	Yamuna river, U/S Paonta Sahib	30.438°N 77.624°E	389 m (1,276 ft)	TC: 32–63	~A	–
S6	Yamuna river, D/S Paonta Sahib	30.438°N 77.624°E	389 m (1,276 ft)	TC: 39–84	~A	–
S7	River Yamuna U/S Ranbaxy	–	–	TC: 33–70	~A	–
S8	River Yamuna D/S Ranbaxy	–	–	TC: 34–70	~A	–

be 8.9. BOD and TC at all the locations is very high, with average ranges 5.25–35.7 and 17,546.5–2,47,45,000 respectively (Table 6, Figs. 14, 15, 16, 17).

In UP, total 13 monitoring stations are authorized by CPCB. Parameters are varying a lot across the state, however pH value is within range (6.5–8.5) at 12 stations, except at Shahpur, where minimum pH is slightly lower (6.2) than the recommended limit. Overall range of DO is 0–16.9; out of 13 stations, only 2 stations (Kesighat, Vrindavan and Etawah) are having DO >4, and only 3 stations, viz. Chambal, Hamirpur and Balua ghat (Allahabad) reflect DO >6. At rest 8 stations (Mazawali, Shahpur, Mathura-03 outlets, Agra-02 outlets, and Bateswar), minimum value of DO $\leq$3.5. BOD shows great variation in the values with an average range of 5.33–16.76. Only one station (Balua ghat, Allahabad) has BOD <3, rest all the 12 stations have it between 2.1 and 47. Existence of TC is extremely bad. All the

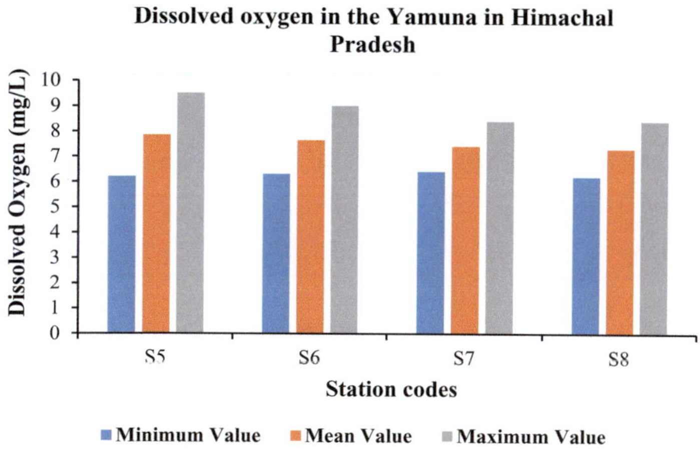

Fig. 6 DO values at 04 stations of HP

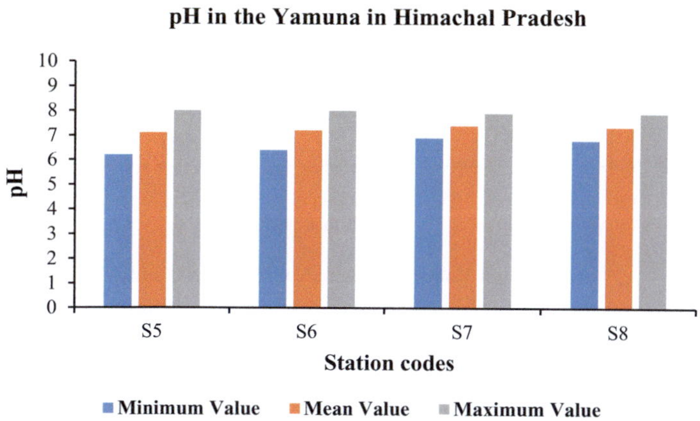

Fig. 7 pH values at 04 stations of HP

stations except Hamirpur (average 620) contain very high amount of TC, average vale ranging from 9300–555,500 (Table 7, Figs. 18, 19, 20, 21).

4 Discussion and Analysis

Most important factor destroying the health of water bodies is increasing population, urbanization and pollution. Our continuous advancement towards urbanization is becoming the major cause of anthropogenic activities, leading to severe levels of pollution. Data indicates that only one station (Yamunotri) in the state of Uttarakhand

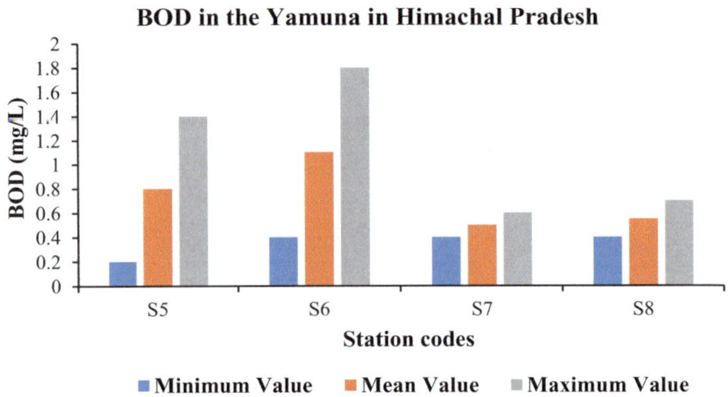

Fig. 8 BOD values at 04 stations of HP

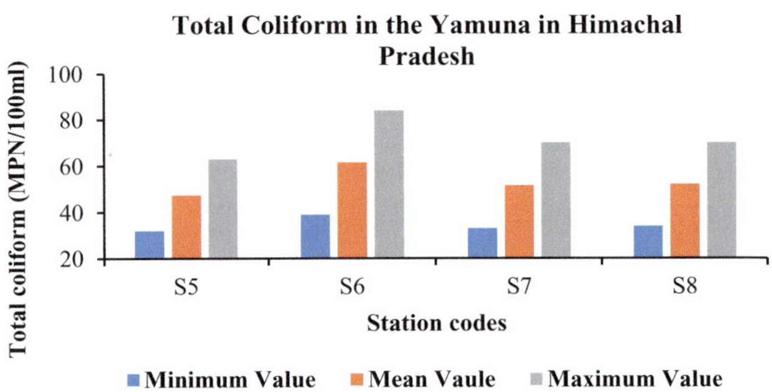

Fig. 9 TC values at 04 stations of HP

comes under category 'A' and is the sources of pure water, which can safely be used for drinking after disinfection. At two stations (Hanumanchatti and Dak Patehr, Dehradun), water falls under category-B, suitable for outdoor bathing purpose, while at one station (Lakhwar dam), water falls under category-C apt for drinking after conventional treatment and disinfection. Overall water quality is good in Uttarakhand primarily due to the absence of any major industrial establishments in this area. Somewhat declining value of the river water on moving down in Uttarakhand is due to the discharge of partly treated effluent in Yamuna besides the municipal waste. Being situated near a residential town in Uttarakhand, Kalsi, at Lakhwar dam, stagnant water and discharge of animal and feacal waste along with untreated effluent is the cause of slightly higher amount of total coliform.

It is clearly observed from the data that water quality of Yamuna in HP is appreciably good, lying under category-A at all the 04 stations. This allows it to be used for drinking purpose, merely after disinfection.

Table 5 Water quality of Yamuna river in Haryana

Station code	Name of the station	Coordinates	Elevation	Abnormal quality parameter	Category	Health hazards
S9	Yamuna at Hathnikund, Yamunanagar	30.313889°N 77.584444°E	–	DO: 4.8–8.1 BOD: 1.6–3.0 TC: 1,300–84,000	D	• Disruption of aquatic ecosystem • Microbial contamination, posing serious health hazards to swimming pools, drinking water, fishing and harvesting zones
S10	Yamuna at Kalanaur, Yamuna Nagar	30.3781868°N 76.7766894°E	–	DO: 4.5–9.2 BOD: 2.0–2.8 TC: 2,700–4,26,000	D	• Disruption of aquatic ecosystem • Microbial contamination, posing serious health hazards to swimming pools, drinking water, fishing and harvesting zones
S11	River Yamuna at Khojipur Panipat	29.3909°N 76.9635°E	219 m (718.504 ft)	DO: 4.6–11.2 BOD: 1.3–14.0 TC: 3,400–4,26,000	Can't be defined	• Disruption of aquatic ecosystem • Microbial contamination, posing serious health hazards to swimming pools, drinking water, fishing and harvesting zones
S12	Yamuna at Sonepat	28.9931°N 77.0151°E	224 m (734.908 ft)	pH: 7.1–8.9 BOD: BDL-11 TC: 23–94,000	Can't be defined	• Salty taste of water • Disruption of aquatic ecosystem • Microbial contamination, posing serious health hazards to swimming pools, drinking water, fishing and harvesting zones

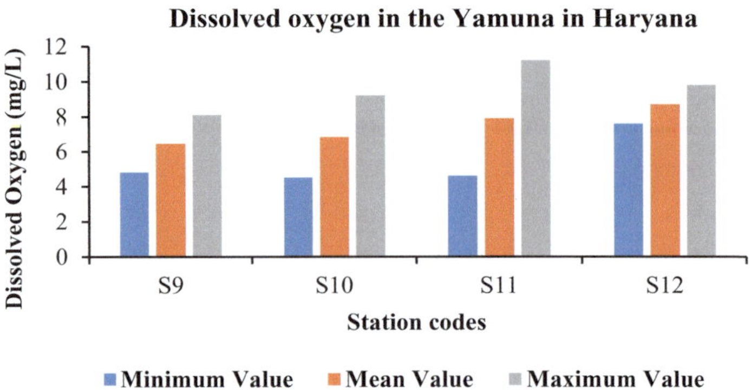

Fig. 10 DO values at 04 stations of Haryana

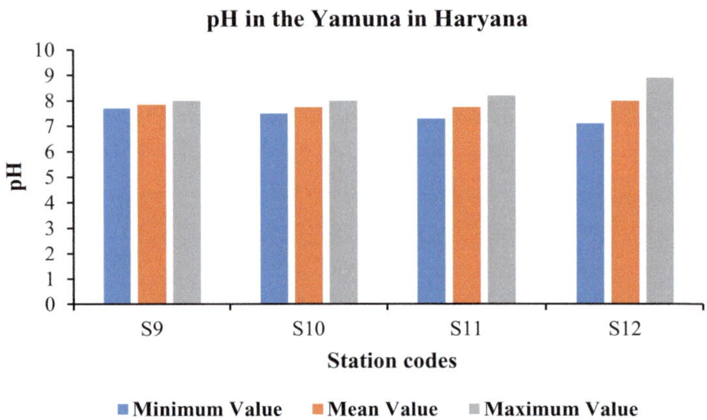

Fig. 11 PH values at 04 stations of Haryana

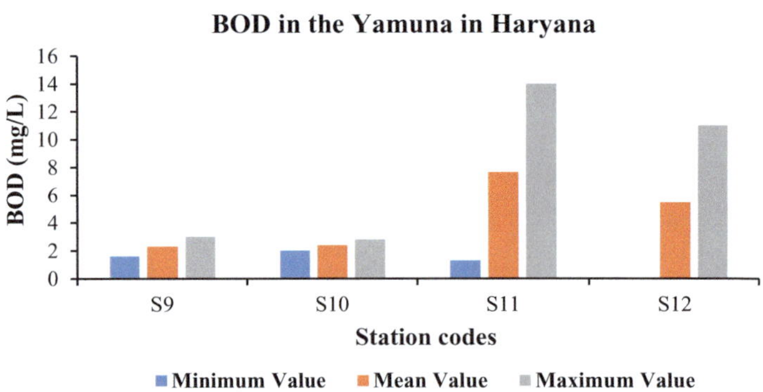

Fig. 12 BOD values at 04 stations of Haryana

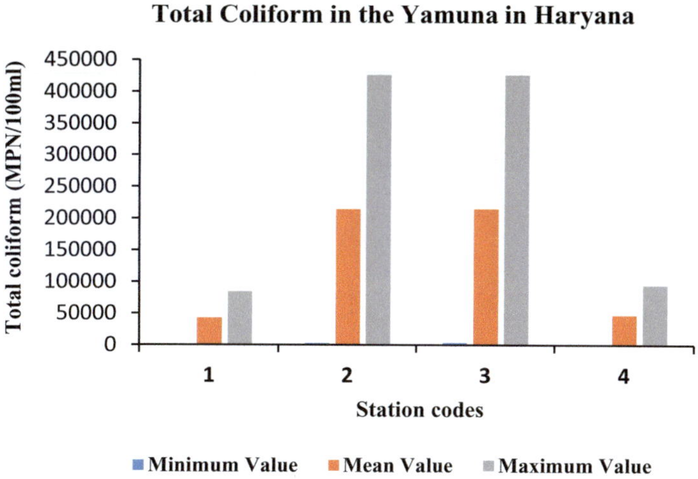

Fig. 13 TC values at 04 stations of Haryana

Beyond HP, Yamuna river quality is getting worsened and pathetic in rest of the three states, viz. Haryana, Delhi and UP. It is very shocking to observe that at two stations only out of four in Haryana (Yamuna Nagar-02 outlets), water parameters belong to category-D, permitting its use for propagation of wildlife and fisheries. At two stations out of thirteen in UP (Hamirpur and Allahabad), water quality belongs to category-C, allowing it to be used for drinking purpose after conventional treatment and disinfection. At rest seventeen stations (02 in Haryana, 04 in Delhi and 11 in UP), water condition is so bad that it can even not be classified under any category (Figs. 22, 23, 24, 25).

High population, establishment of sewage drains, pumping stations and several industries are the main reason behind the dire situation of river in the three states. Main cities contributing to domestic waste to Yamuna include Panipat and Sonepat in Haryana, Delhi, and Ghaziabad, Agra, Mathura, Vrindavan, Etawah and Allahabad in UP. Seasonal non-availability of water is also responsible for the bad health of Yamuna [8]. This is indeed unfortunate to observe that a river of religious significance converts into a drain in its lower stretch carrying agricultural, industrial and domestic effluents with it. As per a report of [2], approximately 359 industrial units are located near Yamuna, out of which 22 units in Haryana, 42 in Delhi and 17 in UP directly discharge their wastes in it. These include industries from almost all the segments, such as paper, sugar, leather, chemical, pharmaceuticals, distillery, etc. Sonepat, Yamuna Nagar, Panipat, Agra, Mathura and Delhi are the main cities around which these industries are situated. Decreased value of DO is an indicative of higher levels of pollution as expressed by higher BOD. In Delhi particularly, 95% of river pollution is due to discharge of waste water through seven key drains (Najafgarh, Yamunapur, Sen Nursing Home, Barathpula, Maharanibag, Kalkaji and Tuglakabad). Overall, three main reason have been identified for poor quality of Yamuna in Delhi by [16].

Table 6 Water quality of Yamuna river in Delhi

Station code	Name of the station	Coordinates	Elevation	Abnormal quality parameter	Category	Health hazards
S13	Yamuna at Palla	28.470951°N 77.318216°E	–	pH: 7.3–8.9 BOD: 1.7–8.8 TC: 93–35,000	Can't be defined	• Salty taste of water • Disruption of aquatic ecosystem • Microbial contamination, posing serious health hazards to swimming pools, drinking water, fishing and harvesting zones
S14	Yamuna at Nizamuddin	28.600469°N 77.262231°E	–	DO: BDL-2.4 BOD: 7.3–35.0 TC: 4,90,000–2,40,00,000	Can't be defined	• Disruption of aquatic ecosystem • Microbial contamination, posing serious health hazards to swimming pools, drinking water, fishing and harvesting zones

(continued)

Table 6 (continued)

Station code	Name of the station	Coordinates	Elevation	Abnormal quality parameter	Category	Health hazards
S15	Yamuna at Okhla bridge (Inlet of Agra canal)	28°34′N 77°18′E	–	DO: BDL-3.1 BOD: 6.7–28.0 TC: 68,000–1,70,00,000	Can't be defined	• Disruption of aquatic ecosystem • Microbial contamination, posing serious health hazards to swimming pools, drinking water, fishing and harvesting zones
S16	Yamuna at Okhla after meeting of Shahdara drain	28.565°N 77.303°E	–	DO: BDL-3.3 BOD: 8.4–63.0 TC: 4,90,000–4,90,00,000	Can't be defined	• Disruption of aquatic ecosystem • Microbial contamination, posing serious health hazards to swimming pools, drinking water, fishing and harvesting zones

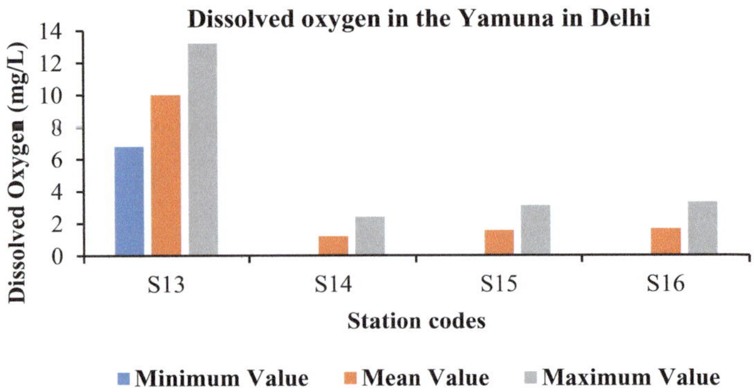

Fig. 14 DO values at 04 stations of Delhi

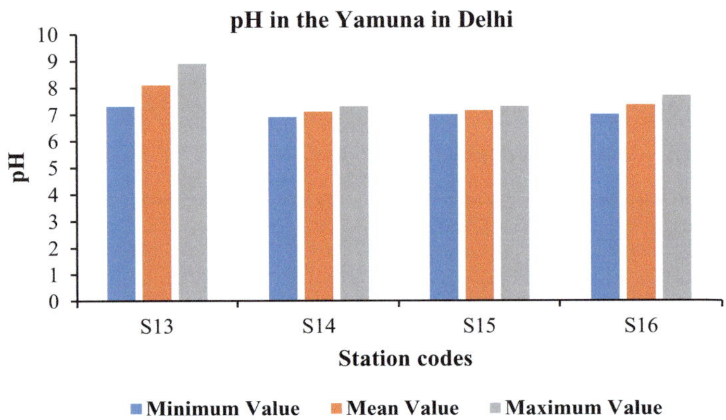

Fig. 15 PH values at 04 stations of Delhi

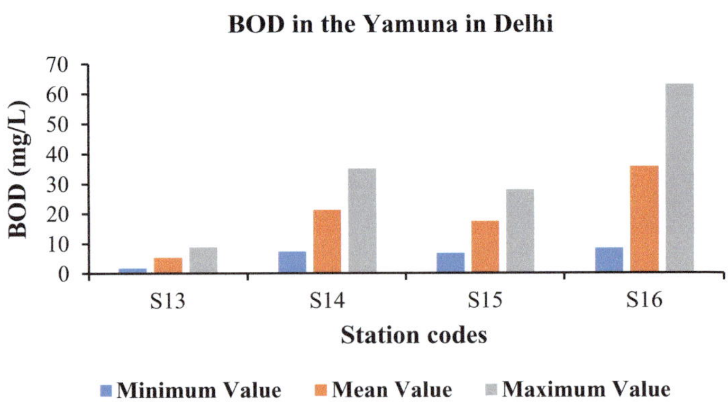

Fig. 16 BOD values at 04 stations of Delhi

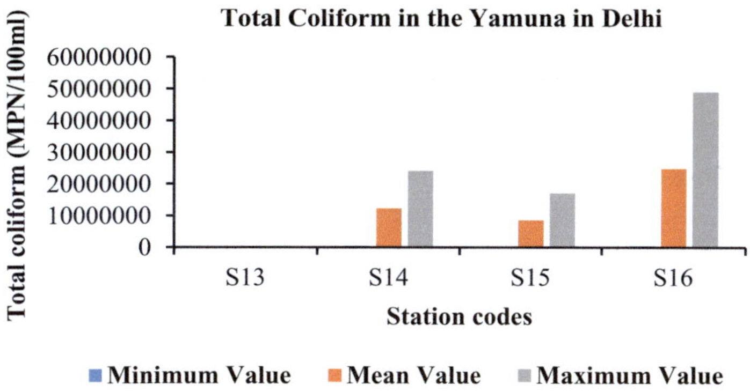

Fig. 17 TC values at 04 stations of Delhi

Insufficient treatment capacity plants for the sewage generated.

Improper establishment of STPs.

Mixing of treated and untreated sewage.

Existence of fresh water is a must to maintain the assimilation capacity of the river; and, this is totally missing in case of Yamuna. Around 92% water is abstracted out of the river for irrigation and domestic purposes, and in return enormous waste is drained into it. Yamuna is in the top list of the most polluted rivers in India with respect to bacterial contamination, especially along Delhi segment where around 22 drains discharge their waste water in the river [3, 7]. High TC resulting in microbial contamination is causing serious health hazards through drinking water, swimming pools, fishing and harvesting zones, where human exposure is very high [1, 6, 17]. Data from present study indicates that > 72% sampling stations across the route of Yamuna are having TC much more than the allowed limit, which advises about the frightening condition of river quality. Higher amounts of TC is also validated by higher BOD values. It reveals the presence of high coliform contamination in water, causing serious health issues in human beings and aquatic organisms. This class of water is not advisable to be used even after chlorination, as disinfection by-products (e.g. trihalomethanes) obtained during the process are toxic in nature. This also damages aquatic life and may result in the deaths of fish and interruption of aquatic ecosystem.

5 Conclusion and Recommendations

Proper utilization and preservation of rivers is our responsibility. Deteriorating water quality of Yamuna is a serious concern, which demands adequate consideration and action. Present study indicates innumerable loss of water quality parameters of Yamuna from Uttarakhand to Delhi, which although is not an unknown fact, but is essential to bring into notice repeatedly. Data presents the situation about the

Table 7 Water quality of Yamuna river in UP

Station code	Name of the station	Coordinates	Elevation	Abnormal quality parameter	Category	Health hazards
S17	Yamuna at Mazawali	–	–	DO: 0–5.9 BOD: 4.1–47.0 TC: 11,000–11,00,000	Can't be defined	• Disruption of aquatic ecosystem • Microbial contamination, posing serious health hazards to swimming pools, drinking water, fishing and harvesting zones
S18	River Yamuna at Shahpur	29.35162°N 77.548416°E	–	DO: 2.8–6.4 BOD: 8.0–18.0 TC: 72,000–1,60,000	Can't be defined	• Disruption of aquatic ecosystem • Microbial contamination, posing serious health hazards to swimming pools, drinking water, fishing and harvesting zones
S19	Yamuna at Kesighat, Vrindavan	27.58°N 77.7°E	170 m (560 ft)	DO: 4.8–6.1 BOD: 7.8–11.6 TC: 64,000–98,000	Can't be defined	• Disruption of aquatic ecosystem • Microbial contamination, posing serious health hazards to swimming pools, drinking water, fishing and harvesting zones
S20	Yamuna at Mathura U/S	27.4925°N 77.673611°E	–	DO: 0.6–10.5 BOD: 5.8–17.0 TC: 1,700–79,000	Can't be defined	• Disruption of aquatic ecosystem • Microbial contamination, posing serious health hazards to swimming pools, drinking water, fishing and harvesting zones
S21	Yamuna at Vishramghat, Mathura	27.4925°N 77.673611°E	–	DO: 3.4–6.2 BOD: 8.6–14.0 TC: 70,000–1,10,000	Can't be defined	• Disruption of aquatic ecosystem • Microbial contamination, posing serious health hazards to swimming pools, drinking water, fishing and harvesting zones
S22	Yamuna at Mathura D/S	27.4925°N 77.673611°E	–	DO: 3.5–6.8 BOD: 7.1–15.0 TC: 22,000–9,20,000	Can't be defined	• Disruption of aquatic ecosystem • Microbial contamination, posing serious health hazards to swimming pools, drinking water, fishing and harvesting zones

(continued)

Table 7 (continued)

Station code	Name of the station	Coordinates	Elevation	Abnormal quality parameter	Category	Health hazards
S23	Yamuna at Agra U/S	27.18°N 78.02°E	170 m (560 ft)	DO: 1.1–7.4 BOD: 4.5–15.0 TC: 1,400–28,000	Can't be defined	• Disruption of aquatic ecosystem • Microbial contamination, posing serious health hazards to swimming pools, drinking water, fishing and harvesting zones
S24	Yamuna at D/S of Agra	27.18°N 78.02°E	170 m (560 ft)	DO: 0.5–12.6 BOD: 5–21 TC: 2,600–1,40,000	Can't be defined	• Disruption of aquatic ecosystem • Microbial contamination, posing serious health hazards to swimming pools, drinking water, fishing and harvesting zones
S25	Yamuna at Bateswar	26.936°N 78.542°E	160 m (520 ft)	DO: 3.0–16.9 pH: 6.9–8.9 BOD: 4.7–21.0 TC: 2.0–3,50,000	Can't be defined	• Salty taste of water • Disruption of aquatic ecosystem • Microbial contamination, posing serious health hazards to swimming pools, drinking water, fishing and harvesting zones
S26	Yamuna at Etawah	26.77°N 79.03°E	197 m (646 ft)	DO: 4.6–16.2 pH: 6.9–8.9 BOD: 4.2–18.0 TC: 49.0–92,000	Can't be defined	• Salty taste of water • Disruption of aquatic ecosystem • Microbial contamination, posing serious health hazards to swimming pools, drinking water, fishing and harvesting zones
S27	Yamuna at Juhika B/C with Chambal, Etawah	22°27′N 75°31′E	843 m (2,766 ft)	BOD: 2.1–12.0 TC: 22.0–9,20,000	Can't be defined	• Disruption of aquatic ecosystem • Microbial contamination, posing serious health hazards to swimming pools, drinking water, fishing and harvesting zones

(continued)

Table 7 (continued)

Station code	Name of the station	Coordinates	Elevation	Abnormal quality parameter	Category	Health hazards
S28	Yamuna at Hamirpur	25.95°N 80.15°E	80 m (260 ft)	BOD: 5.6–5.7 TC: 450–790	~ C (excluding considering BOD)	• Disruption of aquatic ecosystem • Bacterial contamination responsible for water borne diseases like cholera, dysentery, jaundice, etc. So, it is recommended to be used only after conventional treatment and disinfection
S29	Yamuna at Allahabad D/S (Balua ghat)	25.45°N 81.85°E	98 m (322 ft)	BOD: 1.9–2.6 TC: 4,600–14,000	~ C	• Disruption of aquatic ecosystem • Bacterial contamination responsible for water borne diseases like cholera, dysentery, jaundice, etc. So, it is recommended to be used only after conventional treatment and disinfection

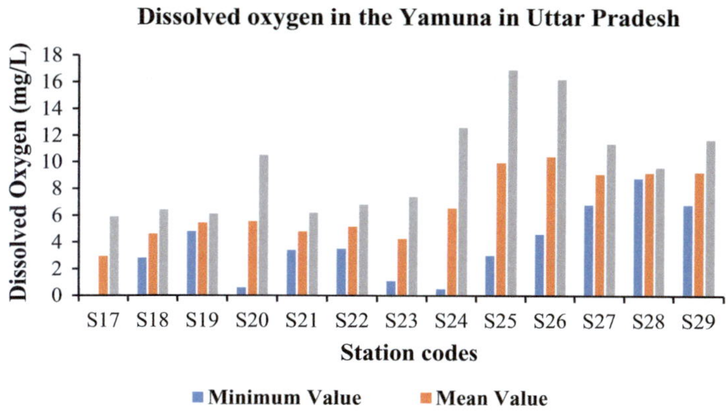

Fig. 18 DO values at 13 stations of UP

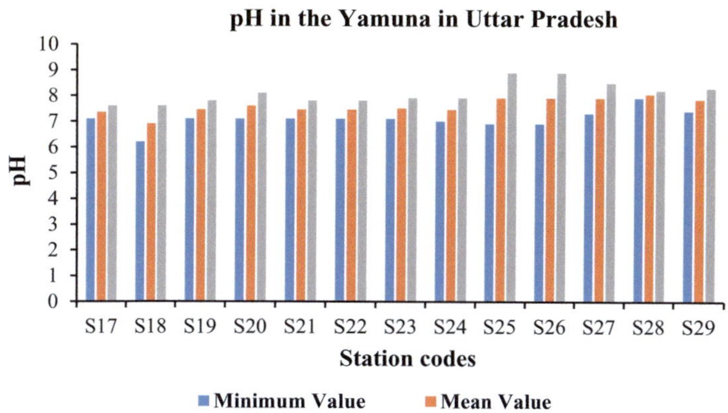

Fig. 19 pH values at 13 stations of UP

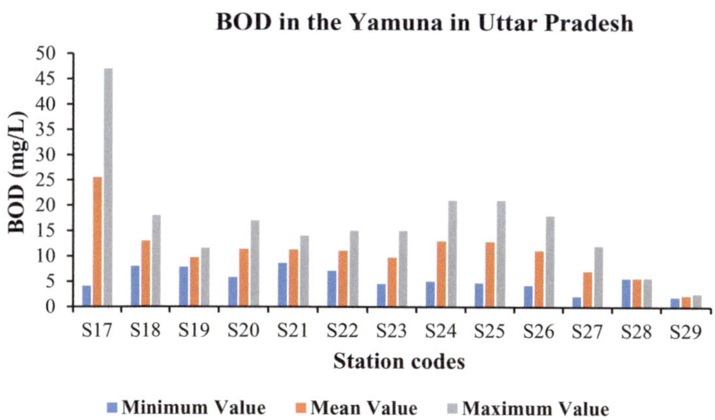

Fig. 20 BOD values at 13 stations of UP

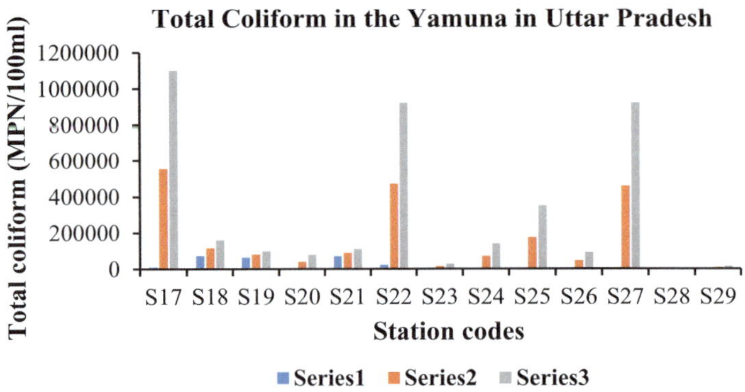

Fig. 21 TC values at 13 stations of UP

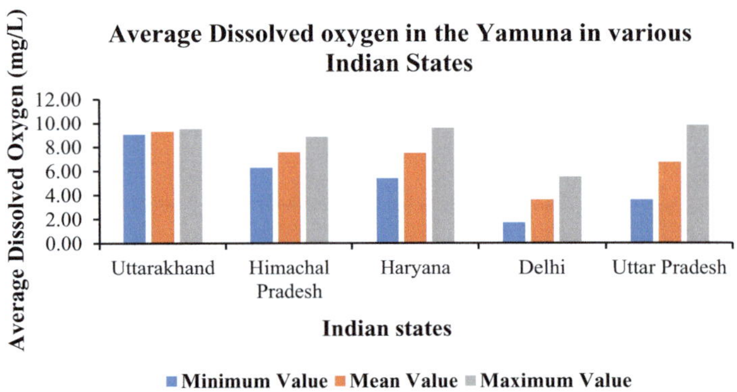

Fig. 22 Comparative DO levels in the five states

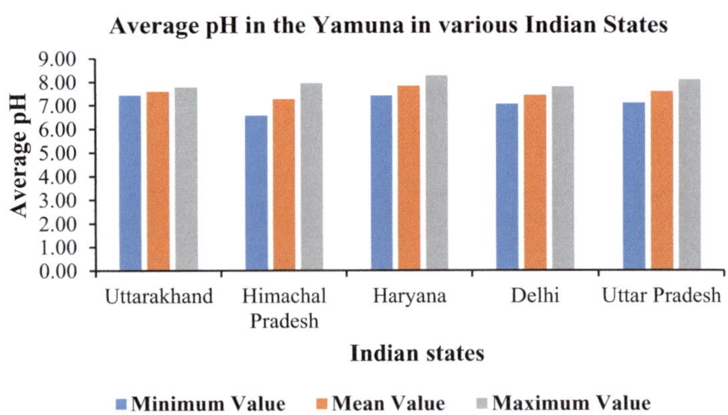

Fig. 23 Comparative pH levels in the five states

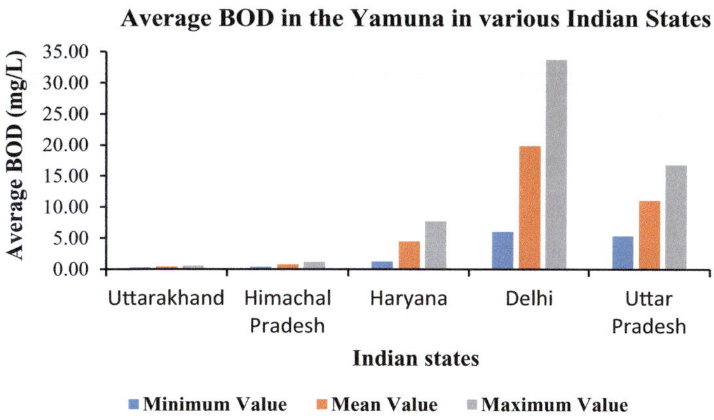

Fig. 24 Comparative BOD levels in the five states

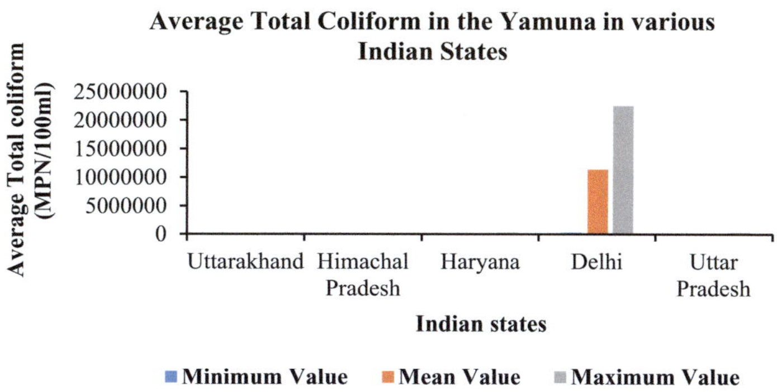

Fig. 25 Comparative TC levels in the five states

fatal levels of pollution, especially in the areas around Haryana, Delhi and UP. It again validates the role of human intervention in the devastating deterioration of nature and its resources. Major contributors of river pollution in these regions are the establishment of high number of industrial establishments, high population density, religious rituals, etc.

Awareness and concern about the issue has been increased in the recent years, and in the same essence, the Government of India has launched Yamuna action plan in 1993, with its second phase launched in April 2003 to restore the water quality across Yamuna river [3]. Without people participation and a holistic approach, the plans also seem not to be sufficient enough to restore the devastated quality of Yamuna river. In fact, few stringent measures need to be followed in order to protect the river, like installing sufficient numbers of sewage/effluent treatment plants (S/ETPs), zero discharge policy of untreated waste in the river, eliminating solid waste disposal in the river, minimizing routine domestic activities and open defecation at river banks, stopping cremation residues and idol immersion in the river. In addition, warning boards should be displayed at severely contaminated stations, and firm environment surveillance should be established for monitoring and checking the compliance of industries established at the banks of the river. Further, as per the guidelines of Hon'ble Supreme Court of India, minimum water flow needs to be maintained throughout the passage of Yamuna at a rate of 10m3/sec [15]. Water abstraction policy also needs to be strengthened to control over unnecessary diversion of river water for different purposes. Development at the cost of environmental health is never advisable. Human intervention are damaging the environment and ecosystem, which is difficult but will have to be minimized for sustainable existence of nature and society.

Conflict of Interest Statement Authors do not have any conflict of interest with anybody concerned.

Data Availability Statement Data supporting the findings of this study are available from the corresponding author upon reasonable request.

References

1. Almeida C, Soares F (2012) Microbiological monitoring of bivalves from the Ria Formosa Lagoon (south coast of Portugal): A 20 years of sanitary survey. Mar Pollut Bull, 64(2), 252–262. Available from: https://doi.org/10.1016/j.marpolbul.2011.11.025
2. CPCB (1999–2000) Water quality status of Yamuna River. ADSORBS/32. Published by Central Pollution Control Board (CPCB), Delhi, India
3. Central Pollution Control Board (CPCB) (2006) Water quality status of Yamuna River (1999–2005). Cent Pollut Control Board Minist Environ & For, Govment India
4. CPCB (2021) Available from: https://cpcb.nic.in/
5. Jain P (2009) Sick Yamuna, Sick Delhi—Searching a correlation. PEACE Institute Charitable Trust, Delhi, India
6. Kistemann T, Claßen T, Koch C, Dangendorf F, Fischeder R, Gebel J, Vacata V, Exner M (2002) Microbial load of drinking water reservoir tributaries during extreme rainfall and runoff. Appl Environ Microbiol, 68(5), 2188–2197. Available from: https://doi.org/10.1128/AEM.68.5.2188-2197.2002
7. Paul MS, Varun M, D'Souza R, Favas PJ, Pratas J (2014) Metal contamination of soils and prospects of phytoremediation in and around river yamuna: a case study from North-Central India. In Maria C. Hernandez-Soriano (eds) InTech, Croatia
8. Qureshi, S., Akolkar, P. & Garg, J.K. (2010) Impact of regulated flow on biological water quality in upper ecological segment of River Yamuna. J Exp Zool India, 13(2), 419–424. Available from: https://www.semanticscholar.org/paper/Impact-of-regulated-flow-on-biological-water-in-of-Ouareshi-Akolkar/9d9f648437ffbe6968de5ddf73b3f8628d02a7f6
9. Sarkar A, Srikanta KP (2007) Water borne diseases in India: environmental health and policy perspectives. Manak Publications Pvt. Ltd., New Delhi
10. Sharma R, Kumar R, Satapathy SC, Al-Ansari N, Singh KK, Mahapatra RP, Agarwal AK, Van Le H, Pham BT (2020) Analysis of water pollution using different physicochemical parameters: a study of yamuna river. Front Environ Sci, 8, 581591–581609. Available from: https://doi.org/10.3389/fenvs.2020.581591
11. Singh B (2020) Assessment of yamuna river water quality at agra: a case study. Int J Eng Res & Technol (IJERT), 105–106. Available from: https://www.ijert.org/assessment-of-yamuna-river-water-quality-at-agra-a-case-study

12. Singh G, Deb M, Ghosh C (2016) Urban metabolism of river yamuna in the national capital territory of Delhi, India. Int J Adv Res, 4(8), 1240–1248. Available from: https://doi.org/10.21474/IJAR01/1323
13. Singhal S, Agarwal S, Bahukhandi K, Sharma R, Singhal N (2014) Bioadsorbent: A cost-effective method for effluent treatment. International Journal of Environmental Science & Research, 3(1), 151–156. Available from: https://www.researchgate.net/publication/342338112_Bio-adsorbent_A_cost-effective_method_for_effluent_treatment
14. Singhal S, Agarwal S, Sharma R, Singhal N, Kumar A (2015) A green route for copper ions removal from textile industry effluent. Asian J Water, Environ Pollut, 12(3), 75–81. Available from: https://doi.org/10.3233/AJW-150010
15. Supreme court of India, Land Mark Judgement of the Hon'ble Supreme Court of India regarding Minimum Flow in River Yamuna (1999). Accessed on 4th Oct. 2017. https://www.cla.auburn.edu/envirolitigators/litigation/yamuna-case-sinha/summary-of-problem-and-court-actions/
16. Upadhyay R, Dasgupta N, Hasan A, Upadhyay SK (2011) Managing water quality of River Yamuna in NCR Delhi. Phys Chem Earth 36(9–11): 372–378. Available from: https://www.sciencedirect.com/science/article/pii/S1474706510000331
17. Wong M, Kumar L, Jenkins TM, Xagoraraki I, Phanikumar MS, Rose JB (2009) Evaluation of public health risks at recreational beaches in Lake Michigan via detection of enteric viruses and a human-specific bacteriological marker. Water Res, 43(4), 1137–1149. Available from: https://doi.org/10.1016/j.watres.2008.11.051
18. Yadav A, Khandegar V (2019) Dataset on assessment of River Yamuna, Delhi, India using indexing approach. Data in Brief, 22, 1–10. Available from: http://creativecommons.org/licenses/by/4.0/

Energy and Sustainability

Parameters and Principles Governing Thermal Comfort in Urban Open Spaces

Jagrati Sehgal, Vratika, and Gireendra Kumar

Abstract Urban heating is one of the most critical urban issues. As a result, the thermal environment in cities is getting hotter every day, intensifying the risk of heat stress. An urban built environment combines diverse microclimatic conditions, including varieties of urban morphological settings that distinctly affect the outdoor thermal environment. It can vary thermal loads and consequently affect the thermal comfort of inhabitants. The thermal comfort in urban open spaces influences urban dwellers' living quality, health, and well-being. Urban open areas are essential for social, cultural, and economic values. Outdoor thermal comfort (OTC) is a complex urban scenario that includes the role of various parameters and defines the perceived quality of urban open spaces. The amalgamation of natural and built environments is critical in controlling OTC. This study investigates the parameters responsible for thermal discomfort in urban open spaces. The study will thoroughly review the existing literature and focus on various factors governing heat exchange between urban built environments, giving rise to dissimilar thermal environments in open spaces. The microclimatic, physical, and personal parameters are recognized, classified, and decomposed. The interrelationship between the parameters is explored and analyzed. This work will cooperate with the scientific community to scrutinize thermal comfort in urban open spaces. The study's outcome will provide insights and scope for future studies examining the quantitative influence of various parameters identified for making resilient future cities.

Keywords Urban heat · Urban heat Island · Outdoor thermal environment · Urban microclimate · Outdoor thermal comfort

---

J. Sehgal (✉) · Vratika · G. Kumar
Malaviya National Institute of Technology, Jaipur, India
e-mail: jagratisehgal@gmail.com

A. K. Sharma and A. Joshi (eds.), *Proceedings of the 3rd International Conference on Opportunities and Challenges for a Resilient Future*, Advances in 21st Century Human Settlements, https://doi.org/10.1007/978-981-96-0931-4_11

1 Introduction

Urban development has changed the built environments of cities to accommodate population growth. The urban population in developing regions is expected to grow 57% by 2030 and 66% by 2050 [1]. The increasing trend of urbanization is imposing a threat to the urban climate. It thus requires serious attention to address the negative environmental impacts of urban areas and the many other problems that urban life brings concerning the health and well-being of people [2, 3]. The devastation of the thermal environment is a critical issue in the present world [4]. A human-made environment is an amalgamation of natural and built environments together. The effect of the urban built environment can give rise to a severe phenomenon of heat, i.e., Urban Heat Islands. Heat islands, Heat Waves, and global warming are overheating the cities [5] and modifying microclimatic conditions depending on urban design characteristics. A high-quality outdoor thermal environment is the solution to improve urban liveability and utilization rate of urban open spaces [6].

Outdoor thermal comfort (OTC) in urban open spaces is crucial and deals with multiple parameters. Urban open spaces are unbuilt spaces between varied built forms. In urban open areas, various geometrical and morphological characteristics of space affect thermal comfort [7]. Potential climatic conditions in urban open spaces are significantly correlated with the layout of cities [8]. Thermal comfort in open spaces can boost opportunities for social, cultural, and economic benefits [9]. OTC is a multi-layered notion. It is not only affected by microclimatic parameters but also by the surrounding built environment and personal state of mind [10]. OTC studies involve both objective and subjective parameters for assessing the thermal environment. The study first discusses the basic principles and reasons behind the formation of varying local climates with increasing heat issues. The study aims to classify and categorize identified parameters and discuss their importance and role in the modification of outdoor thermal comfort conditions.

2 Methodology

This study attempts to provide a detailed overview of various factors affecting the thermal environment in open urban spaces. The first step involved searching and screening various OTC studies dealing with objective and subjective assessments in the field. The second step deals with interpreting various OTC studies in urban areas and segregating them according to the parameters studied and the approach of OTC assessment. The third step involves extracting information and identifying key parameters. Finally, the identified parameters are recognized, classified, and decomposed into three major categories, i.e., microclimatic, physical, and personal. The study methodically groups and arranges the pertinent data under several parts and subsections. It examines and explains parameters determined to be significant in over-the-counter research. The study covers the efficiency of various identified parameters concerning one another under various settings and scenarios. An extensive classification of the factors used to evaluate outdoor thermal comfort is presented in this study. It provides insights into the basic heat exchange principles in urban areas. The literature review systematically recognized, classified, and decomposed several OTC parameters.

3 Results

3.1 Principles Governing OTC

The perception of extreme heat or cold in the surroundings can lead to thermal discomfort. In the current period of climate change and global warming, the primary source of the disruption of the thermal environment is surplus heat trapped in urban areas, which results in a disturbed energy balance. All the heat produced by human activity and solar radiation is reflected or retained in opaque materials like pavement, buildings, and other structures. Heat absorption and dissipation through the built environment govern the thermal environment. It increases the overall temperature of the urban environment, which causes thermal imbalance in the urban area [11]. In an urban microclimate, the energy fluxes pass through multiple reflection points and result in the absorption of some amount of energy in the built environment. An urban built environment combines diverse microclimatic conditions, including varieties of urban morphological settings that differ in dimensions. It affects the outdoor thermal environment distinctly in different open spaces [12] by altering the thermal loads, affecting the thermal comfort of citizens. The three essential aspects of urban infrastructure that influence urban energy balance are shading, ventilation, and reflectivity [11]. Urban geometry primarily controls the heat exchange in outdoor areas by regulating the ability of sun and wind to reach the built environment and regulating energy exchange in urban microclimate. As a result, they affect how much heat is removed from building surfaces, affecting the microclimate and OTC in urban settings [13]. Urban canyons cause cities to retain heat by reducing natural ventilation [14]. Outdoor thermal comfort is overall regulated by surrounding built structures, surface materials, their emissivity and albedo, geometry, and built form [15]. The concentration of human activity, heating, air conditioning, and heat-producing structures produce more heat resources. This phenomenon at the urban scale is termed an Urban Heat Island (UHI), where the temperature of urban areas is increased compared to rural surroundings. The disturbed thermal environment is the paramount effect of UHI. Literature on urban energy balance that compiled the results of several studies in various urban centers around the globe expressed the urban energy balance as follows [16];

$$Q_I + Q_F = Q_H + Q_E + \Delta Q_S + \Delta Q_A$$

where Q_I is net all-wave solar radiation, Q_F is the anthropogenic heat flux (car exhaust, light bulbs, etc.), Q_H is the sensible heat flux, Q_E is the latent heat flux, ΔQ_s is the net storage flux, and ΔQ_A is the net horizontal heat advection.

3.2 Parameters Governing OTC

Thermal comfort within urban settings is influenced by numerous parameters associated with urban heat, which interact with one another. It is primarily affected by climatic conditions, geographical characteristics, the surrounding built environment, and the impact of a particular region's cultural and social aspects. The optimal combination for achieving comfort varies based on individual factors such as clothing choices, level of physical activity, and adaptive measures. In this regard, three broad categories of parameters influencing the urban environment have been identified: microclimatic variations, physical factors of the built environment, and personal factors focusing on the physiological and psychological characteristics of the people.

Microclimate. The difference in environmental conditions in a small region of the earth's surface is called microclimate (MC) [17]. On the other hand, an Urban Microclimate (UMC) is characterized by a unique climate inside a comparatively small geographical area within an urban area, primarily shaped by the built environment surrounding it [18]. The microclimatic characteristics are represented by air temperature, relative humidity, wind speed, and mean radiant temperature.

Air Temperature. Research findings indicate that there is a likelihood of a 1.4–4.8°C rise in global mean surface temperatures in urban areas before the year 2100 [19]. This rising air temperature adversely affects outdoor thermal comfort, increases energy consumption, and impacts human health. For this shift, the summer months are worse than the winter ones. Air Temperature is a paramount factor for urban microclimate and receives significant attention in an empathetic urban environment. The convective heat exchange between the human body and the environment is regulated by air temperature, indirectly affecting respiratory, evaporative, and radiative heat exchange [20]. Several studies have highlighted air temperature as the most influential microclimate factor on outdoor thermal comfort [21].

Relative Humidity. Humidity has traditionally been considered the least significant climatic factor that impacts outdoor thermal comfort. Several variables, such as vegetation cover, proximity to water bodies, air conditioning, etc., impact the humidity levels. The correlation between air temperature and humidity is widely acknowledged. It is represented by indices such as the Wet Bulb Globe Temperature (WBGT), which considers both variables to provide a more precise representation of thermal stress [22]. Typically, relative humidity levels in urban areas fluctuate between 58 and 96% during rainy months, rarely falling below 48%, which is deemed comfortable, and can occasionally reach 100% [17]. By decreasing the body's capacity to release heat through evaporation, high humidity levels can intensify thermal discomfort by making people feel warmer and more uncomfortable. On the other hand, low humidity can cause dry air, irritating the skin and breathing passages [23].

Wind Speed. Wind speed is an essential factor in regulating the thermal environment. It is challenging to predict and manage because it is impacted by several variables, including local and regional landscape typology and global air movement from a high-pressure area to a lower-pressure region [24]. Wind, however, fluctuates rapidly in both direction and intensity (wind gusts) seasonally or yearly. Air temperature and wind speed are significantly correlated in a few studies. Studies have shown that the preferred wind speed decreases with a drop in air temperature [25]. The level of wind may be regarded as beneficial or unpleasant depending on the climate, as during the winters, it creates an unfavorable comfort condition opposite to the summers. Several studies show that higher wind speed helps improve thermal comfort and reduce UHI's impact [26]. The extreme roughness of the buildings in a highly populated urban area also obstructs wind movement [17].

Mean Radiant Temperature. Mean Radiant Temperature (MRT) is another crucial meteorological parameter and is also considered the most reliable indicator that affects human outdoor thermal comfort [27, 28]. The mean radiant temperature indicates the radiant heat the human body receives. All radiative fluxes, such as longwave emissions from surfaces and direct, diffuse, and reflected solar radiation, are included in the radiation [29]. The human body's radiant relationship with the environment happens continuously with sunshine throughout the day. It is challenging to calculate MRT due to multiple reflections of radiant energy in the built environment [12]. Elevated Mean Radiant Temperature (MRT) levels can result in discomfort as individuals absorb additional heat from their environment, regardless of the relatively lower air temperature.

Physical. Urban morphology, which includes building size, spacing, artificial surface characteristics, and amount of green space, are primary physical parameters influencing the micro-scale thermal environment [30, 31]. Human-centered or physical approaches to building environments distinguish suitable Outdoor Thermal Comfort (OTC) solutions from other strategies. Further insights into how research responding to climate change has addressed OTC usage in metropolitan areas can be obtained by looking into these main approaches [13]. The physical characteristics studied are built form, density, geometry, canyon geometry, vegetation, blue spaces, and surface materials.

Built Form. Numerous studies have delved into the capacity of built form to improve microclimates, addressing various aspects such as building shapes, types, densities, and overall urban layout [32]. Amirtham et al. [33] state that changes in urban built form notably impact the micro-scale thermal environment. The thermal climate at the canopy layer is closely linked to street geometry, primarily determined by the specific characteristics of each site [33]. The configuration of buildings influences heat transfer within the urban environment. Many studies reveal that an increase of 10% in the height-to-floor area ratio leads to a 0.3% decrease in the nighttime UHI effect. A building's setback from the street, the distance between buildings with long facades, and the permeability of the building are all critical factors that contribute to improving pedestrian-level wind conditions, which in turn helps to cool the outdoor air during the summer [34, 35].

Building Density. The increase in building density allows the formation of deep-enclosed urban canyons [30]. It prevents solar radiation from escaping, allows less longwave radiation to reflect, and blocks air movement, exacerbating microclimatic conditions [11]. In densely populated areas, courtyards can create cooler microclimates when connected to ground-level streets, facilitating air movement. Kubota et al. discovered that residential neighborhoods with low-density and shallow street canyons can improve ventilation [36]. Yang [37] found that as the distance between buildings exceeds 40–50 m, the building density decreases. The buildings offer less resistance to the wind, resulting in a gradual increase in wind speed [37]. Building density exists concerning built-form configuration. Research suggests that irregularly shaped neighborhoods with varied plot sizes and building heights perform better in thermal comfort than uniformly compact urban areas [38]. Rose [39] discovered that adopting a medium-rise, medium-density built form in coastal cities can notably enhance outdoor comfort conditions [39]. Meanwhile, in Singapore, [40] observed that integrating higher and lower densities in urban areas led to decreased incoming solar radiation, thereby improving outdoor thermal comfort [40].

Built Geometry. Urban-built geometry is typically regulated by development restrictions that precede patterns of economic expansion over human comfort. Based on earlier research, three aspects of urban geometry have been shown to impact urban albedo: plan area density, building height, and building height homogeneity [41]. Increasing building height within a given street width creates deeper urban canyons, intensifying the interaction between building surfaces and reducing urban albedo due to increased solar reflection and absorption [11, 21]. Some research indicates that thoughtfully arranging a few tall buildings inside a canyon can increase wind speed at the pedestrian level, which will reduce the warmth of the surrounding air. Winds that flow parallel to the street have been shown to rise by up to 90%, which can result in a 1°C drop in temperature. The tower impact was even more noticeable in cases with perpendicular wind flow, when air temperatures dropped by 1.1°C and wind speeds increased tenfold [11, 42]. It was discovered that altering building heights was a key tactic for improving the urban wind environment. Based on numerical research, [43] concluded that non-uniformity in building heights could promote better urban airflow and a cooler climate and that towering structures did not necessarily exist as impediments [43].

Canyon Geometry. Canyon geometry, best represented by canyon orientation, the street's height-to-width ratio (H/W), and the cross-section of a typical street canyon can be used to describe it [44, 45]. Studies reveal that the geometry of urban canyons results in reduced energy loss from longwave radiation emitted by building surfaces [46]. Urban canyon surfaces absorb considerable radiation during daylight hours, facilitated by multiple reflections between urban structures. The parameters considered for the canyon geometry include orientation, aspect ratio, and sky view factor (SKF). The orientation of streets is a crucial factor influencing solar access and wind speed within urban canyons, thus playing a significant role in determining the thermal comfort of sidewalks. Research shows that while the orientation of urban canyons affects the overall amount of heat retained in limitation, it affects each urban facet because of variations in solar exposure [30]. Most studies have found that streets oriented in an east–west direction typically experience the most unfavorable thermal conditions during the daytime. As a solution, they suggest that deeper canyons enable mutual shading of buildings, enhancing the thermal performance along sidewalks. Since direct solar radiation can increase the radiant temperature by up to 25.1°C, several studies recommend shading through deeper canyons or galleries [47].

The aspect ratio is determined by the ratio between the average height (H) of the canyon walls and the width (W) [45]. A canyon is deemed uniform when its aspect ratio is approximately 1, indicating minimal wall openings. It is considered shallow if the aspect ratio is below 0.5 and deep if the aspect ratio equals 2 [48, 49]. The aspect ratio is critical to energy transfer in urban canyons. Greater net radiation, a more significant share of stored solar flux, and a longer time lag between peak radiation and maximum storage rate are all caused by a higher aspect ratio, which suggests a deeper canyon [30]. In Greece, a significant rise in median, maximum, and minimum cooling rates was observed with a decreased aspect ratio [50]. Furthermore, research from Algeria indicated that deeper canyons exhibited lower air and surface temperatures [51]. The increased shading resulting from a high aspect ratio leads to lower daytime air temperatures and PET values.

Sky View Factor (SVF) denotes the proportion of the sky visible in the overhead hemisphere. It is often determined using fisheye lens cameras to capture hemispheric images. It is a valuable tool for assessing street layout and land use attributes. It offers a systematic method to quantify urban geometric characteristics, particularly in diverse and complex environments. The value of this dimensionless factor varies between 0 and 1 [49, 11, 52]. A decreased SVF indicates increased shading during the daytime, resulting in a cooler environment. However, more radiation is trapped at nighttime. Seasonally, during the summers, a lower SVF limits solar radiation, reducing energy consumption for space cooling, whereas, during the winters, it leads to higher energy usage for lighting and space heating [53]. However, there is no specific standard for the SVF values as radiation varies for low, medium, and high-altitude areas. Many studies have revealed the correlation between the surface and air temperature and the SVF [49, 11]. Research conducted on the influence of SVF on the thermal environment of urban canyons has indicated that lower daytime air temperatures (cool island) and higher nighttime air temperatures (heat island) are commonly associated with reduced SVF values. This correlation is explained by the diminished penetration of incoming solar radiation into urban canyons with lower SVF during the day, consequently impacting air and mean radiant temperatures, influencing the comfort level in outdoor spaces [54, 55].

Vegetation. Urban vegetation has been suggested as a primary method for enhancing outdoor thermal conditions and air quality. Vegetation influences urban climate by promoting evapotranspiration, which aids in heat dissipation. Additionally, it provides shading for building surfaces and intercepts longwave radiation. Strategically placed vegetation can also help regulate airflow in urban areas [56]. Studies have shown that vegetation can reduce its temperature by 8–12°C compared to the surroundings. It can exist in parks, street trees, roof gardens, vertical gardens, etc., from macro to micro scale [14]. The cooling effect of vegetation depends on factors like species, canopy, size and coverage, density, complexity, quality, and quantity of trees. The most significant cooling effect has been demonstrated by parks, which also lessen the impact of urban heat islands. Regarding thermal comfort, trees are superior to grasslands, providing several other benefits, like reducing pollution and noise levels and stabilizing the soil. The MRT can drop by 18 °C with a tree height increase of 0 m to 12 m. According to [57], in a high-density setting, a dense tree design and a uniform distribution of vegetation can provide a noticeably colder microclimate and improved thermal comfort [57].

Blue Spaces. Blue planning, which includes the water bodies or water courses, is one of the effective strategies for improving outdoor thermal comfort because of evaporation and high heat capacity, which tend to lower land temperatures [13]. Water bodies serve as heat sinks in urban areas because they absorb more solar radiation than typical building and pavement materials without appreciably raising their temperature. As a result, the temperature at the water's surface and longwave radiation decreases [58]. Fewer studies examine the cooling effects of water bodies compared to green spaces, as greenery strategies tend to have more significant mitigation effects. However, studies indicated that blue spaces contribute to a 2–6°C cooling effect in surrounding environments [59]. The cooling effect of water bodies depends on their size, shape, location, distance of the water body, fluid flow characteristics, etc [60]. The presence of water bodies also increases the humidity in the surroundings, limiting the thermal impact of water bodies. However, studies have revealed that it significantly reduces the MRT and air temperature and positively influences thermal comfort's behavioral or psychological parameters [61, 21].

Surface Materials. The thermal characteristics of surface materials are crucial for absorbing, retaining, and releasing radiant energy, encompassing thermal conductivity and heat capacity [30]. High albedo materials and hard paved surfaces have high reflectivity to solar radiation and enhanced emissivity across the spectrum, increasing thermal stress [62]. Certain crucial factors like material composition, color, texture, and shading influence the heating of surface materials [32]. The surface albedo of building materials and urban geometry influences the reflection of incoming solar radiation. Studies reveal that by strategically choosing building materials, the surface albedo in urban areas could surpass natural surfaces, leading to increased reflection of short-wave radiation [30]. According to the studies, reflecting materials can effectively reduce the heat island effect, depending on the characteristics of the building, the surrounding urban environment, and the site's climatic and geographical conditions. Furthermore, results indicate that whereas reflecting materials have a significant cooling potential on surface temperatures, their effect on air temperature is less consistent and much smaller [63, 64]. Although highly reflecting materials lower the air temperature in urban open spaces, they also cause pedestrians to receive higher levels of solar heat [14].

Personal. Thermal comfort is related to human satisfaction concerning the thermal conditions of the surroundings. Therefore, in addition to the physical and microclimatic factors, it is essential to understand the personal factors that influence the outdoor environment [65, 66]. The personal characteristics include physiological, psychological, behavioral, and socio-cultural factors.

Physiological. Physiological factors affect the human body's sensations that regulate the thermal state of mind. However, many researchers have not studied the relationship between physiological parameters and outdoor thermal comfort [20]. Temperature variations in the human body are caused by heat exchange between the body and the environment, which is the first step in the relationship between physiological parameters and outdoor thermal comfort. Several physiological parameters significantly impacting thermal comfort include the metabolic rate and the clothing value. Clothing insulation decreases as the outside temperature increases. Nevertheless, when the air temperature exceeded 25°C, the decline slowed, probably because people dressed more modestly to fit in with society. Further observation reveals that the increase in clothing insulation again reaches a limit below 10°C in the ambient temperature [67]. The equilibrium between the rate of heat production from metabolism and the losses resulting from exchange with the surrounding environment determines human thermal comfort. However, in outdoor settings, people's metabolic rates vary [68]. Some research demonstrates that sensitivity to warm and cold climates varies depending on the activities performed [69]. Intense activities such as exercise result in increased metabolic heat production, leading individuals to be less sensitive to cold and more sensitive to heat. Conversely, individuals engaged in moderately demanding tasks experienced a more rapid decline in sensitivity when the thermal environment was colder compared to warmer conditions [20].

Psychological. From a psychological perspective, thermal comfort represents the mind's assessment of satisfaction with the thermal conditions present in the outdoor environment [70]. People typically experience more neutral thermal perception during cooler months than in summer [32]. The human perception of thermal comfort is derived from experience, exposure duration, environmental adaptation, and physiological acclimatization. It can be further observed that providing a pleasant outdoor microclimate and thermal comfort might motivate individuals to spend more time outside. Creating a comfortable outdoor microclimate can lower the death and morbidity rate associated with heat-related health hazards and increase the productivity of people working in open spaces. The high level of occupant acceptability of the outdoor thermal environment and the climatic patterns of outdoor thermal comfort are frequently attributed to adaptation [20]. Research has indicated that people in different locations respond differently to international OTC models and indexes because they have distinct thermal preferences due to the adaptation process [68].

Behavioral. The studies show that the behavioral parameter differs with gender, age, and associated activities [32]. Behavior is a way for people to express their satisfaction with their outdoor surroundings. Compared to research on subjective thermal perception, behavioral investigations are significantly rarer. Studies reveal that the impact of the thermal environment on children may differ from that of other populations due to the differences in their physiology and behavioral patterns from adults [71].

Socio-Cultural. Social and cultural factors are primarily examined based on the societal context and cultural distinctions across various regions, encompassing regional attire, lifestyle philosophies, national temperament, religious convictions, and more. These factors significantly influence individuals' everyday experiences and cognitive processes, shaping their perception and adjustment to the outdoor thermal environment. Diverse social and cultural backgrounds may result in varying preferences and behaviors related to sunlight exposure, clothing choices, activity preferences, and more, ultimately impacting individuals' sensitivity and tolerance to climate-related factors [72]. In addition to the individual's socioeconomic status, educational attainment can influence their subjective assessment of the outdoor thermal environment. Generally, those with higher economic means and remarkable educational achievements tend to be more discerning [67]. It has also been demonstrated that cultural influences on behavior affect outdoor thermal comfort [20].

4 Discussion

Analyzing outdoor thermal comfort in urban open settings is challenging, as several parameters influence it. The concept of outdoor thermal comfort is complex and cannot be fully explained by a single aspect of the environment or human state of mind. The knowledge of all the parameters affecting OTC is primarily in the domain. Many studies deal with OTC but discuss mostly one category of parameters or a few related ones. The variability of microclimatic parameters is the main factor to consider. Microclimatic parameters are crucial for OTC, and at the same time, they are highly dynamic in nature. It is challenging to modify these parameters in an outdoor environment. Various psychological and behavioral factors and the built environments' physical characteristics impact OTC. It is determined that both individual and physical characteristics are pertinent and equally significant for an extensive understanding of thermal comfort. Physical factors can be changed to manage wind circulation and the exchange of solar radiation. The combination of physical elements in urban open spaces formulates varied thermal environments (Fig. 1).

OTC studies have found that design conditions with balanced geometrical and vegetation properties result in better OTC conditions. However, few studies have found one physical parameter better than another. There is no rule or standard for designing physical elements; it primarily depends on the region's climatic conditions.

The importance of air and mean radiant temperature is justified in the OTC domain. The impact of humidity and air velocity is comparatively less explored. Relative humidity and air velocity variation are also highly variable and dependent on weather characteristics; therefore, it is difficult to make general observations. There is scope to explore the relevance of relative humidity and air velocity in the OTC domain. Moreover, the relationship of microclimatic parameters can be explored in varied conditions.

It has been recognized that vegetation and building geometry impact OTC conditions. Very little knowledge is available on the effect of surface materials and water

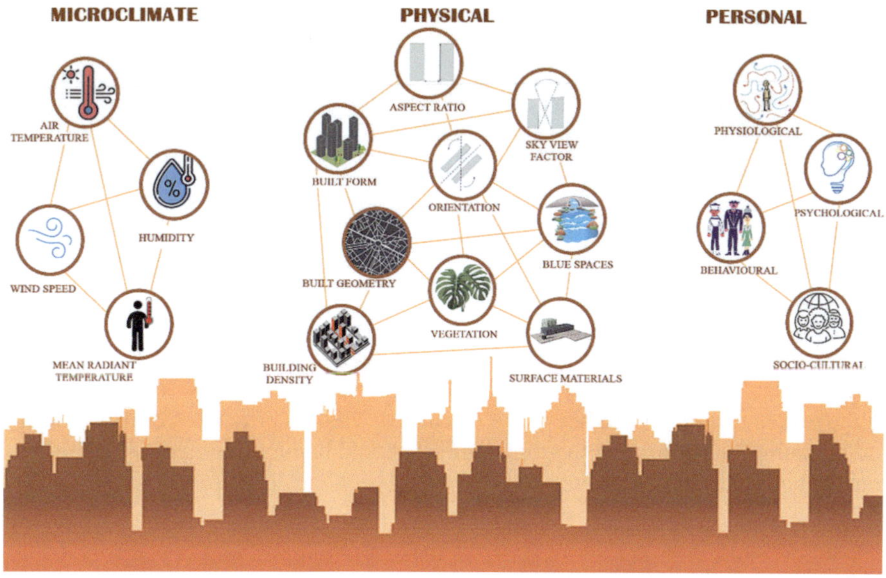

Fig. 1 Interrelationships of parameters governing outdoor thermal comfort

bodies on OTC conditions. There is future scope to explore the effect of varied physical parameters and their sub-factors on OTC conditions. Furthermore, a combined assessment of physical parameters together and comparison with each other is still a gap in the OTC domain.

Personal parameters are essential to include in OTC studies because of the impact of acclimatization and adaptability. The physiological and psychological parameters depend highly on cultural and climatic differences and must always be included. There are limited studies that include the behavioral aspect of OTC. Social and cultural aspects are majorly neglected in OTC studies. Future research on over-the-counter medications should address subjective factors in greater detail to determine the significance of each element.

Microclimatic parameters are the basis of every OTC study. In addition, most studies deal with physical or personal parameters while assessing OTC conditions. The difficulty lies in navigating the relationships between the three categories, each supported by several sub-factors. Even though each element affects thermal comfort independently, when combined, they create environments that vary significantly from location to location. Together, if all these parameters are in synchronization, they can profoundly impact thermal comfort conditions. More OTC studies are needed in the future to discuss various parameters with each other with multiple sub-factors. OTC is dynamic and varies based on climatic conditions. It is found that the interrelation of numerous parameters and an in-depth understanding of each one of them is required at varied locations and climatic conditions to address the importance and role of each parameter geographically within a different cultural context.

5 Conclusion

The researchers are interested in growing knowledge of OTC studies. There is a need to have a chronic state of knowledge of all the OTC parameters. This study provides an outlined understanding of heat exchange principles and parameters responsible for OTC. This study is an effort to unfold various layers of microclimatic, physical, and personal parameters affecting OTC. Various OTC studies are reviewed for different parameters, and relevant information is extracted from other sources. Each parameter is further classified and explained by all the sub-factors responsible. All of the combined information is systematically represented in the study. The importance and role of various parameters and their factors are discussed for regulating the thermal environment. The scope and future dimensions of OTC studies focusing on the role of different parameters are discussed. Various principles of urban design and planning for optimizing thermal comfort should be incorporated in the future to create a healthier microclimate.

References

1. UN (2022) World population prospects 2022: summary of results. Population Division. https://www.un.org/development/desa/pd/content/World-Population-Prospects-2022
2. De B, Mukherjee M (2018) Optimisation of canyon orientation and aspect ratio in warm-humid climate: case of Rajarhat Newtown, India. Urban Clim 24:887–920. https://doi.org/10.1016/j.uclim.2017.11.003
3. Ibrahim Y, Kershaw T, Shepherd P, Elwy I (2021) A parametric optimisation study of urban geometry design to assess outdoor thermal comfort. Sustain Cities Soc 75:103352. https://doi.org/10.1016/j.scs.2021.103352
4. Akbari H, Kolokotsa D (2016) Three decades of urban heat islands and mitigation technologies research. Energy Build 133:834–842. https://doi.org/10.1016/j.enbuild.2016.09.067
5. Santamouris M (2020) Recent progress on urban overheating and heat island research. Integrated assessment of the energy, environmental, vulnerability and health impact. Synergies with the global climate change. Energy Build 207:109482. https://doi.org/10.1016/j.enbuild.2019.109482
6. Sun C, Lian W, Liu L, Dong Q, Han Y (2022) The impact of street geometry on outdoor thermal comfort within three different urban forms in severe cold region of China. Build Environ 222:109342. https://doi.org/10.1016/j.buildenv.2022.109342
7. Ma X, Leung T, Chau C, Yung EH (2022) Analyzing the influence of urban morphological features on pedestrian thermal comfort. Urban Climate 44:101192. https://doi.org/10.1016/j.uclim.2022.101192
8. Mahmoud H, Ghanem H, Sodoudi S (2021) Urban geometry as an adaptation strategy to improve the outdoor thermal performance in hot arid regions: Aswan University as a case study. Sustain Cities Soc 71:102965. https://doi.org/10.1016/j.scs.2021.102965
9. Gaber N, Ibrahim A, Rashad AB, Wahba E, El-Sayad Z, Bakr AF (2020) Improving pedestrian micro-climate in urban canyons: city center of Alexandria, Egypt. Urban Clim 34:100670. https://doi.org/10.1016/j.uclim.2020.100670
10. Schnabel MA, Perera K, Donn M, Maddewithana H (2015) Addressing human thermal adaptation in outdoor comfort research–a literature review. ResearchGate. https://www.researchgate.net/publication/287208087_Addressing_Human_Thermal_Adaptation_in_Outdoor_Comfort_Research_-_A_Literature_Review.

11. Lin P, Gou Z, Lau S, Qin H (2017) The impact of urban design descriptors on outdoor thermal environment: a literature review. Energies 10(12):2151. https://doi.org/10.3390/en10122151
12. Johansson E, Emmanuel R (2006) The influence of urban design on outdoor thermal comfort in the hot, humid city of Colombo, Sri Lanka. Int J Biometeorol 51(2):119–133. https://doi.org/10.1007/s00484-006-0047-6
13. Aghamolaei R, Azizi MM, Aminzadeh B, O'Donnell J (2022) A comprehensive review of outdoor thermal comfort in urban areas: effective parameters and approaches. Energy & Environ 34(6):2204–2227. https://doi.org/10.1177/0958305x221116176
14. Taleghani M (2018) Outdoor thermal comfort by different heat mitigation strategies-a review. Renew Sustain Energy Rev 81:2011–2018. https://doi.org/10.1016/j.rser.2017.06.010
15. Peng M, Huang H (2022) The synergistic effect of urban canyon geometries and greenery on outdoor thermal comfort in humid subtropical climates. Front Environ Sci 10. https://doi.org/10.3389/fenvs.2022.851810
16. Golden J, Carlson J (2009) Climate, energy and urbanization: a guide on strategies, materials and technologies for sustainable development in the desert. https://www.researchgate.net/publication/269764329_Climate_Energy_and_Urbanization_A_Guide_on_Strategies_Materials_and_Technologies_for_Sustainable_Development_in_the_Desert.
17. Bherwani H, Singh A, Kumar R (2020) Assessment methods of urban microclimate and its parameters: a critical review to take the research from lab to land. Urban Clim 34:100690. https://doi.org/10.1016/j.uclim.2020.100690
18. Wang Y, Ni Z, Peng Y, Xia B (2018) Local variation of outdoor thermal comfort in different urban green spaces in Guangzhou, a subtropical city in South China. Urban For & Urban Green 32:99–112. https://doi.org/10.1016/j.ufug.2018.04.005
19. Climate Change 2021: The Physical Science Basis. IPCC (2021). https://www.ipcc.ch/report/ar6/wg1/
20. Lai D, Lian Z, Liu W, Guo C, Liu W, Liu K, Chen Q (2020) A comprehensive review of thermal comfort studies in urban open spaces. Sci Total Environ 742:140092. https://doi.org/10.1016/j.scitotenv.2020.140092
21. Liu Z, Li J, Xi T (2023) A review of thermal comfort evaluation and improvement in urban outdoor spaces. Buildings 13(12):3050. https://doi.org/10.3390/buildings13123050
22. Byrne MP, O'Gorman PA (2018) Trends in continental temperature and humidity directly linked to ocean warming. Proc Natl Acad Sci USA 115(19):4863–4868. https://doi.org/10.1073/pnas.1722312115
23. Huang C, Tsai H, Chen H (2020) Influence of weather factors on thermal comfort in subtropical urban environments. Sustain 12(5). https://doi.org/10.3390/su12052001
24. Nikolopoulou M (2004) Designing open spaces in the urban environment: a bioclimatic approach. Cent Renew Energy Sources (CRES)
25. Kim H, Macdonald E (2016) Measuring the effectiveness of San Francisco's planning standard for pedestrian wind comfort. Int J Sustain Dev World Ecol/ Int J Sustain Dev World Ecol 24(6):502–511. https://doi.org/10.1080/13504509.2016.1232319
26. Lai D, Chen C, Liu W, Shi Y, Chen C (2018) An ordered probability model for predicting outdoor thermal comfort. Energy Build 168:261–271. https://doi.org/10.1016/j.enbuild.2018.03.043
27. Ridha S, Ginestet S, Lorente S (2018) Effect of the shadings pattern and greenery strategies on the outdoor thermal comfort. Int J Eng Technol 10(2):108–114. https://doi.org/10.7763/IJET.2018.V10.1043
28. Peng S, Piao S, Ciais P, Friedlingstein P, Ottle C, Bréon FM, Nan H, Zhou L, Myneni RB (2012) Surface urban heat island across 419 global big cities. Environ Sci Technol 46(2):696–703. https://doi.org/10.1021/ES2030438
29. Crawford RH, Stephan A, Jamei E, Rajagopalan P (2015) Urban growth and pedestrian thermal comfort, pp 907–918. https://www.melbourne.vic.gov.au
30. Erell E, Pearlmutter D, Williamson T (2012) Urban microclimate : designing the spaces between buildings. Urban Microclim. https://doi.org/10.4324/9781849775397

31. Tsoka S, Tsikaloudaki K, Theodosiou T, Bikas D (2020) Urban warming and cities' microclimates: investigation methods and mitigation strategies—a review. Energies 13(6):1414. https://doi.org/10.3390/en13061414
32. Banerjee S, Chattopadhyay S (2020) A meta-analytical review of outdoor thermal comfort research: applications, gaps and a framework to assess low-income settlements in Indian megacities. Urban Clim 33:100641. https://doi.org/10.1016/j.uclim.2020.100641
33. Amirtham LR, Horrison E, Rajkumar S (2015) Impact of urban morphology on Microclimatic conditions and outdoor thermal comfort—a study in mixed residential neighbourhood of Chennai, India
34. Giridharan R, Lau S, Ganesan S, Givoni B (2007) Urban design factors influencing heat island intensity in high-rise high-density environments of Hong Kong. Build Environ 42(10):3669–3684. https://doi.org/10.1016/j.buildenv.2006.09.011
35. Yuan C, Ng E (2012) Building porosity for better urban ventilation in high-density cities–a computational parametric study. Build Environ 50:176–189. https://doi.org/10.1016/j.buildenv.2011.10.023
36. Kubota T, Miura M, Tominaga Y, Mochida A (2008) Wind tunnel tests on the relationship between building density and pedestrian-level wind velocity: development of guidelines for realizing acceptable wind environment in residential neighborhoods. Build Environ 43(10):1699–1708. https://doi.org/10.1016/j.buildenv.2007.10.015
37. Yang G, Xuan Y, Zhou Z (2022) Influence of building density on outdoor thermal environment of residential area in cities with different climatic zones in china—taking Guangzhou, Wuhan, Beijing, and Harbin as examples. Buildings 12(3):370. https://doi.org/10.3390/buildings12030370
38. Sharmin T, Steemers K, Matzarakis A (2015) Analysis of microclimatic diversity and outdoor thermal comfort perceptions in the tropical megacity Dhaka, Bangladesh. Build Environ 94:734–750. https://doi.org/10.1016/j.buildenv.2015.10.007
39. Rose L, Horrison E, Venkatachalam LJ (2011) Influence of built form on the thermal comfort of outdoor urban spaces. In: National University of Singapore, Department of Architecture & Sathyabama University, Sholinganllur, Chennai, India, The 5th International Conference of the International Forum on Urbanism (IFoU). http://globalvisions2011.ifou.org/Index/Group%202/FOUA00137-00244P2.pdf
40. Yang W, Wong NH, Jusuf SK (2013) Thermal comfort in outdoor urban spaces in Singapore. Build Environ 59:426–435. https://doi.org/10.1016/j.buildenv.2012.09.008
41. Kondo A, Ueno M, Kaga A, Yamaguchi K (2001) The influence of urban canopy configuration on urban albedo. Bound-Layer Meteorol/Bound-Layer Meteorol 100(2):225–242. https://doi.org/10.1023/a:1019243326464
42. Priyadarsini R, Wong NH (2005) Parametric studies on urban geometry, air flow and temperature. Int J Archit Sci 6(3)
43. Chan AT, So ES, Samad SC (2001) Strategic guidelines for street canyon geometry to achieve sustainable street air quality. Atmos Environ 35(32):5681–5691. https://doi.org/10.1016/s1352-2310(01)00483-6
44. Arnfield AJ (1990) Canyon geometry, the urban fabric and nocturnal cooling: a simulation approach. Phys Geogr 11(3):220–239. https://doi.org/10.1080/02723646.1990.10642404
45. Oke T (1988) Street design and urban canopy layer climate. Energy Build 11(1–3):103–113. https://doi.org/10.1016/0378-7788(88)90026-6
46. Emmanuel R (2012) An urban approach to climate sensitive design. Taylor & Francis eBooks. https://doi.org/10.4324/9780203414644
47. Watkins R, Palmer J, Kolokotroni M (2007) Increased temperature and intensification of the urban heat Island: implications for human comfort and urban design. Built Environ 33(1):85–96. https://doi.org/10.2148/benv.33.1.85
48. Ahmad K, Khare M, Chaudhry K (2005) Wind tunnel simulation studies on dispersion at urban street canyons and intersections—a review. J Wind Eng Ind Aerodyn 93(9):697–717. https://doi.org/10.1016/j.jweia.2005.04.002

49. Jamei E, Rajagopalan P, Seyedmahmoudian M, Jamei Y (2016) Review on the impact of urban geometry and pedestrian level greening on outdoor thermal comfort. Renew Sustain Energy Rev 54:1002–1017. https://doi.org/10.1016/j.rser.2015.10.104
50. Giannopoulou K, Santamouris M, Livada I, Georgakis C, Caouris Y (2010) The impact of canyon geometry on intra urban and urban: suburban night temperature differences under warm weather conditions. Pure Appl Geophys 167(11):1433–1449. https://doi.org/10.1007/s00024-010-0099-8
51. Bourbia F, Boucheriba F (2010) Impact of street design on urban microclimate for semi arid climate (Constantine). Renew Energy 35(2):343–347. https://doi.org/10.1016/j.renene.2009.07.017
52. Oke TR (1981) Canyon geometry and the nocturnal urban heat Island: comparison of scale model and field observations. J Climatol 1(3):237–254. https://doi.org/10.1002/joc.3370010304
53. Yan H, Fan S, Guo C, Wu F, Zhang N, Dong L (2014) Assessing the effects of landscape design parameters on intra-urban air temperature variability: the case of Beijing, China. Build Environ 76:44–53. https://doi.org/10.1016/j.buildenv.2014.03.007
54. Johnson DB (1985) Urban modification of diurnal temperature cycles in birmingham, U.K. J Climatol 5(2):221–225. https://doi.org/10.1002/joc.3370050208
55. Yamashita S, Sekine K, Shoda M, Yamashita K, Hara Y (1986) On relationships between heat island and sky view factor in the cities of Tama River basin, Japan. Atmos Environ 20(4):681–686. https://doi.org/10.1016/0004-6981(86)90182-4
56. Picot X (2004) Thermal comfort in urban spaces: impact of vegetation growth. Energy Build 36(4):329–334. https://doi.org/10.1016/j.enbuild.2004.01.044
57. Duarte DH, Shinzato P, Gusson CDS, Alves CA (2015) The impact of vegetation on urban microclimate to counterbalance built density in a subtropical changing climate. Urban Clim 14:224–239. https://doi.org/10.1016/j.uclim.2015.09.006
58. Robitu M, Musy M, Inard C, Groleau D (2006) Modeling the influence of vegetation and water pond on urban microclimate. Sol Energy 80(4):435–447. https://doi.org/10.1016/j.solener.2005.06.015
59. Völker S, Baumeister H, Classen T, Hornberg C, Kistemann T (2013) Evidence for the temperature-mitigating capacity of urban blue space—a health geographic perspective. Erdkunde 67(4):355–371. https://doi.org/10.3112/erdkunde.2013.04.05
60. Gunawardena K, Wells M, Kershaw T (2017) Utilising green and bluespace to mitigate urban heat island intensity. Sci Total Environ 584–585:1040–1055. https://doi.org/10.1016/j.scitotenv.2017.01.158
61. Lehnert M, Tokar V, Jurek M, Geletič J (2020) Summer thermal comfort in Czech cities: measured effects of blue and green features in city centres. Int J Biometeorol 65(8):1277–1289. https://doi.org/10.1007/s00484-020-02010-y
62. Pisello AL, Santamouris M, Cotana F (2013) Active cool roof effect: impact of cool roofs on cooling system efficiency. Adv Build Energy Res 7(2):209–221. https://doi.org/10.1080/17512549.2013.865560
63. Synnefa A, Karlessi T, Gaitani N, Santamouris M, Assimakopoulos D, Papakatsikas C (2011) Experimental testing of cool colored thin layer asphalt and estimation of its potential to improve the urban microclimate. Build Environ 46(1):38–44. https://doi.org/10.1016/j.buildenv.2010.06.014
64. Taha H (2008) Meso-urban meteorological and photochemical modeling of heat island mitigation. Atmos Environ 42(38):8795–8809. https://doi.org/10.1016/j.atmosenv.2008.06.036
65. Chan S, Chau C, Leung T (2017) On the study of thermal comfort and perceptions of environmental features in urban parks: a structural equation modeling approach. Build Environ 122:171–183. https://doi.org/10.1016/j.buildenv.2017.06.014
66. Potchter O, Cohen P, Lin T, Matzarakis A (2018) Outdoor human thermal perception in various climates: a comprehensive review of approaches, methods and quantification. Sci Total Environ 631–632:390–406. https://doi.org/10.1016/j.scitotenv.2018.02.276

67. Aljawabra F, Nikolopoulou M (2018) Thermal comfort in urban spaces: a cross-cultural study in the hot arid climate. Int J Biometeorol 62(10):1901–1909. https://doi.org/10.1007/s00484-018-1592-5
68. Fang Z, Lin Z, Mak CM, Niu J, Tse KT (2018) Investigation into sensitivities of factors in outdoor thermal comfort indices. Build Environ 128:129–142. https://doi.org/10.1016/j.buildenv.2017.11.028
69. Huang J, Zhou C, Zhuo Y, Xu L, Jiang Y (2016) Outdoor thermal environments and activities in open space: an experiment study in humid subtropical climates. Build Environ 103:238–249. https://doi.org/10.1016/j.buildenv.2016.03.029
70. Antonini E, Vodola V, Gaspari J, De Giglio M (2020) Outdoor wellbeing and quality of life: a scientific literature review on thermal comfort. Energies 13(8):2079. https://doi.org/10.3390/en13082079
71. Havenith G (2007) Metabolic rate and clothing insulation data of children and adolescents during various school activities. Ergon 50(10):1689–1701. https://doi.org/10.1080/00140130701587574
72. Tung C, Chen C, Tsai K, Kántor N, Hwang R, Matzarakis A, Lin T (2014) Outdoor thermal comfort characteristics in the hot and humid region from a gender perspective. Int J Biometeorol 58(9):1927–1939. https://doi.org/10.1007/s00484-014-0795-7

Analysis of Noise Levels in Cities of Uttarakhand and its Compliance to UN Sustainable Development Goal (SDG)-11

Tarumay Ghoshal, Naveen Singhal, Shailey Singhal, and Jabrinder Singh

Abstract Sustainability is the key of life. As we understand the earth is the only known space suitable for habitat in the whole universe it is our prime responsibility to protect and nourish this beautiful living world in harmony with nature so that present and future generations can live in a healthy and sound environment. Unfortunately, in most of the cases this basic principle of sustainability is being deviated due to rapid urbanization and industrialization. The environmental resources: Air, water and soil being polluted at a faster rate than possible recovery. Noise Pollution is such type of a problem being aggravated due to heavy traffic, industries, and non-compliance in the designated zones. Noise Pollution comes with various adverse auditory and non-auditory effects on living beings which are leading to health issues these days. Uttarakhand is also facing increase in noise pollution level due to migration, tourist activities, heavy traffic flow, congested areas with non-compliance of noise rules. The Ambient noise level is largely being deviated from standards. This is a great concern for cities like Kahsipur, Haridwar and of course Dehradun which is announced as smart city in Uttarakhand. United Nations Sustainable Development Goal (SDG) 11 says about sustainable cities and communities. In this context, it very much important to have check and balance in noise level and its compliance to SDG for maintaining a heathy environment. Therefore, this study represents a detailed analyses of Noise level data from various locations of Uttarakhand and their correlation with Noise Pollution rules and SDG 11.

Keywords Noise · SDG · Sustainability · Pollution

T. Ghoshal (✉) · N. Singhal · J. Singh
DIT University, Dehradun 248009, India
e-mail: tarumay.ghoshal@dituniversity.edu.in

S. Singhal
UPES, Dehradun, Uttarakhand 248007, India

A. K. Sharma and A. Joshi (eds.), *Proceedings of the 3rd International Conference on Opportunities and Challenges for a Resilient Future*, Advances in 21st Century Human Settlements, https://doi.org/10.1007/978-981-96-0931-4_12

1 Introduction

Sustainability is way of life. It is the principle that ensures human is taking care of surrounding environment which comprises both living and non-living beings through a holistic approach. Also sustainable principles reduce conflicts between nature and human and affirm healthy and prosperous surrounding where present and future generations can live. Modern cities which are backbones of societal development and progress must be built following all such principles to ensure healthy surroundings where every living being can stay peacefully. Unfortunately, it is observed that human induced activities are the sources of many environmental and health problems which pose threat to existence of life forms as well as cause harm to diverse ecosystems. Pollution of environmental aspects like air, water and soil are prime examples which are directly or indirectly related to human activities. One such example is Noise Pollution which conventionally considered as part of air pollution. It is basically any undesirable or discomforting sound that negatively affects the health and well-being of humans and other living beings. Noise is measured in Decibels (dB) unit. 1 dB is the lowest sound that the human ear can detect. As per the advisory by the World Health Organization (WHO) noise exposure levels must not cross 70 dB for a 24-h period, and 85 dB for a 1-h period to avoid hearing problems [1]. Noise is characterized by (i) Intensity, (ii) frequency, (iii) Variation, (iv) Duration and (v) Personal Attitude. Intensity refers loudness or levels of Noise. Noise level more than 10dB causes irritation. Similarly, higher frequency, abrupt variation and longer duration noise cause more disturbances [2]. Increasing noise levels are becoming problems for modern cities which are primarily attributed to operation of heavy machines in Industry, vehicles, loud speakers, crackers burning in festivals, construction work etc. Especially cities with narrow roads and heavy traffics are encountering these problems more where traffic jam is common. Noise has several adverse effects on health which are primarily divided in auditory and non-auditory categories. Auditory effects are (i) Acoustic Trauma, (ii) Tinnitus, (iii) Temporary and permanent hearing losses [3]. Acoustic trauma represents sudden hearing loss due to loud noise whereas Tinnitus is perception based buzzing in ear even in the absence of real noise. Temporary and permanent hearing losses are results of prolong noise beyond human capacity due to temporary or permanent shifting of threshold levels of hearing capacity. Non auditory effects of noise include (i) irritation, (ii) Heartbeat changes, (iii) Sleep disturbances etc.

Hence, for a sustainable and smart city these problems should be addressed. United Nations (UN) sustainable development goal (SDG) 11 advocates for sustainable cities and communities where human induced problems such as noise pollution need to be mitigated with technological solutions and city planning. SDG 11 pertaining to noise level reduction is very much related to health and well-being (UN SDG 3), resilient infrastructure, sustainable transportation. Therefore, city planners must focus on these issues in the context of smart city.

The state of Uttarakhand of India situated in the foothills of Himalaya has always been popular to travellers, nature lovers for the natural diversity in its rivers, mountains, forests and ecosystems. Moreover, the state is known as "the land of the gods" (Devbhumi) because of its number of Hindu pilgrimage sites. The holy rivers: Ganga, Yamuna, Alakananda and many more are flowing through the state always attract religious people to visit to the cities and towns situated near those rivers. Due to being gifted with such wealth of Natural diversity and serenity, people from other states have always tried to relocate to the cities of Uttarakhand for residential purpose or to spend weekend destination. Over the years, cities of the state have grown with infrastructure development, public transport connectivity, academic institutions, industries. However, with this positive side of development, points of concerns are also increasing with respect to air and water pollution and capacity management. The growing human activities in the cities like Dehradun, Haridwar are leading to these problems multi fold. Noise pollution is one among such problem in the cities of Uttarakhand. In a recent study by [4] Noise levels were found quite high throughout the year near SIDCUL industrial zone and Siwalik Nagar residential area of Haridwar which were responsible for auditory and non-auditory effects on people. Semwal et al. [5] took observations at various locations around clock tower of Dehradun and found noise levels quite alarming in both residential zones and commercial zones whereas silence zones showed wide variations especially near hospital area which was great concern. Chauhan [6] observed that at various prominent locations of Dehradun, noise level was above prescribed noise standard of CPCB, India. They considered both day and night time sound levels in the city for the study and non-compliance was a great concern. Madan et al. [7] studied the noise levels in Haridwar city during 2010 'Kumbh Mela' a famous religious festival in India and found that in all the bathing 'ghats' near Ganga the noise levels ranged 50–90 dB which were above the ambient noise standards. Also people felt headache, annoyance, lack of concentration due to impacts of increased noise level as per their study. A similar study by [8] also indicated the same results before. Not only Dehradun and Haridwar, but other cities like Haldwani, Nainital and Srinagar are also facing same. Research and media reports recently have expressed concerns of noise levels in these cities also [9, 10]. So, with increasing urbanization and modernizations these cities are also being engulfed by Noise pollution which is very much concern for all city dwellers.

The UN defined sustainable development goal SDG 11 focuses on SDG 11.2: Affordable and Sustainable Transport Systems, 11.3: Inclusive and Sustainable Urbanization, 11.6: Reduce The Environmental Impact of Cities, 11.8: Strong National and Regional Development Planning [11]. Government of India has adopted the SDG 11 policy framework along with other sixteen SDGs in overall planning and development and to achieve target of SDG by 2030 as specified by UN [12].

Therefore, the cities which are continuously growing in various states of the country need to apply the framework appropriately for achieving the SDG target and to ensure a sound and sustainable society for all citizens. Same is expected for the cities of Uttarakhand. Being close proximity with wide natural resources and diversity, the cities and towns of Uttarkhand needs more focus as any negative effect from human activities will have serious consequences in future. Especially

Table 1 Ambient air quality standards with respect to noise

Area code	Category of area/zone	Limits in dB(A) Leq	
		Day time	Night time
(A)	Industrial area	75	70
(B)	Commercial area	65	55
(C)	Residential area	55	45
(D)	Silence zone	50	40

Dehradun which has been announced as smart city in the state needs more planning and implementation in more scientific ways.

Hence this study has focussed on noise level data of several cities of the state, identification of the problems and gaps and recommendation for appropriate steps. Also in this study an attempt has been taken to correlate the solutions and policies needed for noise level management with SDG framework (SDG 11).

2 Data and Methodology

The Noise level data year wise have been collected from the website of Uttarakhand Pollution Control Board. The website regularly publishes data which have been collected and analysed during festival and non-festival time. For analysis in this study the equivalent continuous sound pressure level termed as Leq has been used which is measured in dB(A) that indicates the time weighted average of the level of sound in decibels on Scale [13]. The data are analysed primarily for following sites: (i) Doon Hospital Dehradun, (ii) Clock Tower, Dehradun, (iii) Race course, Dehradun, (iv) Mallital, Nainital, (v) Tikonia Chowk, Haldwani, (vi) Pentagon Mall Chowk, Sidcul, Haridwar, (vii) Govt. Hospital, Rudrapur, (viii) M.P. Chowk Kashipur. For all these sites, six-year monthly data have been compared with the Noise Level standards given by Noise Pollution rules 2000 in India [14]. The standards are given below in Table 1.

Lastly, statistical analysis has been performed on these datasets such as standard deviation and Trend Analysis.

3 Results and Discussion

The monthly Noise level data for consecutive five years: 2017–2023 is provided in Fig. 1a–d. Figure 1a gives alarming situation as the Doon Hospital area is under silence zone and it is expected that Noise level will be between 40–50 dB. On the contrary, continuous and year to year non-compliance being observed in the area where health related sensitive situation prevails. Figure 1b shows the Noise levels at Clock Tower region where 55–65 dB (Commercial zone) is expected. Although the

Fig. 1b clearly indicates that except in initial phase of 2021 none of the instances in five years have met the standards. Clock tower region is one of the most significant position in Dehradun city where traffic load is huge due to proximity of railway station, main market area and connecting roads to schools and various offices. The area often becomes congested especially during office time and school time. Figure 1c represents noise level for Mallital, Nainital region (commercial zone). The results show more or less satisfactory as per the standards for commercial zone is concerned. However, the significant variation is observed during winter time as traffic is expected more during this time for pool of tourist activities. The variation slows down but continues during summer and reduces during rainy seasons. The variation of noise primarily responsible due to tourist activities. Figure 1d represents Noise level for Tikonia Chowk (commercial zone), Haldwani where except 2023 the other years show values near the limit however with abrupt variations. The analysis also reveals huge variations in Noise levels between in pre and post 2022. This is most prominent in Fig. 1a-c and less in 1d. These variations are undoubtedly attributed to increase in traffic volume in Dehradun City as well as increased tourist activities in hill city Nainital [15–18]. These factors are increasing multi-fold during 2022 to 2023 due to enhancement in road conditions and communications [19]. Similar situation is observed near Race course (Residential Zone), Dehradun where after the year 2021 significant rise in noise levels are found (figure not shown). Due to unavailability of data before 2021, a six-year comparison could not be made for (i) Pentagon Mall Chowk, SIDCUL, Haridwar, (ii) M.P. Chowk Kashipur and (iii) Govt. Hospital, Rudrapur. However, it was found the noise levels during late 2021 to 2023 were quite high as compared to ambient standards. Figure 2 represents trend analysis of average Leq versus Year for the area near Doon Hospital. It can be easily found that the Noise level is continuously increasing in this silence zone which is not at all desirable. Table 2 summarizes the Trend Analysis results for the all the sites. The data show that positive slopes of the trend. It indicates over the time noise level is increasing drastically all over the Uttarakhand. Data also indicates that silence zones like hospital area needs serious planning because of rapid increase in noise pollution level. This observation is further strengthened by standard deviation values for the two silence zones: Doon Hospital and Govt Hospital Rudrapur. The standard deviations for the two sites were found quite high (not shown) compared to other sites that indicate abrupt sound level changes which can only be possible due to heavy vehicle activities during peak time of hospitals or diverted traffics near the area. Table 3 gives more clarity to the noise level non-compliance with respect to ambient standards. It is observed that in recent two years 2022–2023 all the sites are experiencing high deviation from the standards which is alarming. Although Mallital area of Nainital shows compliance with respect to standards but the value indicates soon it will enter in the non-compliance zone. Same trend is for Tikonia Chowk, Haldwani.

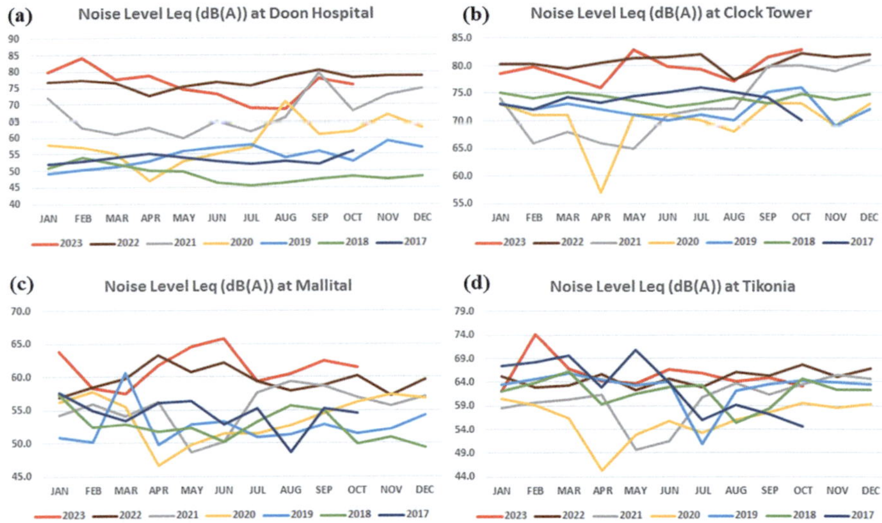

Fig. 1 Monthly plot of noise level Leq at various sites for 2017–2023

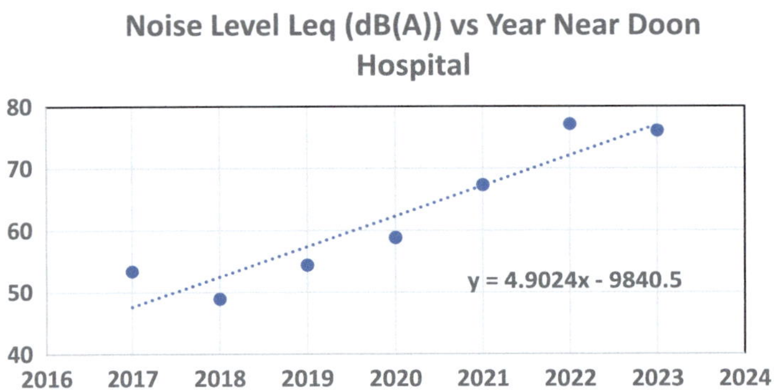

Fig. 2 Trend analysis-noise level Leq versus year at near Doon hospital

Table 2 Trend analysis results of Leq versus year

Area	Trend analysis results	
	Slope	Intercept
Doon hospital, Dehradun	4.9024	−9840.5
Clock tower, Dehradun	1.1377	−2223.4
Race course, Dehradun	2.736	−5467.5
Mallital near NPP office Nainital	1.3563	−2684.1
Tikonia Chauraha Nainital Road, Haldwani	0.392	−729.91
Govt. Hospital, Rudrapur	8.2567	−16,639
Pentagon mall Chowk, SIDCUL, Haridwar	1.0006	−2.2922
M.P. Chowk Kashipur	6.0142	−12,077

Table 3 Percentage deviation from standards (Leq)

Area	Percentage deviation from standards (Leq)						
	2023	2022	2021	2020	2019	2018	2017
Doon hospital, Dehradun	52	54	35	18	9	−2	7
Clock tower, Dehradun	22	24	12	8	11	14	13
Race course, Dehradun	24	28	10	−7	1	−3	1
Mallital near NPP office Nainital	−5	−8	−15	−17	−19	−19	−16
Tikonia Chauraha Nainital Road, Haldwani	1	0	−8	−14	−3	−5	−3
Govt. Hospital, Rudrapur	27	14	−6	NA	NA	NA	NA
Pentagon mall Chowk, SIDCUL, Haridwar	15	12	23	NA	NA	NA	NA
M.P. Chowk Kashipur	40	38	25	NA	NA	NA	NA

4 Conclusion and Recommendation

The Noise level data analysis indicated a very distinctive pattern for pre and post 2022 year scenarios which have marked significant rise in Noise level as compared to year 2017. This indicates that within last six years the human activities have become Multi-fold which are attributed to rapid urbanization and increase in heavy traffic in various cities. Remarkably Silence zones were found to be more disturbed compared to other sites. Government agencies must focus on these areas as these are highly sensitive from health perspectives. More than industrial activities abrupt vehicle sound, high pitch horns, mics and speakers are being increased in the cities [20]. The add-ons are crackers burning and construction activities being carried at the sites [21, 22]. Narrow roads are some of the reasons having less vehicle capacity and create traffic jam which lead to sudden rise in Noise level [23]. The study is also indicating that although human activities are increasing with the pace of urbanization in the cities of Uttarakhand but the principles of sustainability has somewhere been overlooked. In order to improve the situation and to comply the expectations of sustainability following steps are further recommended in respect to Noise level management.

4.1 SDG 11.2: Affordable and Sustainable Transport Systems

The state government has initiated electric Bus service in Dehradun which is a great step as this public transport system do not emit pollutants and having vehicular noise level sufficiently low in comparison to conventional fuel driven transit system. However, the electric bus service is available in few roots only which need to be operated in other roots of Dehradun as well as in other cities of the state. Electric four wheelers and two wheelers are to be promoted in the cities. Regular maintenance records of the vehicles need to checked to avoid sound from faulty parts. The state of

Uttarakhand is still waiting for metro rail connectivity which needs to be accelerated. Provisions for cyclists and pedestrians to be made on the roads wherever possible to reduce dependency on vehicles. Vehicle speeds need to be checked by deploying sensors as part of the policy which is partially enabled but to be made widely available.

4.2 SDG 11.3: Inclusive and Sustainable Urbanization

The cities of the state needs noise barriers to be installed at the sites facing severe non-compliance. The new offices, schools, universities, industrial units, health care units are to be set up in well planned manner to get less exposure to noise. Roads need to be widened for avoiding traffic jams. Illegal encroachment of Government areas need to be dealt with legal provisions. Wherever possible parallel and bye pass roads need to be constructed to divert traffic especially near silence zones. The roads need to be equipped with smart traffic light with sensor to stop unnecessary horns of vehicles with the concept: 'more horn more waiting' [24].

4.3 SDG 11.6: Reduce the Environmental Impact of Cities

The cities need to progress towards zero emission and reduce dependency on personal vehicles and increase use of public transit system. Unplanned construction activities need to be stopped. Public gathering or any type of debate or use of speakers near school-college area and silence zone need to be absolutely prohibited. Ambient standards at respective zones given in Table 1 need to be implemented in letter and spirit.

4.4 SDG 11.8: Strong National and Regional Development Planning

Expansion and cross-over of ambient standard zones need to be crucially monitored. During construction activities, time restriction, temporary instalment of noise barriers, battery powered energy storage systems need to be used in a smart city as part of planning [25]. Promotion of sound insulation material utilization in building construction need to be adopted in policy. Legal enforcement for Non-Compliance to Noise Pollution Rules to be enacted strictly.

The study was limited to few selected sites. The authors will communicate more on the Noise contour map focussed on specific zones in coming future.

Acknowledgements The authors are extremely grateful to Uttarakhand Pollution Control Board for utilizing the noise pollution data available from its website.

References

1. Centers for Disease Control and Prevention (2018) Public Health and Scientific Information. https://www.cdc.gov/nceh/hearing_loss/public_health_scientific_info.html
2. Dreamtech Engineering Textbooks (eds) (2015) Energy, environment, ecology, and society
3. Canadian centre for occupational health and safety (2023) Noise-auditory effects. https://www.ccohs.ca/oshanswers/phys_agents/noise/noise_auditory.html
4. Saini P, Sharma N, Kumar A, Tyagi S, Sharma S (2023) Analysis of noise pollution level in and around SIDCUL area in district Haridwar (Uttarakhand) India. Environ Conserv J 24(4):264–268
5. Semwal S, Upreti BM (2019) Durational assessment and impact of noise levels in different localities in Dehradun Valley: a capital city of Uttarakhand. J Emer Tech Inn Res 6(6):725–729
6. Chauhan A, Pande KK (2010) Study of noise level in different zones of Dehradun city, Uttarakhand. Rep Opin 2(7):65–68
7. Madan S, Pallavi P (2010) Assessment of noise pollution in Haridwar city of Uttarakhand state, India during Kumbh Mela 2010 and its impact on human health. J Appl Nat Sci 2(2):293–295
8. Chauhan A (2008) Study of noise pollution level in different places of Haridwar and Dehradun city (India). Environ Conserv J 9(3):21–25
9. Panwar M, Thapliyal S (2018) Estimation of noise pollution in the hill town Srinagar (Uttarakhand): a geographical study. Ann Natl Assoc Geogr, India 38(1)
10. Verma R (2022) Pollution in Uttarakhand, Jagran. https://www.jagran.com/uttarakhand/nainital-noise-pollution-is-high-in-haldwani-tikonia-23122743.html
11. The global goals (n.d.) Sustainable cities and communities. https://www.globalgoals.org/goals/11-sustainable-cities-and-communities/
12. Aayog N (2023) An overview of SDGs. https://www.niti.gov.in/overview-sustainable-development-goals
13. Uttarakhand Pollution Control Board (n. d.) Noise quality, Uttarakhand pollution control board. https://ueppcb.uk.gov.in/pages/display/97-noise-data
14. Environment Law Research And Guidance Foundation (n.d.) The noise pollution (regulation and control) rules. 2000. https://wwfin.awsassets.panda.org/downloads/the_noise_pollution__regulation_and_control__rules__2000.pdf
15. Alam J, Joshi A, Ahmed R (2020) Impact of seasonality in tourism of Nainital. J Uttarakhand Acad Adm (JUAAN) 1(1)
16. Bond (2020) Increase in Nainital's tourist inflow is a sign that tourism activities are getting back to normal: UTDB secretary. Hotelier India. https://www.hotelierindia.com/business/12443-increase-in-nainitals-tourist-inflow-is-a-sign-that-tourism-activities-are-getting-back-to-normal-utdb-secretary
17. Srivastava AK, Bisht BS, Uniyal RS (2022) Building smart and sustainable cities: a case study of Dehradun City, Uttarakhand, India. Orient Anthropol 22(1):41–50
18. Tyagi T (2021) Traffic volume 'thrice' the infra capability in Doon, year-end sees sharp increase. The Times of India. https://timesofindia.indiatimes.com/city/dehradun/traffic-volume-thrice-the-infra-capability-in-doon-year-end-sees-sharp-increase/articleshow/88597367.cms
19. Press Trust of India (2022) Roads being built in Uttarakhand will open its doors to prosperity: Gadkari, business standard. https://www.business-standard.com/article/current-affairs/roads-being-built-in-uttarakhand-will-open-its-doors-to-prosperity-gadkari-122010401183_1.html
20. Chauhan P, Sanotra S (2017) Assessment of noise pollution level in different sites of Dehradun city, Uttarakhand. Environ & Ecol 35(3D):2433–2436
21. Ahuja A (2019) Pollution levels shoot up in US Nagar on Diwali. Times of India. https://timesofindia.indiatimes.com/city/dehradun/pollution-levels-shoot-up-in-us-nagar-on-diwali/articleshow/71810067.cms
22. Shukla S (2019) Pollution increased due to fireworks. Jagran. https://www.jagran.com/uttarakhand/nainital-fire-and-air-pollution-increased-due-to-fireworks-kashipur-to-dissolve-poison-in-air-and-rudrapur-topped-in-noise-pollution-19707025.html

23. Mishra S, Jain A (2023) Long weekend rush chokes main tourist towns in Uttarakhand. Times of India https://timesofindia.indiatimes.com/city/dehradun/long-weekend-rush-chokes-main-tourist-towns-in-uttarakhand/articleshow/99349534.cms
24. Kumar A, Agarwal K, Kulraj K, Roy S (2021) Smart traffic light using sound sensor. Int J Creat Res Thoughts 9(5):449–460
25. Atlas Copco (2024). How to reduce noise in construction operations? https://www.atlascopco.com/en-in/construction-equipment/resources/green-solutions-guide/how-to-reduce-noise-in-construction-operations

Assessing the Influence of Urban Development on the Urban Heat Island Through Remote Sensing and Geospatial Techniques in Jhansi, India

Rubika Singh and Neetu Kapoor

Abstract In the context of increasing urbanization and climatic change, the global phenomenon of Urban Heat Island (UHI) has become a critical concern, with urban surface temperatures steadily increasing across cities worldwide. UHI has intensified over recent decades, consequently exerting a substantial impact on the local climate and affecting the quality of human life. UHI is a phenomenon in which urban regions experience higher temperatures compared to the adjacent outlying areas, as a result of urbanization, population pressure, industrialization, insufficient vegetation, and transportation systems. The present study is focused on Jhansi, aiming to comprehensively analyze the formation of UHIs in the region. By employing a multidimensional approach, the research investigates the dynamic changes in land use/land cover and fluctuations in urban surface temperature in Jhansi as well as correlation analysis is determined through remote sensing and meteorological data, and Geographic Information System (GIS) analysis. The findings of the study indicate that there is an overall increase in the intensity of UHI from 2001 to 2021 and this is because of the conversion of permeable surfaces like open spaces, water bodies, and vegetation towards impermeable surfaces like buildings, roads, and other infrastructure. Higher temperatures were observed in urban areas characterized by dense built-up structures and areas devoid of vegetation, but lower temperatures were seen in regions with vegetation and waterbodies. The findings also reveal that LST demonstrates a negative correlation with NDVI, signifying the cooling impact of green spaces, and a positive association with NDBI, underscoring the warming effect of built-up areas on the urban microclimate. These findings are crucial for urban planners and policymakers to devise effective mitigation and adaptation strategies while dealing with increasing urbanization and climate change. The study helps in developing an in-depth analysis of UHI formation, with implications extending beyond Jhansi to urban centers globally, which can become a foundation for evidence-based decision-making for more resilient and sustainable urban environments.

R. Singh (✉) · N. Kapoor
National Institute of Technology, Hamirpur 177005, India
e-mail: 22MAR002@nith.ac.in

A. K. Sharma and A. Joshi (eds.), *Proceedings of the 3rd International Conference on Opportunities and Challenges for a Resilient Future*, Advances in 21st Century Human Settlements, https://doi.org/10.1007/978-981-96-0931-4_13

Keywords Urban heat island · Urbanization · Land use and land cover · Land surface temperature

1 Introduction

Cities are significant contributors to greenhouse gas (GHG) emissions; in the last 60 years, the pace of urbanization worldwide has increased by 21% [1, 2]. Numerous environmental concerns have been exacerbated by temperature fluctuations carried on by changes in the physical landscape that are caused by urbanization [3]. Due to urbanization, a naturally occurring landscape with an abundance of vegetation cover and pervious space is converted into an impervious area. This process changes the patterns of Land Use Land Cover (LULC) and has an impact on the quantity and quality of urban greenspace [4, 5]. The effects of increasing urbanization on the environment are now evident, as cities exhibit greater daytime and nighttime temperatures than outlying rural areas [1]. This phenomenon is known as the "urban heat island (UHI) effect." The primary contributing factors to the formation of UHI are changes in the physical properties of the surface (thermal capacity, albedo, and heat conductivity) that result from the replacement of vegetation with concrete and asphalt; a reduction in the amount of surface moisture available for evapotranspiration; modifications to the near-surface flow and irradiative fluxes resulting from complicated street and tall building geometry; and anthropogenic heat [3]. UHI is divided into two categories: Surface Urban Heat Island (SUHI) and Atmospheric Urban Heat Island (AUHI). SUHI calculations depend on Land Surface Temperature (LST), while calculations for AUHI rely on air temperature [6]. The study mainly focuses on SUHI. The effects of UHI can be particularly severe, leading to issues like deteriorating living conditions, rising energy costs, elevated ground-level ozone, and even higher rates of mortality [7].

According to previous studies, LST and LULC are key variables for the study of UHI [8]. Understanding the effects of LULC changes on urban climates requires an accurate study of these changes and mapping of continuing land use changes, which can assist policymakers in environmental management [9].

Studies showed that because of the world's fast urbanization, industrialization, and population growth, the study of the UHI effect has become increasingly important. Recent research based on the combination of GIS and temperature data from remote sensing indicates that urbanization is having an adverse impact on Indian cities. [5, 6, 10–14]. The purpose of the current study, which focuses on Jhansi, is to comprehensively examine the way UHI developed in the area.

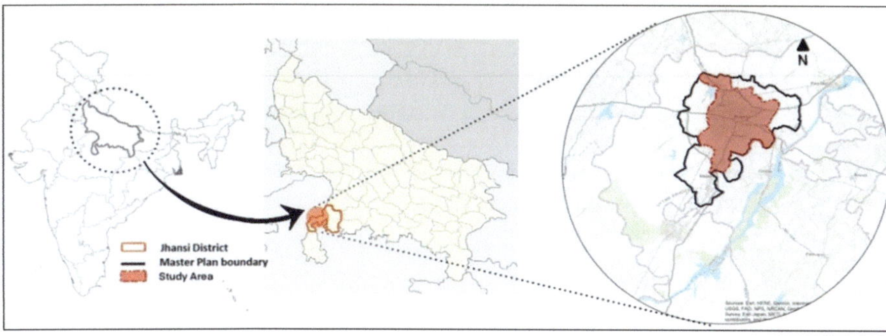

Fig. 1 Location map

2 Material and Methods

2.1 *Study Area*

Jhansi a historic city in Uttar Pradesh state of India (Fig. 1), is located at 25.44 N 78.56 E. The divisional and district administrative centers are located in Jhansi. The study area of Jhansi city is 163 km^2. The city experiences hot, dry summers and cold winters due to its mild subtropical to semi-arid climate. The mean daily minimum temperature of 9.20°C and the mean daily maximum temperature of 24.10 °C. January is the coldest month of the year. The warmest month is May, with a mean daily maximum temperature of 46 °C and a mean daily lowest temperature of 28.80°C. The district experiences 850 mm of annual rainfall on average, with the monsoon months of June, July, August, and September accounting for 91% of this total. The Jhansi district is located in Uttar Pradesh's belt of regions susceptible to drought. For the month, the relative humidity is 41% on average. The wind speed on average is 4.8 km/h. Jhansi has a significant rail and road intersection. The North–South Corridor, which connects Kashmir to Kanyakumari, passes through Jhansi. Since this city is situated along the East–West route, the development of its real estate and infrastructure has accelerated recently.

2.2 *Methodology*

The current study was carried out using remote sensing data that was downloaded from the USGS Earth Explorer website at 30 m resolutions, i.e., for the years 2001 and 2021, with zero cloud cover. Table 1 lists the specifications of the satellite data that was acquired for this study. Landsat 7 ETM + , and Landsat 8 OLI/TIRS satellite data are used to prepare LULC and LST as shown in Fig. 2. Similarly, Landsat satellite data is also used to construct the Normalized Difference Vegetation Index (NDVI)

Table 1 Data Specifications of different years

Satellite	Sensor	Date	Path & Row
Landsat 7	ETM +	18/09/2001	145, 042
Landsat 8	OLI/TIRS	3/10/2021	145, 042

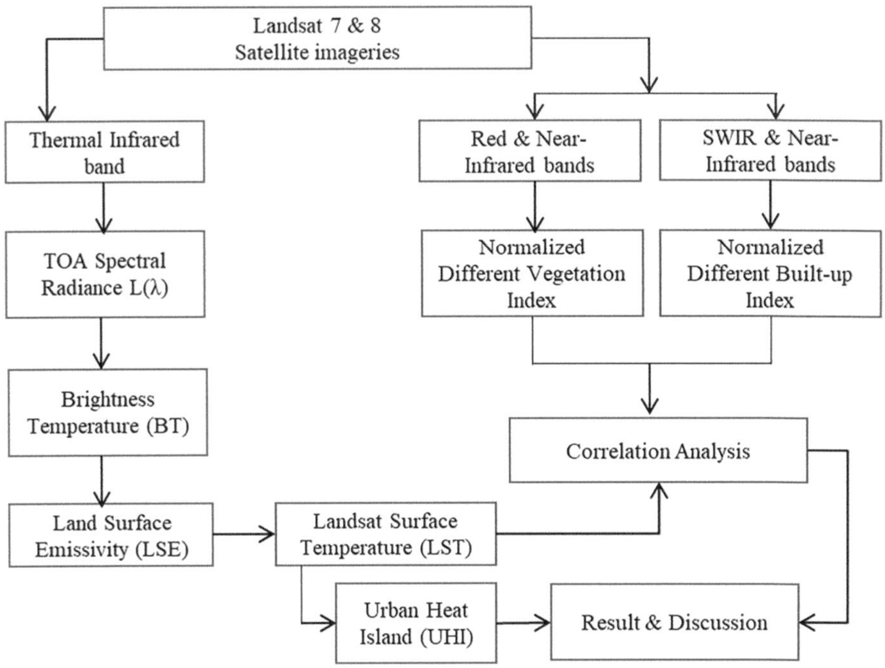

Fig. 2 Methodology

and Normalized Difference Built-up Index (NDBI). Additionally, UHI is computed using LULC and LST data, and its relation is associated with NDVI and NDBI. The boundary shape file, which is shown in Fig. 1, was clipped from the satellite data to define the study region. The boundary shape file was obtained from the Survey of India (SOI).

2.3 Image Pre-Processed and Classification

Landsat 8 products have eleven bands total, of which band 10 is a thermal band; and Landsat 7 ETM + has eight bands total, of which band 6 is a thermal band. Using ArcGIS 10.7 and Environment for Visualizing Images (ENVI 5.3) software, layer stacking, and geometric and atmospheric correction were applied to Landsat 7 and 8 satellite data images. For image classification, a supervised classification

Table 2 Description of LULC classes

Sr. no	LULC classes	Description
1	Built-up land	All infrastructure, including roads, pavements, man-made structures, communities, settlements, mixed-use and industrial regions, and residential and commercial buildings
2	Vegetation	Trees, grasslands, mixed forests, gardens, parks, and play areas, as well as natural vegetation
3	Cropland	Agriculture area
4	Bare land & Fallow land	Open space, bare soils, developed land, excavation sites, fallow land, earth and sand land infills, building sites, and the other land cover types
5	Water bodies	Rivers, lakes, ponds, canals, marshy ground

technique was used. Training sets were chosen for classification in this process using a Maximum Likelihood Classifier (MLC). Pixels are categorized into classes using a statistical decision-making procedure through MLC. Image classification was used to define the LULC types into five classes: built-up, vegetation, cropland, bare and fallow land, and water as shown in Table 2.

2.4 Spatial Metrics and Validation

Assessment of classification accuracy was carried out on the resulting classified imagery using the kappa index and error matrix, which allow for the distinction between expected and ground-truth classification. The distinct mixture seen in urban areas was comprised of residential property, roadways, railroads, water bodies, and flora. The Kappa Index (I), which can be computed using Eq. (1) is used to assess the degree of agreement between the classifier's result and the reference dataset while accounting for the accuracy of a random classifier. For the accuracy assessment, it is advised that at least 50 samples be obtained for each LULC class in the error matrix to reduce the possibility of a biased sample [15].

$$K = (Observed\ accuracy - Chance\ accuracy)/(1 - Chance\ accuracy) \quad (1)$$

2.5 Calculation of LST, NDVI and NDBI

The process of estimating LST using ArcGIS 10.7 requires converting digital digits (DN) to radiances (Lλ) and measuring surface temperature using brightness maps using emissivity adjustments and radiance brightness temperatures. Thus, the LST inversion bands were determined to be band 6 of Landsat 7 and band 10 of Landsat 8.

Digital numbers to spectral radiance conversion; The digital number (DN) values of the TIR pixel data are first converted into radiance. To obtain the TIR band radiation for the Landsat 7 TM data [9], Eq. (2).

$$L_\lambda = \{(L_{max} - L_{min}).(Qcal - Qcal_{max})/(Qcal_{max} - Qcal_{min})\} + L_{min} \quad (2)$$

The Landsat 8 TIR band's radiation is derived using Eq. (3)

$$L_\lambda = ML^* \, Qcal + AL \quad (3)$$

where L_λ = spectral radiance, L_{max} = spectral radiance to $Qcal_{max}$, L_{min} = spectral radiance to $Qcal_{min}$, Qcal = Level 1 pixel in DN, $Qcal_{max}$ = maximum quantized calibrated pixel value (1), $Qcal_{min}$ = minimum quantized calibrated pixel value (255), ML = band-specific multiplicative rescaling factor from the metadata (RADIANCE_MULT_BAND_n from the metadata), AL = band-specific additive rescaling factor from the metadata.

$$T_b = K_2 / \, ln \, (K_1 / L_\lambda + 1) \; - 273.15 \quad (4)$$

Calculation of satellite Brightness temperature: Following the conversion of the DN to radiance, the radiance values were transformed using Eq. (4) into brightness temperature.

where the calibration constants K1 and K2 are used, and T_b is the brightness temperature in °C.

Conversion from Kelvin to Celsius: Following that, Eq. (5) was used to estimate the LST using corrected Land surface emissivity (LSE) measurements [10, 16].

$$LST = T_b / \{1 + (\lambda Tb/\rho) * In(LSE)\} \quad (5)$$

where ρ is a constant that can be determined by calculating h · c/σ (1.438 · 10 − 2 mK), where h is Planck's constant = 6.626 · 10 − 34 Js and c = light velocity = 2.998 · 108 m/s, and Tb is brightness temperature (°C) derived from Eq. (4).

Calculation of NDVI: The NDVI, a standardized vegetation index that is useful for monitoring changes in vegetation coverage due to natural occurrences like drought or human activities like construction, was computed to evaluate the quantitative distribution of vegetation cover and its density. It is estimated using Near Infra-red (NIR) and Red (R) bands The NDVI has values ranging from -1 to + 1, dense vegetation is indicated by increasing positive NDVI values, while non-vegetation surfaces, such as water and barren ground, are indicated by values near zero or decreasing negative values [7, 17]. Equation (6) is used to calculate NDVI [14, 18,

19].

$$NDVI = (NIR - R) / (NIR + R) \quad (6)$$

Calculation of NDBI: In previous research, NDBI the most common and widely recognized technique for identifying built-up areas exhibited a strong association with surface temperature [9, 20]. NDBI was computed using Eq. (7) to detect urban areas, where the NIR region is characterized by lower reflectance than that of Landsat 8 OLI shortwave infrared (SWIR) [11].

$$NDBI = (SWIR - NIR) / (SWIR + NIR) \quad (7)$$

NDBI values ranges from −1 to + 1, with values near + 1 typically denoting densely built-up regions.

2.6 UHI Intensity

The UHI map is estimated using Eq. (8) in order to assess the heat variance of Jhansi and the surrounding area.

$$UHI = (T_s - T_{mean}) / SD) \quad (8)$$

where T_s is defined as the LST in kelvin, T_{mean} represents the mean LST in kelvin and SD denotes the standard deviation of the calculated LST map.

3 Results and Discussions

3.1 LULC Change Analysis

Two classified LULC maps of Jhansi City for the years 2001 and 2021 have been cre-

ated as part of the present study. As seen in Fig. 3, the LULC maps of the years 2001 and 2021 are divided into five LULC classes: built-up area, vegetation, cropland, bare and fallow land, and water bodies. It should be noted that the total accuracy was estimated to be 91% in 2001 and 94% in 2021. The values of the Kappa Index for 2001 and 2021 were found to be 0.87 and 0.91, respectively. As Table 3 illustrates, it was noted that there is 20% increase in built-up area between the years 2001 and 2021, rising from 15.18 km^2 to 48.23 km^2. All of the central areas have been set aside for human settlement. Agricultural practices are common in the periphery regions. Also, it was observed that the majority of the area of cropland and bare and fallow land has been transformed into built-up areas from the year 2001 to 2021 as shown in Table 4. The primary causes of the land conversion process are population growth and extensive urbanization. As shown in Table 3, it was observed that there is a 5% decrease in vegetation from the year 2001 to 2021, it was 25.68 km^2 in 2001 and 17.69 km^2 in 2021. It was noted that from the year 2001 to 2021, there was a 6% increase in cropland and a 1% decrease in waterbodies. Also in 2021, there was a 20% decrease in bare and fallow land. The data analysis revealed that over the previous 20 years, Jhansi has experienced substantial growth. The urban area's built-up area has increased through both internal densification and external expansion.

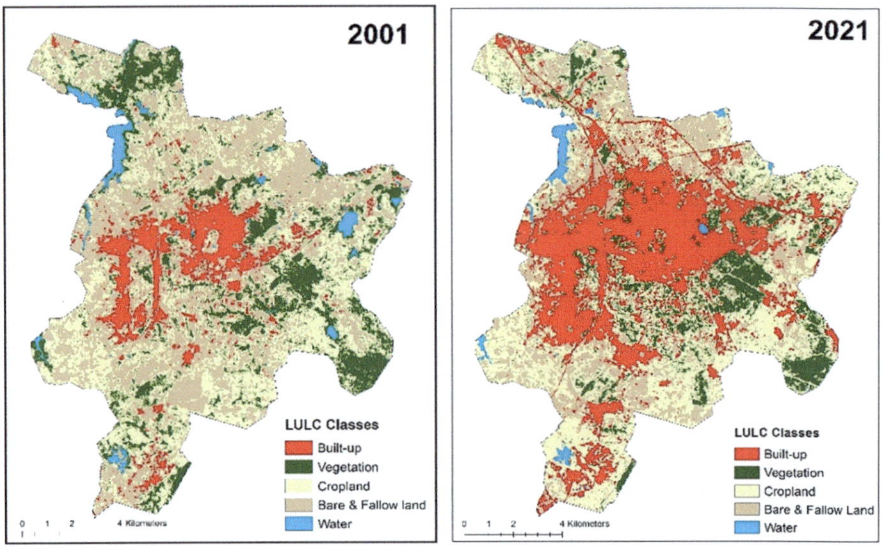

Fig. 3 LULC maps of years 2001 and 2021

Table 3 LULC change dynamics in 2001 and 2021

LULC Classes	2001		2021		2001–2021	
	Km^2	%	Km^2	%	Km^2	Change %
Built-up	15.18	9%	48.23	30%	33.05	20%
Vegetation	25.68	16%	17.69	11%	-7.99	−5%
Cropland	45.94	28%	55.64	34%	9.71	6%
Bare and fallow land	72.42	44%	39.24	24%	−33.18	−20%
Water	4.14	3%	2.55	2%	−1.59	−1%
Total	163.36		163.36			

Table 4 LULC change matrix 2001–2021

LULC Classes		2021 (Area Km^2)					
		Built-up	Vegetation	Cropland	Bare & Fallow Land	Water	Total
2001 (Area Sq.km)	Built-up	14.37	0.07	0.45	0.27	0.03	15.18
	Vegetation	2.92	9.41	10.10	2.88	0.36	25.68
	Cropland	6.19	5.23	22.17	12.29	0.06	45.94
	Bare & Fallow Land	24.25	2.91	21.32	23.66	0.28	72.42
	Water	0.49	0.07	1.61	0.15	1.82	4.14
	Total	48.23	17.69	55.64	39.24	2.55	163.3

3.2 LST Variation

The LST change detection was carried out throughout two temporal periods: 2001 and 2021. The study area observed severe temperature changes in different parts due to urban growth. The LST map for 2001 and 2021 is displayed in Fig. 4. In general, the LST was higher in built-up and bare soil regions than in vegetated and water-covered areas. The maximum and minimum LST in 2001 and 2021 are shown in Table 5. In the years 2001 and 2021, the highest average temperature was recorded in bare and fallow land. Due to low reflectivity in barren areas, more solar radiation causes the soil to be warmer than on urban surfaces. Additionally, the results demonstrated that the central region, which is linked to Jhansi City's core urban setup and has a high population density, contained the majority of urban built-up regions. Additionally, after bare and fallow land, places with high LST values were identified in the city's Centre [21]. The maximum LST values recorded in the built-up area were 28°C in 2001 and 30°C in 2021. Roads, pavement, and the widespread use of tiles in urban areas are examples of impervious surfaces that may contribute to higher LST in certain regions [21]. Furthermore, the lower LST was observed on water bodies, which were followed by vegetation and agriculture. Overall, the findings supported the idea that vegetation has a decreasing effect on changes in LST and that built-up regions and bare ground have an increasing effect.

3.3 Correlation Analysis Between LST, NDVI, and NDBI

To examine the visible and NIR bands of the electromagnetic spectrum, the NDVI, a simple graphical or numerical indicator, is used for remote sensing measurement analysis. The NDVI value for a given pixel is always calculated to be between minus one (-1); no green leaves, however, produce a value that is very close to zero. A value close to $+ 1$ (0.8–0.9) denotes the maximum density of green leaves that can exist,

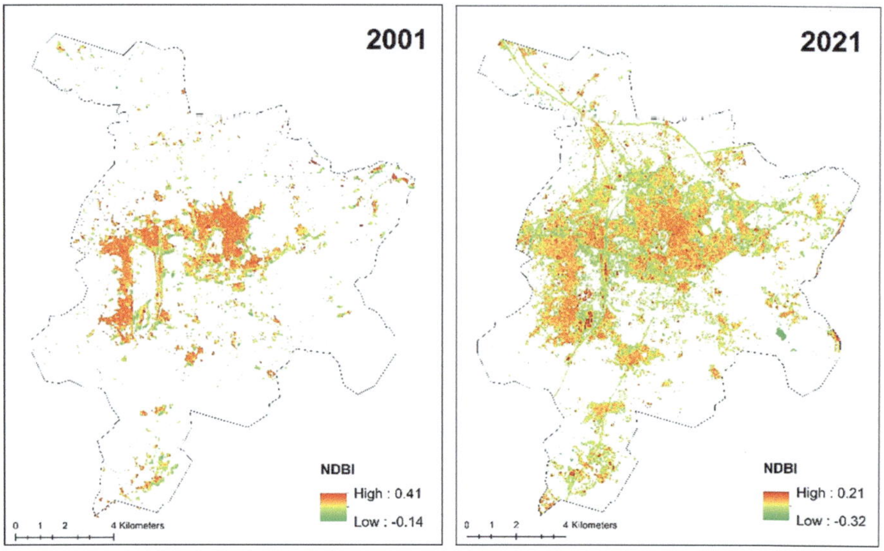

Fig. 4 LST maps of year 2001 and 2021

Table 5 LST over the years 2001 and 2021

Year	LST °C (min)	LST°C (max)	LST °C (mean)	SD
2001	19.57	34.8	27	1.99
2021	22.6	30.8	25.56	0.99

while a zero denotes no vegetation. Figure 5a illustrates the spatial distribution of the NDVI for the years 2001 and 2021 in Jhansi. Built-up area is extracted from urban areas using NDBI, a common indicator. Figure 5b illustrates the spatial pattern of NDBI in Jhansi city. Using ArcGIS 10.7 software, the correlation between LST and NDVI and LST and NDBI was assessed for the years 2001 and 2021. The relation between the LST and NDBI showed a positive correlation and the R2 values for 2001, and 2021 are, respectively, 0.17 and 0.45. the correlation between the LST and NDVI showed a negative relationship because of construction damage to the green area and the resulting high-temperature range at this location. The LST and NDVI correlation shows that the R2 values for 2001 and 2021 are 0.0002 and 0.24, respectively.

3.4 *Evaluation UHI Intensity*

According to Eq. (8), which estimates UHI intensity, Jhansi City recorded a maximum UHI intensity of 5.2 °C and a minimum of −2.3 °C in 2021. In the year 2001, the

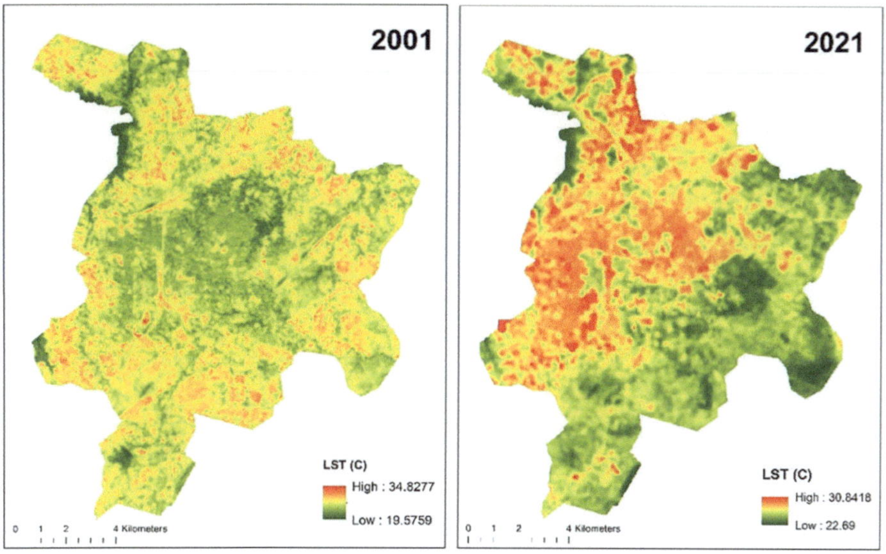

Fig. 5 **a** NDVI of years 2001 and 2021 **b** NDBI of years 2001 and 2021

maximum UHI intensity was 4.1 °C and the minimum was −3.5 °C respectively. As seen in Fig. 6, the city's urban core experienced the maximum UHI intensity in 2021. Urban development, geological processes, air pollution, human heat, impervious surfaces, reduced vegetation, and reduced wind speeds are some of the variables that may cause this kind of change in UHI conditions. Regardless of other environmental considerations, vegetation zones are generally cooler than other places, The increased heat absorption due to the higher thermal conductivity slows down the rate of cooling, which raises the UHI. Therefore, it is most reasonable to assume that having vegetation in an urban environment can lower the UHI value.

3.5 Role of Vegetation in Reducing UHI

The present study shows that due to bare land, increased built-up, and lack of vegetation, there is formation of UHI effect. Adaptation and mitigation techniques help in minimizing the UHI effect. Increasing vegetation cover is one of the easiest ways to mitigate the UHI effect in urban areas. The most effective way to reduce heat island effects is to strategically plant trees and other vegetation to shade pavement and buildings and reduce runoff. For instance, adding more mid-size parks and open spaces for leisure and recreation in urban areas, establishing green parking lots, and putting a green roof, a vegetative layer at the top, vertical greenery, and plant components on the building's façade. The benefits of green roofs are that they reduce nearby air temperature by 2–5°C energy consumption is reduced by 10% [22], and park and

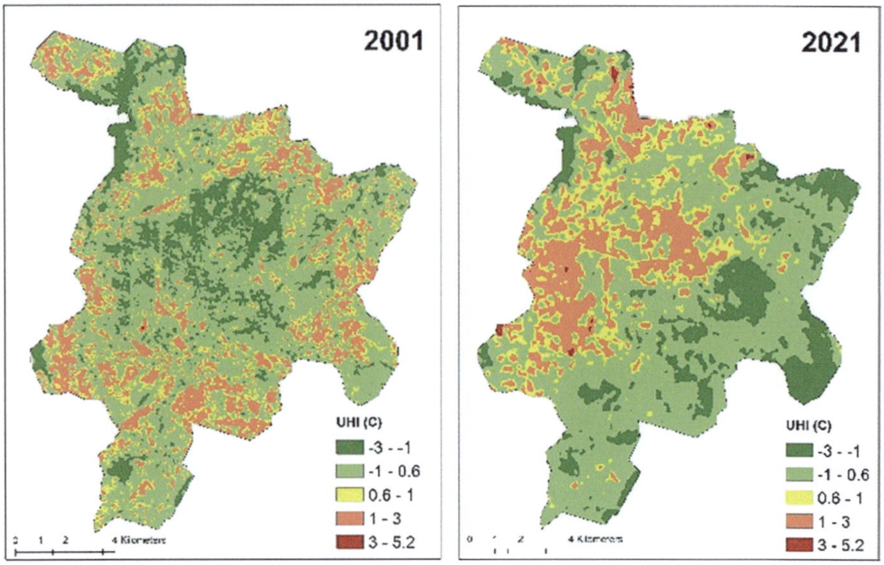

Fig. 6 UHI maps of year 2001 and 2021

open spaces increase the thermal comfort of the urban area. Through evapotranspiration, vegetation enhances the cooling effects and reduces the effect of UHI. In addition, planting trees contributes to the restoration of the ecosystem by enhancing air quality, lowering greenhouse gas emissions, consuming less energy, and creating more green space. During the design stages of urban planning, it is essential to implement methods intended for reducing the effect UHI.

4 Conclusion

The study employs GIS and remote sensing techniques to investigate in detail the way LULC and urbanization affect the UHI in Jhansi. The pattern and distributions of UHI in Jhansi City have been investigated and results reveal an increase in built-up area and reduced greenery. By using ArcGIS software in the research, the LST, LULC, NDVI, NDBI, and UHI of the research region were extracted. For the research, Landsat ETM + and OLI were employed. The findings showed that urbanization affected the spatial distribution of LST and that, as a result of urban development, an urbanized structure, and a decrease in green spaces, the core region of Jhansi had an increase in temperature between 2001 and 2021. The study also determined the intensity of UHI over Jhansi city. According to the analysis, the intensity of UHI in 2021 was 5.2 °C greater than in 2001. The development of built-up land, the presence of barren or rocky surfaces, the presence of fallow land, and the reduction of agriculture, vegetation, and waterbodies are the primary causes of UHI formation. This study also analyses the correlations between the LST and the NDVI as well as the NDBI. There was a negative relation between LST and NDVI, with the UHI effect being less evident in regions with vegetation. However, because built-up areas are warmer than vegetated and farmland, there was a positive connection observed in the relationship between LST and NDBI. However, because the diurnal and seasonal changes of LST and UHI effects are not included, this study has certain limitations. Consequently, seasonal, and diurnal changes can be added to this study in the future to make it even more comprehensive. The findings of the study will support decision-makers and urban planners in managing urban areas sustainably, enhancing green spaces, and lessening the detrimental consequences of rising UHI on the urban environment.

References

1. El-Hattab M, Amany SM, Lamia GE (2018) Monitoring and assessment of urban heat islands over the Southern region of Cairo Governorate, Egypt. Egypt J Remote Sens Space Sci 21(3):311–323
2. Fu P, Weng Q (2018) Responses of urban heat island in Atlanta to different land-use scenarios. Theor Appl Climatol 133(1–2):123–135
3. Faizan M (2020) Assessment of urban heat island using GIS and remote sensing-a case study of Chennai City, India. Available: https://www.researchgate.net/publication/355814388
4. Chen C, Wang L, Yang G, Sun W, Song Y (2023) Mapping of ecological environment based on google earth engine cloud computing platform and landsat long-term data: a case study of the Zhoushan Archipelago. Remote Sens (Basel), 15(16)
5. Mathew A, Khandelwal S, Kaul N (2016) Spatial and temporal variations of urban heat island effect and the effect of percentage impervious surface area and elevation on land surface temperature: Study of Chandigarh city, India. Sustain Cities Soc 26:264–277
6. Shukla A, Jain K (2021) Analyzing the impact of changing landscape pattern and dynamics on land surface temperature in Lucknow city, India. Urban For Urban Green, 58
7. Al-Saadi LM, Jaber SH, Al-Jiboori MH (2020) Variation of urban vegetation cover and its impact on minimum and maximum heat islands. Urban Clim, 34
8. Barat A, Kumar S, Kumar P, Parth Sarthi P (2018) Characteristics of Surface Urban Heat Island (SUHI) over the Gangetic Plain of Bihar, India. Asia Pac J Atmos Sci, 54(2), pp 205–214
9. Lee K, Kim Y, Sung HC, Kim SH, Jeon SW (2022) Surface urban heat island in South Korea's new towns with different urban planning. Environ Monit Assess, 194(5)
10. Sharma R, Pradhan L, Kumari M, Bhattacharya P (2021) Assessing urban heat islands and thermal comfort in Noida City using geospatial technology. Urban Clim, 35
11. Halder B, Bandyopadhyay J, Banik P (2021) Monitoring the effect of urban development on urban heat island based on remote sensing and geo-spatial approach in Kolkata and adjacent areas, India. Sustain Cities Soc, 74
12. Singh P, Kikon N, Verma P (2017) Impact of land use change and urbanization on urban heat island in Lucknow city, Central India. A remote sensing based estimate. Sustain Cities Soc 32:100–114
13. Mohan M, Kikegawa Y, Gurjar BR, Bhati S, Kolli NR (2013) Assessment of urban heat island effect for different land use-land cover from micrometeorological measurements and remote sensing data for megacity Delhi. Theor Appl Climatol 112(3–4):647–658
14. Singh P, Verma P, Chaudhuri AS, Singh VK, Rai PK (2023) Evaluating the relationship between Urban Heat Island and temporal change in land use, NDVI and NDBI: a case study of Bhopal city, India. Int J Environ Sci Technol

15. Congalton RG (1991) A review of assessing the accuracy of classifications of remotely sensed data
16. Weng Q, Lu D, Schubring J (2004) Estimation of land surface temperature-vegetation abundance relationship for urban heat island studies. Remote Sens Environ 89(4):467–483
17. Chuai XW, Huang XJ, Wang WJ, Bao G (2013) NDVI, temperature and precipitation changes and their relationships with different vegetation types during 1998–2007 in Inner Mongolia, China. Int J Climatol 33(7):1696–1706
18. Avdan U, Jovanovska G (2016) Algorithm for automated mapping of land surface temperature using LANDSAT 8 satellite data. J Sens, 2016
19. Roy DP et al (2014) Landsat-8: Science and product vision for terrestrial global change research. Remote Sens Environ 145:154–172
20. Sang L, Zhang C, Yang J, Zhu D, Yun W (2011) Simulation of land use spatial pattern of towns and villages based on CA-Markov model. Math Comput Model 54(3–4):938–943
21. Reisi M, Ahmadi Nadoushan M, Aye L (2019) Remote sensing for urban heat and cool islands evaluation in semi-arid areas. Glob J Environ Sci Manag, 5(3), pp 319–330, 2019
22. Refahi AH, Talkhabi H (2015) Investigating the effective factors on the reduction of energy consumption in residential buildings with green roofs. Renew Energy 80:595–603

Energy and Cost Analysis of Building Envelope for the Warm and Humid Climatic Regions of India

Kishor S. Kulkarni, Garima Singh, and Ankita Paul

Abstract Energy consumption by buildings and their capacity for saving energy are the most widely discussed topics worldwide in the context of a deteriorating energy crisis and global warming. Energy consumption can be decreased by adding or improving thermal insulation to building components. Building envelope optimization is extremely important for energy savings. This study focused on the optimum building envelope from the viewpoint of value engineering, considering both the energy saving and life cycle cost. The study considered a total of six walling and two roofing envelopes. The study examined 12 building envelope assemblies while considering various walling and roofing envelopes. The findings indicate that insulating building envelopes will contribute to reducing energy use. The study's findings support the use of value engineering to obtain the appropriate building envelope.

Keywords Building energy consumption · Life cycle cost analysis · Value engineering · Insulation · Optimum building envelope

1 Introduction

The construction industry in India is growing rapidly. The construction sector consumes about 32% of India's energy production, the residential building consumes about 24% and commercial real estate consumes about 8%. Residential Buildings make up 75% of the construction sector. Residential buildings are one of the major energy consumers so, it is important to reduce the amount of energy consumed by the buildings. Buildings use energy during construction, operation, and demolition throughout their life cycle. The building energy consumption usually accounts

K. S. Kulkarni (✉) · G. Singh · A. Paul
Architecture Planning and Energy Efficiency, CSIR-Central Building Research Institute, Roorkee, Uttarakhand, India
e-mail: kishorsk@cbri.res.in

A. Paul
Vellore Institute of Technology, Katpadi, Tamil Nadu, India

A. K. Sharma and A. Joshi (eds.), *Proceedings of the 3rd International Conference on Opportunities and Challenges for a Resilient Future*, Advances in 21st Century Human Settlements, https://doi.org/10.1007/978-981-96-0931-4_14

for about 40% of the global energy consumption [1]. The largest primary energy consumers and contributors to carbon dioxide emissions and global warming are residential structures. Hence, more attention needs to be paid to energy conservation because of the growing urbanisation, energy crises, and global warming.

Building envelopes can use thermal insulations to lower the energy used by buildings for heating and cooling. The walls, roofs, and foundations of structures, which regulate the passage of heat, humidity, and air, have an impact on how much energy is used in a building. As a result, the building envelope is crucial in regulating heat intake or loss. Passive technology, HVAC-related technologies (heating, ventilation, and air-conditioning), and renewable energy-related technologies are the three divisions of building energy technology. Through passive technologies, it is challenging to encourage the development of building envelopes due to high energy savings and high prices, whereas low energy savings and low costs make the design of building envelopes more prevalent and less competitive. However, it has been demonstrated that when insulation thickness grows, costs for both energy and insulation rise. Determining the ideal and cost-effective thickness of the building envelopes is therefore essential. Optimising thermal insulation in building materials contributes to energy conservation and lower greenhouse gas emissions. The majority of studies on energy usage are either solely focused on energy savings or life cycle costs. The goal of the study is to identify the ideal building envelope for the warm and humid climate of India while taking into account life cycle costs and energy savings. Hence, in the present study value engineering method is adopted to determine the optimum thickness of the envelope. Value engineering is a method of making decisions that will assist in selecting the optimal envelope for the specific climatic conditions while taking into account both energy efficiency and envelope Life Cycle Cost (LCC).

1.1 Background Study

Axaopoulos et al. [2] explained that the most important factors affecting the calculation of optimum thickness are the climatic conditions, insulation material, consumed energy and its cost, level of indoor comfort, the efficiency of cooling and heating systems and the orientation of external walls [2]. Study shows that the wall with the largest heating load yields the greatest economic benefits and has the shortest payback periods. Geng et al. [3] suggested that the Structural Insulation Panels (SIPs) have become an important prefabricated building component, and they have two layers of rigid materials as the skin and core [3]. Studies say that Vacuum Insulation Panels (VIPs) can serve as a cost-effective alternative for insulation of buildings such as SIP in colder regions. A SIP building with SIP-VIP wall combinations is more suitable for colder regions. Al-Homoud [4] study addresses the impact of heat envelope design on the thermal performance of buildings in tropical areas [4]. The study concludes that the efficient use of thermal insulation in buildings improves heating with low operating costs. Amir and Alibaba [5], conducted a study to choose between

Extruded Polystyrene (XPS) and Expanded Polystyrene (EPS) providing a comparative measure in the selection of input assets as we will compare both your properties and real-life simulations for a comprehensive numerical analysis of two items [5]. According to Liu et al. [6] for a country like China has four different climatic zones, it is important to reduce energy consumption by changing the insulation thickness of the external walls of the residential buildings [6]. Nematchoua et al. [7] conclude from the study that, in countries falling under the northern hemisphere, it is advised to insulate east and west-facing walls to decrease the cooling energy of the buildings [7].

Rad, and Fallahi [8], conducted studies on the determination of material and optimum thickness of building insulations and explored other parameters including solar shading, the position of insulation in the wall, the percentage of glazing in the wall, and the type of fuel, and type and efficiency of air conditioning system. The study outcome infers that; the initial cost of the insulation will increase linearly with the increase in its thickness because the cost of the insulation is proportional to its thickness when the size of the cross-section of the insulation doesn't change [8]. Basińska et al. [9] conducted economic analysis by using the Global Cost Method (GCM) and Simply the Payback Period Method (SPBT) [9]. The results after the economic analysis showed that the internal wall insulation systems for low-energy buildings become less effective and do not help in energy saving. Ozel [10] studied the thermal, economic, and environmental effects on the walls are investigated quantitatively under the changing temperatures of two different wall structures and two different installation materials in a cold climate [10]. The results show that high downtime, low cost, and low heat load are found in a brick wall with XPS, and shorter duration, greater downtime, and higher heat load are found in the concrete wall. Wei and Chen [11] studied and stated that value engineering can be defined as a functional analysis of a product or service to implement it at the lowest life cycle cost [11]. Energy-saving effect of the building along with the cost of materials and structure of energy design needs to be considered for value engineering. Yuan et al. [12] proposed that Value Engineering technology integrating life cycle cost with energy-saving is used to assess the envelope scheme and select the optimal one [12]. In most studies, the optimum insulation thickness computations were performed based mainly on the heating and cooling loads and other parameters such as the costs of the insulation material and energy, efficiencies of the heating and cooling systems, the lifetime, and the current inflation and discount rates. For this reason, the annual heating and cooling energy requirements of a building were the main inputs required to analyse the optimum insulation thickness.

2 Materials and Methods

In the current study, a residential building in the warm and humid city of Thiruvananthapuram, India, was taken into consideration as per the National Building Codes 2005. The two-story residential structure has a ground floor built-up area of 146 m^2 and additional space of 52 m^2 on the first-floor level with a wall area of 110 m^2 and roof areas of 280 m^2 respectively. The structure faces south in its orientation. A Reference Building (RB) model was used for the analysis, consisting of a traditional envelope made of RCC slab for the roof and laterite block for the walls. The different insulated envelopes were considered to determine the optimum building envelope for the building. The details of the building envelope considered are presented in Table 1. The twelve building cases were made with different walling and roofing building envelopes. The combination of walling and roofing envelopes for different building cases is presented in Table 2.

Table 1 Building envelope details

Type	Envelope details	Thermal transmittance (U) W/m^2K
Wall-1	External plaster 12.5 mm + Laterite block 200 mm + Internal plaster 12.5 mm	2.07
Wall-2	External plaster 18 mm + Autoclaved Aerated Concrete (AAC) block 200 mm + Gypsum Plaster- 3 mm	0.97
Wall-3	Polymer Gypsum plaster 10 mm + EPS Insulation 70 mm + AAC 200 mm + Gypsum Plaster 3 mm	0.27
Wall-4	Polymer Gypsum plaster 10 mm + XPS Insulation 70 mm + AAC 200 mm + Gypsum Plaster 3 mm	0.23
Wall-5	Fibre cement board 12 mm + EPS beads lightweight concrete 65 mm + Fibre cement board 12 mm	0.60
Wall-6	Shotcrete concrete 30 mm + EPS Insulation 70 mm + Shotcrete concrete 30 mm	0.45
Roof-1	External plaster 20 mm + RCC slab 125 mm + Internal plaster 12.5 mm	4.00
Roof-2	External plaster 25 mm + PUF Insulation 120 mm + 125 mm RCC slab + 12.5 mm plaster Inside	0.20

Table 2 Building model details

Building Model	Envelope details	Building Model	Envelope details
RB	Wall-1 + Roof-1	IEB-6	Wall-1 + Roof -2
IEB-1	Wall-2 + Roof-1	IEB-7	Wall-2 + Roof -2
IEB-2	Wall-3 + Roof-1	IEB-8	Wall-3 + Roof -2
IEB-3	Wall-4 + Roof-1	IEB-9	Wall-4 + Roof -2
IEB-4	Wall-5 + Roof-1	IEB-10	Wall-5 + Roof -2
IEB-5	Wall-6 + Roof-1	IEB-11	Wall-6 + Roof -2

RB-Reference Building; IEB-Insulated Envelope Building

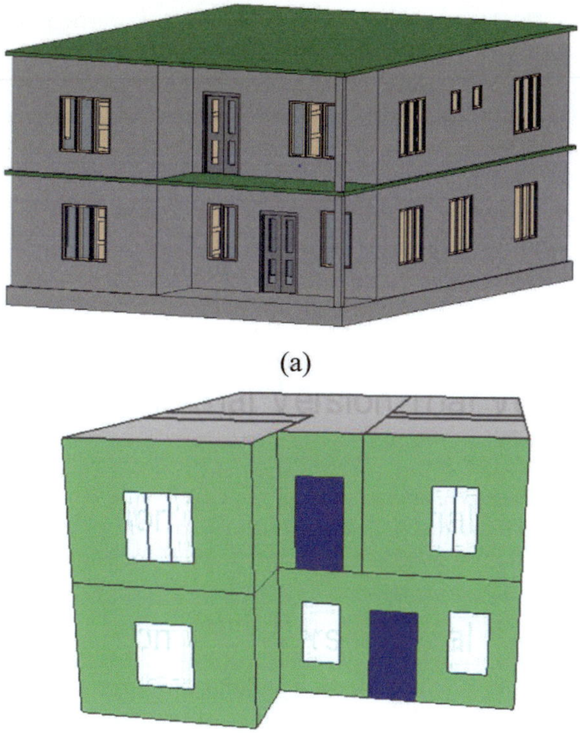
(a) (b)

Fig. 1 3-D Model of **a** Residential building **b** Energy analysis

Using Revit Architecture 2020, the residential building's 3D model was created. The building's 3D model was converted to gbxml format and imported into the Design Builder program for energy analysis as shown in Fig. 1.

The annual energy consumption for heating and cooling is determined via simulation, and the energy-saving function is derived using eq. 1. The energy-saving rate of an Insulated Envelope Building (IEB) is calculated using eq. 2.

$$F = (E_{RH} + E_{RC}) - (E_{IH} + E_{IC}) \tag{1}$$

where,

F—Energy saving function for the respective IEB in kWh
E_{IH}—Annual heating energy consumption of the respective IEB in kWh
E_{IC}—Annual cooling energy consumption of the respective IEB in kWh
E_{RH}—Annual heating energy consumption of the RB in kWh
E_{RC}—Annual cooling energy consumption of the RB in kWh

$$\text{Energy saving rate (in \%)} = \text{Energy Consumption of } ((\text{RB} - \text{IEB})/\text{RB}) \times 100 \tag{2}$$

The Life Cycle Cost analysis (LCC) is carried out for twelve building envelope types considering the cost of construction, materials, operation, and demolition. The construction costs for various materials and finishes were obtained from the Delhi Schedule Rates. The annual operation cost is obtained from the building energy simulation. As per the literature review, the demolition cost is considered 10% of the total construction cost for the building. In eq. 3, the time value of money, the Present Value Interest Factor of the Annuity (PVIFA) and the Present Value Interest Factor (PVIF) are adopted for operation and demolition costs respectively for the LCC calculation. The interest rate (i) is considered 8% and the life of the building envelope (n) is considered 40 years.

$$\mathrm{C} = \mathrm{C}_1 + \mathrm{C}_2\left(\frac{i}{n} - \frac{1}{i(1+i)^n}\right) + \mathrm{C}_3\left(\frac{1}{i(1+i)^n}\right) \tag{3}$$

where,

C_1—Construction cost of the envelope in Rs.

C_2—Annual operation cost in Rs.

C_3—Demolition cost of the building envelope in Rs.

I—Interest rate in %

N—Life of the building in Years

Following detailed studies of energy conservation, life cycle costs, and Value Engineering (VE) for each building envelope, the inquiry discerns the ideal building envelope based on its economic worth. It is important to consider both the energy saving and the life cycle cost of an envelope for the optimization of the building envelope. Hence, the best-valued envelope is selected based on the Value engineering evaluation model as presented in eq. 4. The envelope with a maximum 'V' value is considered the optimal envelope for the selected building. VE analysis helps in unifying the energy-saving function of the envelope assembly

$$\mathrm{V} = \frac{\mathrm{F}}{\mathrm{C}} \tag{4}$$

where,

V—Value of an envelope kWh/Rs.

F—Energy-saving function of the envelope in kWh

C—Life cycle cost of the envelope in Rs.

3 Results and Discussion

The entire annual energy consumption of buildings is shown in Fig. 2 together with the energy-saving function (F) for insulated building envelopes concerning the building. The IEB 9 with a roof envelope (made of external plaster 25 mm + PUF insulation 120 mm + 125 mm RCC slab + 12.5 mm plaster inside) and a walling envelope (made of Polymer Gypsum plaster 10 mm + XPS insulation 70 mm + AAC 200 mm + Gypsum Plaster 3 mm) demonstrates the greatest energy savings compared to the other insulated buildings. The energy savings rate and life cycle cost for various building envelope assemblies are shown in Fig. 3, where IEB-9 (34.66%) and IEB-8 (34.34%) have the highest rates of energy conservation. These building envelopes have PUF on the roof and XPS or EPS insulation in the walls, respectively. The energy-saving rate makes it clear that adding insulation to the wall and roof assembly will aid in lowering energy usage. The IEB-1 has the lowest life cycle cost of all the models.

The value engineering study for various envelope assemblies is shown in Fig. 4. IEB-8 has a high VE value (4.53 kWh/thousand rupees), whereas IEB-9 has a lower value (3.95 kWh/thousand rupees). The combination of EPS insulation with AAC

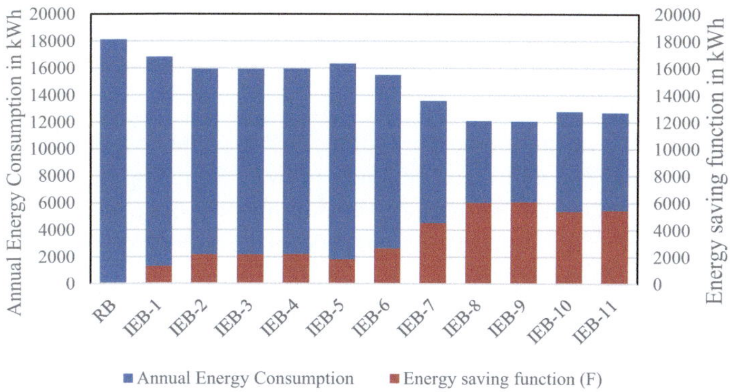

Fig. 2 Annual Energy Consumption and Energy saving function(F) for different envelope assembly

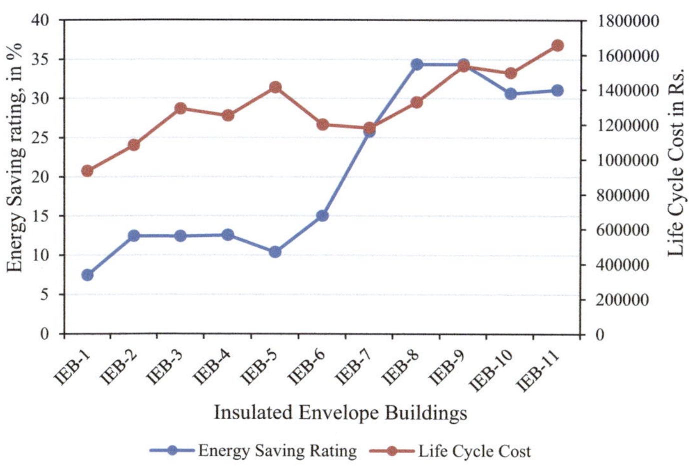

Fig. 3 Energy saving rating and life cycle cost for different envelope assembly

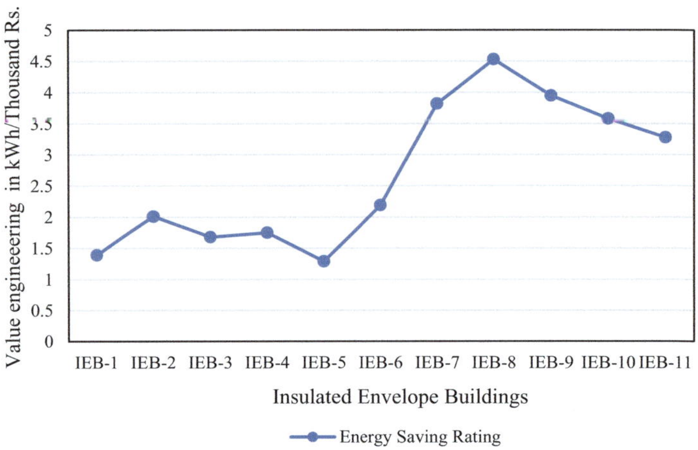

Fig. 4 Value engineering analysis for different envelope assembly

in the walls and PUF insulation on the roofing provides the best optimum envelope for the chosen research condition. The study utilizes value engineering analysis to identify the most ideal insulation strategy, considering both life cycle cost and energy-saving function, thus facilitating the integration of goals aimed at reducing energy consumption and life cycle costs.

4 Conclusion

In the current study, value engineering analysis is used to identify the ideal insulated envelope design. This approach made it easier to combine the goals of energy efficiency and building envelope life cycle costs. According to the value engineering analysis, IEB-8 (with Gypsum plaster 10 mm, EPS 70 mm, and AAC 200 mm make up the walls of this envelope, while external plaster 25 mm, PUF 120 mm, RCC slab 125 mm, and 12.5 mm internal plaster make up the roof) is the best envelope for the selected city's warm and humid climate. The study supports the Value engineering method, stating that choosing an envelope scheme that takes life cycle cost and energy efficiency into account is significantly better than concentrating only on one of these factors, highlighting the rational and scientific benefits of this all-encompassing method. However, because time and market value will have an impact on the life cycle cost of every envelope, it will be required to do ongoing study in the future to improve the technique used in this paper.

References

1. Kaynakli O (2012) A review of the economical and optimum thermal insulation thickness for building applications. Renew Sustain Energy Rev 16(1):415–425
2. Axaopoulos I, Axaopoulos P, Gelegenis J (2014) Optimum insulation thickness for external walls on different orientations considering the speed and direction of the wind. Appl Energy 117:167–175
3. Geng Y, Han X, Zhang H, Shi L (2021) Optimization and cost analysis of thickness of vacuum insulation panel for structural insulating panel buildings in cold climates. J Build Eng 33:101853
4. Al-Homoud MS (2004) The effectiveness of thermal insulation in different types of buildings in hot climates. J Therm Envelope Build Sci 27(3):235–247
5. Amir AH, Alibaba HZ (2018) Comparison between heat conductivity of EPS (expanded polystyrene) and XPS (extruded polystyrene). Int J Recent Res Civ Mech Eng (IJRRCME) 4(2):24–31
6. Liu X, Chen X, Shahrestani M (2020) Optimization of insulation thickness of external walls of residential buildings in hot summer and cold winter zone of China. Sustain 12(4):1574
7. Nematchoua MK, Raminosoa CR, Mamiharijaona R, René T, Orosa JA, Elvis W, Meukam P (2015) Study of the economical and optimum thermal insulation thickness for buildings in a wet and hot tropical climate: Case of Cameroon. Renew Sustain Energy Rev 50:1192–1202
8. Rad EA, Fallahi E (2019) Optimizing the insulation thickness of external wall by a novel 3E (energy, environmental, economic) method. Constr Build Mater 205:196–212
9. Basińska M, Kaczorek D, Koczyk H (2021) Economic and Energy Analysis of Building Retrofitting Using Internal Insulations. Energies 14(9):2446
10. Ozel M (2013) Thermal, economical and environmental analysis of insulated building walls in a cold climate. Energy Convers Manag 76:674–684
11. Wei T, Chen Y (2020) Green building design based on BIM and value engineering. J Ambient Intell Humaniz Comput 11(9):3699–3706
12. Yuan Z, Zhou J, Qiao Y, Zhang Y, Liu D, Zhu H (2020) BIM-VE-based optimization of green building envelope from the perspective of both energy saving and life cycle cost. Sustain 12(19):7862

Governance and Policies

A National Policy for Street Vendors and Its Implication in Delhi

Monika Yadav, Pulipati Mounica, Chaluvadi Mounika, and Khushboo Khanna

Abstract Indian markets are one the guide to explore the chronology of the country. They widely contribute to the pronunciation of cultural significance. The gradual change in these markets describes the economic timeline of the country. Through time these market streets (the bazaars) have invited a lot of informal setups like hawkers and street vendors. On the other hand, the change in marketing infrastructure also occurred. The current perception of markets originates from the social and economic contexts of the West particularly united states and Europe. A typical bazar of India can be defined as an order in that chaos. They are many such bazaars built gradually in old cities of the country. To retain the texture of these markets it is essential to understand the importance of informal markets and simultaneously work on enhancing the cultural significance. A specific research of the transformations occurred in Chandni Chowk, Delhi has been studied to suggest the strategies to preserve the essence of Indian markets by enhancing the cultural significance and also the quality of the informal sector. This paper will be a consolidation of understanding the working of the informal sector, but the heritage and architectural conservation are not addressed.

Keywords Cultural significance · Bazaar · Chandni chowk · Urban markets

1 Introduction to the Study Area

Who Are Street Vendors?

A street vendor is broadly defined as a person who offers goods or services for sale to the public without having a permanent built up structure but with a temporary static structure or mobile stall (or head load).

M. Yadav (✉) · P. Mounica · C. Mounika · K. Khanna
JBR Architecture College, Hyderabad, India
e-mail: lncs@springer.com; monika.yadav@jbrgroup.org.in

A. K. Sharma and A. Joshi (eds.), *Proceedings of the 3rd International Conference on Opportunities and Challenges for a Resilient Future*, Advances in 21st Century Human Settlements, https://doi.org/10.1007/978-981-96-0931-4_15

Street vendors may be **stationary** by occupying space on the pavements or other public/private areas, or may be **mobile** in the sense that they move from place to place carrying their wares on push carts or in cycles or baskets on their heads, or may sell their wares in moving bus etc.

In this policy document, the term urban vendor is inclusive of both traders and service providers, stationary as well as mobile vendors and incorporates all other local/region specific terms used to describe them, such as, **hawker**, ***pheriwalla***, ***rehri-patri walla***, **footpath *dukandars*, sidewalk traders, etc. (NPSV).**

A detailed questionnaire was developed for vendors which were used to get details of the vending and related problems in selected location like Chandni Chowk Market & Tilak Nagar Market. In totality, the size of the population studied for the survey was 200 (Chandni Chowk—135 & Tilak Nagar Market—65). The qualitative information was collected by conducting few Focused Group Discussions (FGD) and also some case studies.

1.1 Focal Group Discussion

During the field visits to different markets in Delhi, the following observations were made:

- **Social aspects**

It was observed that the level of literacy among street vendors is very low. However, this is not universal as even graduates are vending machines on the streets due to lack of employment opportunities in the formal sector.

1.2 Working Conditions

Sellers have faced atrocities in day to day government officials (municipal and police officials).

1.3 Financial Conditions

They earn enough to satisfy your hunger, however, less to save something for the future. He came to realize that most of the vendors have large families (number of dependents) to feed and save nothing or very less.

1.4 Political Consciousness

Vendors have been observed to have good political awareness, as in they know about your district or the representatives of the Legislative Assembly.

1.5 Observations Specific to Women Vendors

All the above conditions also apply to women, but women far more disastrous situations specifically face.

1.6 Take on Government Officials

As the National Policy on Urban Street Vendors. Vendors are now being implemented in Delhi, it is important to see how government agencies and stakeholders are doing this work.

1.7 Strategies

Need for a Holistic Approach

According to my view, a broad and comprehensive approach is needed to solve the problem of street vendors. For example, to make urban plans, it is necessary to take into account the right of providers to public spaces.

1.8 Conceptual Model

Various actors who has been playing the major role in the space provision or policy making of street vendors are (Fig. 1):

National Policy:

(1) The Street Vendors (Protection Of Livelihood & Regulation of Street Vending)Bill, 2012
(2) National Policy on Urban Street Vendors, 2009
(3) National Policy on Urban Street Vendors, 2004

Model Act/Bills:

(1) Model Street Vendors (Protection of Livelihood & Regulation of Street Vending) Bill, 2009

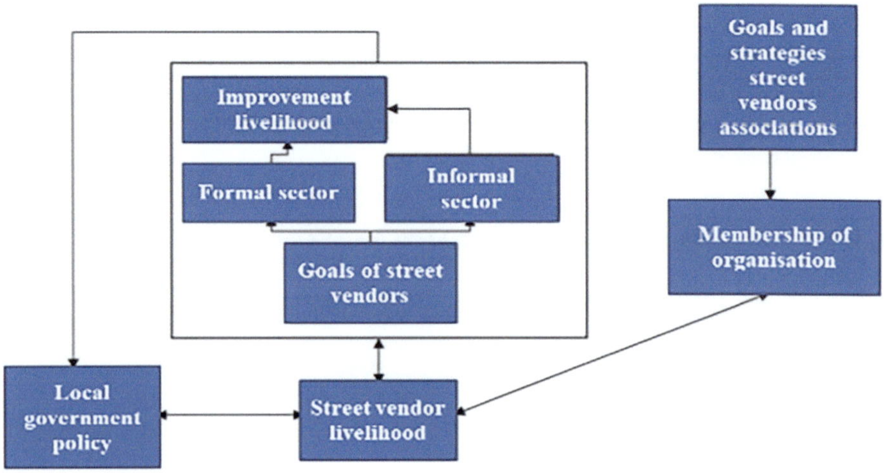

Fig. 1 Conceptual model

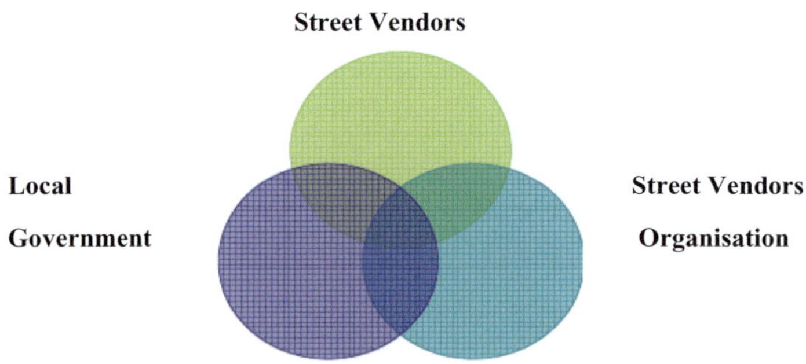

Fig. 2 Conceptual diagram

Municipal Corporation Policies/Schemes:

(1) New Delhi Municipal Council (NDMC)
(2) Municipal corporation of Delhi (MCD)

1.8.1 Conceptual Diagram of Mutual Influence of Main Actors in My Research

See (Fig. 2).

1.8.2 Unit of Analysis

The analysis is based on two geographical levels:

- Micro level: Street vendors
- Macro level: Street vendors organizations and local government

1.8.3 Aim

The aim of the research is to understand the problem & issues of the street vendors & why they are not continuing vending in the formal markets.

1.8.4 Objectives

To find out how vendors make their choices on vending location.
To examine the status of implementation of National Policy for Urban Street Vendors for vending provision.
To make the strategy for inclusion of the street vending following the National policy.

1.8.5 Selection of Study Area: Criteria

Criteria for selection of case studies:

Presence of high pedestrian population – since they are usually the target consumer group of vendors products
Presence of MRTS & high traffic flow – ensures high connectivity and a large inflowing pedestrian population
An existing Tehbazari licensing system – the only hawking initiative successfully implemented in Delhi, so far.

On the basis of these criteria, 3 commercial centers were chosen representing 3 representative.

markets in the hierarchical commercial set up of NDMC:

a. **Chandni Chowk Market**
b. **Non-heirarchy Commercial Centre:** Tilak Nagar Market.

2 Methodology

Stage 1: Literature Search

The concept of vending, merits and demerits of vending activity, vis—vis conflicts with other interest groups, can be evaluated by secondary source data. Policy initiatives by Government and non—Government organizations for vending profession were collected via secondary sources.

Stage 2: Primary Source Information

A sample survey of all vending establishments in the case study area was carried out with the help of a brief questionnaire. The questionnaire collected information about the nature, characteristics of vending activities, vendors dependence on the site context, and the related problems and issues.

Stage 3: Analysis of Primary and Secondary Source Information

Stage 4: Issue Identification, Proposals & Recommendations

The specific outputs of the terminal project to be produced in form of recommendations for the policies and implementation, and accommodating hawking in selected case study.

2.1 Scope and Limitations of the Study

The focus of the study is essentially planning and policy measures and not social activism. The study focuses only on vendors & vendors, and not the entire informal sector trade in Delhi. The scope is limited to urban street vendors only. It focuses on the broad policy frameworks for vending provision of NDMC.

About the NPUSV Policy

3 National Policy for Urban Street Vendors, 2004

The Ministry of Urban Development and Poverty Alleviation, Government of India formulated this policy in the year 2002 and was approved on January 20th, 2004. It aims to ensure that urban street vendors find recognition for their contribution to society and economy and is supportive of their livelihood.

Basic objectives of the National Policy for Urban Street Vendors:

1. To provide facilities for the proper use of space identified as creating vending zones in the Urban Development Plan / Zoning.
2. To stop or discretionary licenses and instead moving to nominal fee—based access regulation. When market forces such as price, quality and demand determine the number of vendors that can be sustained.
3. To make street vendors a special component plans for urban / Zoning treating development as an integral and legitimate part of the urban distribution system. To promote self—fulfillment among street vendors.
4. To establish participatory mechanisms with representation from urban vendors 'organizations' (Union/Co—operatives/Associations), voluntary organizations,

Table 1 Planning norms

S.No	Use zone/premise	No. of informal sector shops/units
I	Retail Trade: Metropolitan City Center District Center Community Center Convenient Shopping Center	3–4 Units per 10 Formal Shops (To be provided in Informal Bazaar/ Service Market Components)
II	Government & Commercial Offices	5–6 Units per 1000 Employees
III	Wholesale Trade & Freight Complexes	3–4 Units per 10 Formal Shops
IV	Hospital	3–4 Units per 100 Beds
V	Bus Terminal	1 Unit per 2 Bus Bays
VI	Schools Primary/ Secondary/Senior Secondary/Integrated	3–4 Units 5–6 Units
VII	Parks District parks Neighbourhood Parks	8–10 Units at Each Major Entry 2–3 Units
VIII	Residential	1 Unit/ 1000 Population
IX	Industrial	5–6 Units per 1000 employees
X	Railways Terminus/ MRTS Stations	To be based on surveys at the time of preparation of project

local authorities, police, residents Welfare Association (RWA) and others the orderly conduct of urban vending activities.

5. To take measures for promoting a better future for child vendors by making appropriate interventions for their rehabilitation and schooling.
6. To facilitate/ promote social security (pension, insurance etc.) and access to credit for street vendors through promotion of self—help groups/co—operatives/ federations/Micro Credit Institutions etc. (Table 1)

3.1 Commercial Areas

The commercial areas of NDMC are in the following hierarchy:

(a) **Chandni Chowk Market:** The area of market is 141.2 ha.
(b) **Non-heirarchy Commercial Centres**: Tilak Nagar market (Fig. 3).

Fig. 3 View of Chandni chowk

3.2 Introduction to the Survey Sites

3.2.1 Chandni Chowk Market

The Moonlit Square!

Location Central Delhi.

- It is one of the oldest and busiest markets.
- It is located in north-central Delhi, India.
- It is the specialty is the variety of traditional Indian items or products and their markets.
- In Indian food, delicacies and sweets (over 1000 types), with sarees with Chikan work Zaree. Lots of narrow streets with many shops selling books, clothing, shoes and leather goods, electronic and consumer products and what not.
- This is also good for shopping window.
- Is the location of the original Haldiram (Figs. 4, 5, 6).

List of Areas in Chandni Chowk

- Ajmeri Gate
- Mori Gate
- Nai Sarak
- Sadar Bazaar

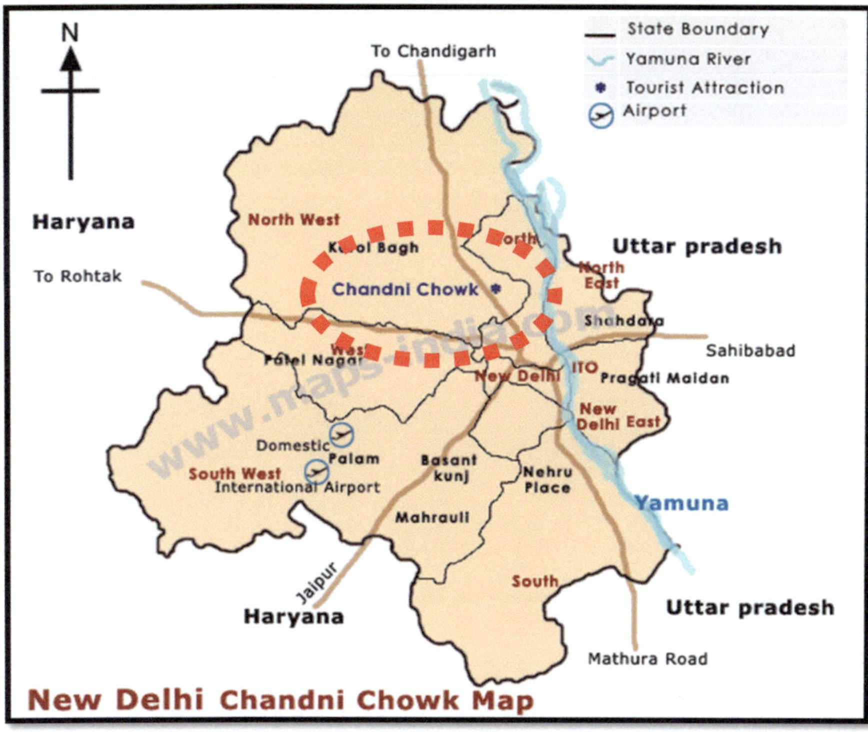

Fig. 4 Location map of Chandni Chowk

4 Structure Type of the Street Vendors

See Figs. 7, 8, 9.

4.1 Tilak Nagar Market

- Tilak Nagar Central Market (closed Wednesday) in west Delhi can be considered as the chicken curry for desi soul.
- Congestion and welcoming - has the comfort of small town - where shops sell Sikh turbans, tailors sew denim and mothers buy wedding dresses for their daughters. Photo 5.3: View of Market Tilak Nagar Pedestrain
- The market has Fabindia, no McDonalds and no multiplexes. It retains the feel of a neighborhood bazaar.
- Shop owners generally come at 10 am and close its stores at 7 pm (Fig. 10).

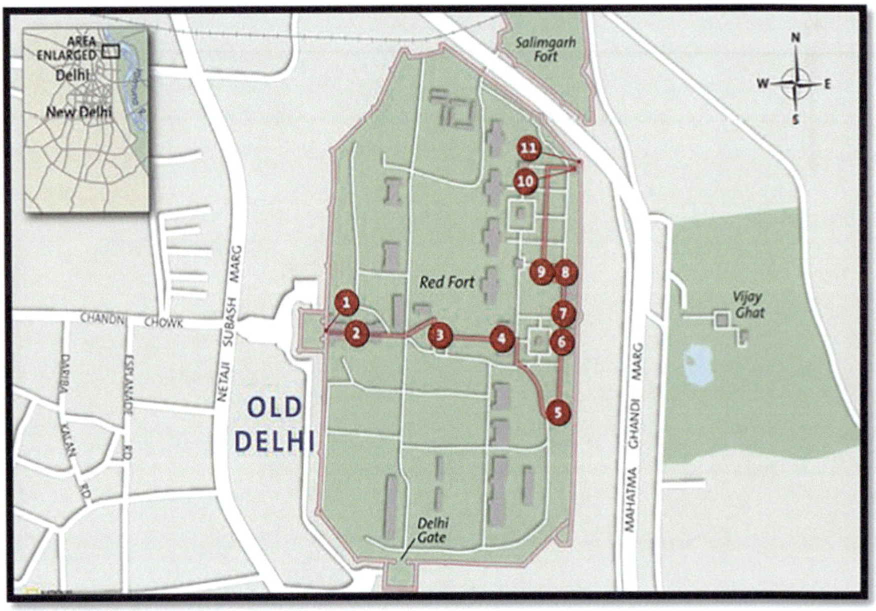

Fig. 5 Surveyed points in the market

Fig. 6 Chandni chowk google image shows how the survey has been done & red ones shows the vendors location on the main road in front of the Red-Fort gate

Fig. 7 Road for Whom? for Vehicles, for Pedestrains or for Vendors

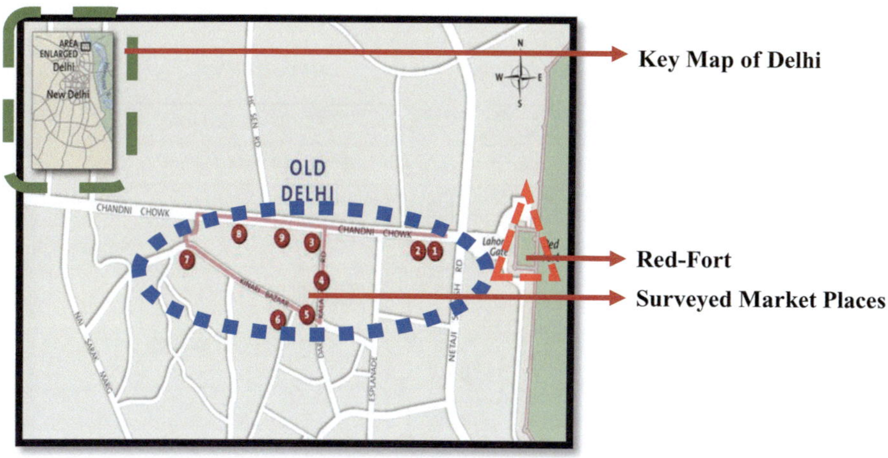

Fig. 8 Red points showing the Surveyed areas in Chandni Chowk site

Fig. 9 Structure type of the street vendors

4.2 Tilak Nagar Market Connecting the Other Wards

See Fig. 11.

Fig. 10 Structure type of vendors

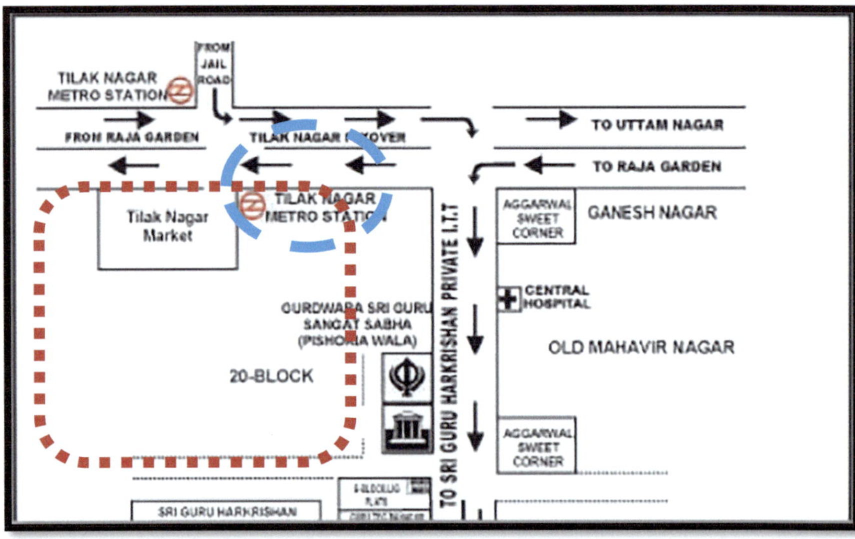

Fig. 11 Tilak nagar metro station and the adjacent roads

5 Distribution of Vending Units, as Per the Kind of Stalls Owned by Them

Vending type	Markets					
	Chandni chowk		Tilak nagar		Total	
	Units	%	Units	%	Units	%
Mobile	95	70	21	32	121	60
Stationary	35	26	44	68	79	40
Total (in %)	135	100	65	100	200	100

Table 2 The spatial parameters of the 2 survey sites

Parameters	Chandni chowk	Tilak nagar
Total Area	230.2 Ha	65.3 Ha
Total Floor Area	22, 44, 386 m^2	4, 03, 226 m^2
Floor Area under Commercial Establishments	5, 22, 472 m^2 (65.6%)	2, 45, 234 m^2 (41.8%)
Total number of vending units	4500	1500
Total number of licensed vending units	2000 (48.42%)	150 (26.99%)
Floor space area occupied	9290 ft^2 (752.37 m^2) (0.06%)	1112 ft^2 (134.11 m^2) (0.11%)
Land Rate	Rs. 15/ ft^2	Rs. 12/ ft^2

(*Source: Primary Survey, Jan –Feb. 2024*).

5.1 Vending Space and Perception Analysis

5.1.1 Spatial Aspects of Vending

See Table 2.

On the basis of the concept of the retail floor area, there can be 1–2 (1.10 precisely) vending units per 1000m^2 of retail floor space in Chandni Chowk, 3—4 (3.66 precisely) vending units per 1000m^2 of retail floor space in a Non—Hierarchical Commercial center as Tilak Nagar, 5—6 (5.48 precisely).

5.2 Problems Faced While Carrying Out Vending Are

- Fear of various authorities.
- Harassment by police officials
- Exposed to the environmental hazards.
- Long hours of work without rest.
- Rainfall damages goods and affect vending.
- Customer's misbehavior
- Extreme tiredness
- Absence of toilet facilities
- Absence of shade to protect from various weather conditions.
- Wastage of unsold goods such as flowers, vegetable, fruits, food items etc. especially in summer.
- Difficulty in procuring vending license and the fear of vending without vending license.
- Problems due to plying vehicles and their drivers especially the heavy vehicles where the drivers behave rudely.
- High rates of fee charged by RWA for permitting vending.

- Not allowing vending by RWAs
- Customers take advantage of being a woman vendor and try to bargain a lot.

6 Conclusion

In conclusion, hawkers can be active in the urban system if given the opportunity to contribute to its development. Being part of marginalized urban poor are treated as troublemakers whose sole purpose is to create chaos in the streets. Whenever the issue of citizens' initiatives to improve cities arises, the reference is invariably to the upper middle and middle class. The vast majority of the urban population is maintained not only of these initiatives, but they are, more often than not, seen as the main problem. The fact is that there is a plan for improving the city can succeed without the participation of the urban poor. They need to be integrated into the planning process and campaigns for a better environment.

Bibliography

1. Anjaria, Jonathan Shapiro (2006) Street hawkers and public space in Mumbai", in EPW, 41(21), pp 2140–2146
2. Aziz, Abdul (1984) Urban poor and urban informal sector. Ashish Publishing House
3. Bhowmik, Sharit K (2000) Improving regulatory framework for income enhancement of the urban poor: a study of Delhi (India) for UNHABITAT, Nairobi
4. Bhowmik, Sharit K (2003) National policy for street vendors. EPW, 38(16), pp 1543–1546
5. Bhowmik, Sharit K (2005) Street vendors in asia: a review. EPW, 40(22–23), pp 2256–2264
6. Bhowmik, Sharit K (2006) Social security for street vendors. Seminar, 568, pp 49–57
7. Chiswick CU (1974) Identification of urban poor. Urban Poverty Task Force Report, Mimeo
8. Cohen M (1974) Designing Anti—Poverty Strategies: Political obstacles and opportunities. Urban Poverty Task Force Report, Mimeo

Decoding Rural Housing: A Local Governance Perspective from Sitapur, U.P

Pragya Singh and Bhanu M. Marwaha

Abstract The research paper examines the existing conditions of rural area- Sitapur district in Uttar Pradesh and highlights the existing local governance issues for present state of Housing in Sitapur District, U.P. Assessing local governance for holistic development of rural areas is crucial for its socio-economic progress, as it promotes sustainability at the community level. Due to rapid urbanization, the transitional areas between urban and rural are on the verge of expansion and typically experience a unique set of challenges and opportunities. Study area Ismailpur located in outskirts of Sitapur District, is in the early stages of urban expansion. The present condition of housing stock in Ismailpur area is unsustainable due to issues like poor drainage system, improper solid waste disposal, insufficient access of electricity, inadequate rural roads, and poorly designed houses, worsened by the absence of building regulations. Methodology for this study involves a questionnaire-based field survey of 40 households in the study area along with the interviews and telecommunications-based interactions with the government officers from the District Rural Development Authority. Related literature review highlights the government and technical reports and policies in detail. The findings suggest that insufficient drainage leads to road waterlogging as the roads are earthen partially, causing transportation disruptions, while waste disposal occurs in open plots. Rural electrification relies on temporary solutions that result in irregular electricity supply. Additionally, poorly designed houses lack ventilation and natural light which fails to meet the standard of living. The present differences in legal and regulatory frameworks between urban and rural areas has led to these unsustainable development practices in Ismailpur. The schemes implemented in the study area include Pradhan Mantri Awaas Yojana—Urban (PMAY-U). Other initiatives, such as the Swachh Bharat Mission (SBM) and the Atal Mission for Rejuvenation and Urban Transformation (AMRUT), could address the needs of these transitional areas if properly implemented.

P. Singh (✉) · B. M. Marwaha
Department of Architecture, National Institute of Technology Hamirpur, Himachal Pradesh 177005, India
e-mail: 22mar004@nith.ac.in

B. M. Marwaha
e-mail: bhanu@nith.ac.in

A. K. Sharma and A. Joshi (eds.), *Proceedings of the 3rd International Conference on Opportunities and Challenges for a Resilient Future*, Advances in 21st Century Human Settlements, https://doi.org/10.1007/978-981-96-0931-4_16

Keywords Local governance · Sustainable development · Housing · Sitapur district

1 Introduction

Land is a natural resource which is provided to human in abundance, the earth constituting 30% of the whole. Is it enough or not? Human settlements are experiencing growth at a very high rate, which has divided this natural resource into two major categories as per the level of development and population density: Urban and Rural. The area lying in between the two is known as transitional area, where the slow shift in land usage is taking place. These transitional areas are often, raw lands which is not prepared to handle the development growth on its own. The actors (resource users and owners) are less active in these areas, which leads to uncoordinated distributions and eventually unsustainable usage.

Urbanization is the shift from rural to urban life-style, since the urban lands are already packed, the human tends to rural lands. Thus, the unplanned and uncoordinated settlements are taking place in the outskirts of the highly dense areas. With the increase in the population in such transitional area, after a time, problems of water-disposal, poor air and water quality, irregular electricity starts occurring eventually.

The roads in these areas are predominantly earthen, resulting in water accumulation during periods of rainfall, leading to flooding. Consequently, the residents in these areas face challenges navigating through waterlogged earthen roads. Additionally, household drainage systems often lead to open ground areas. Stagnant, untreated water in these open areas poses health risks as it becomes a breeding ground for disease-carrying mosquitoes and other pests. Once this water is absorbed, it can lead to water pollution and a negative impact on the quality of drinking water, posing serious health concerns. Constructing drains or gutters near the foundation can also be problematic, potentially causing structural damage or weakening.

The elimination of open defecation in rural India is a complex and multifaceted challenge that involves various social, economic, cultural, and infrastructure factors. While significant progress has been made in recent years through government initiatives and awareness campaigns, complete eradication of open defecation has not been achieved for several reasons, which may include lack of sewage systems, long standing habits, low finances, etc. Similarly, waste management infrastructure is also poor, which leads to the different causes of air and water pollution.

Another issue in transitional areas pertains to electricity in rural areas which is an unfinished and challenging task, with uninterrupted electricity access being a distant goal. Unfortunately, residents in transitional areas are often overlooked in terms of service provision. To cope, many resort to illegal electricity connections, which are temporary solutions.

Approval of plans by the local body, in transitional areas poses a distant challenge for governance, given their existing array of responsibilities. Consequently, overlooked building codes and residents' lack of awareness result in the planning of enclosed spaces with no windows which restricts natural air ventilation, impacting Indoor Air Quality (IAQ) and consequently affecting the well-being of occupants.

This paper is a work-in-progress to study the issues thoroughly and giving a future scope to find the strategies that promote sustainable housing development in transitional areas, incorporating self-sustaining infrastructure for sanitation, drainage, and roads. To gain a comprehensive understanding of these challenges, a survey was undertaken in the Ismailpur study area, situated on the outskirts of Sitapur District in Uttar Pradesh.

Thus, the objectives of this research are:

- To identify **existing practices** in the study area
- To conduct **interviews** of Government Officials to collect **on-going status of policies** in study area
- To check the **implementation status of existing policies** in the study area.

2 Literature Review

2.1 Local Governance

Local governance refers to the way a community or region is governed at the local level, often through elected representatives, local administrative bodies, or community organizations. It involves decision-making, planning, and the provision of public services within a specific geographical area.

It is almost 30 years since the 73rd and 74th Constitutional Amendment Acts, creating the new Local Governance Framework in India, were made operational in April 1993. The Acts established a three-tier system of Panchayati Raj in every state—at the village, intermediate and district levels [1].

Panchayati Raj Institutions (PRIs) and local governance bodies in India are vital for local development, democratic representation, and the empowerment of local communities. They are at the forefront of addressing the unique challenges and opportunities faced by rural and urban areas, making them essential components of India's governance system.

Panchayats are expected to play an effective role in the planning and implementation of functions related to 29 subjects enlisted in the Eleventh Schedule of the Constitution. Many SDG targets are within the purview of these subjects [2].

Gram panchayats (GPs) across the country have begun preparing gram panchayat development plans (GPDP), presenting an opportunity for the GPs to synchronize their plans with SDGs. Resources from various centrally and state sponsored schemes can be leveraged and converged at the GP level. It is important to set GP-level targets with measurable indicators that will have vertical and horizontal linkages, convergence possibilities, resource mobilization potential and feasible action by the GPs [2].

The report sponsored by Ministry of Panchayati Raj (MoPR) which examine the utilization and effectiveness of the Fourteenth Finance Commission (FFC) funds received by a sample of 120 GPs in 20 districts spread across 16 Indian states which includes decision making process, activities carried out by GPs utilizing FFC, which consists majorly road constructions and their maintenance along with water supply facilities, office Infrastructure of record keeping, perception and knowledge of beneficiaries, timely release of grants and their effects [3]. Further, the report states that though people's participation is appreciable in Gram Sabha meetings, their participation in plan preparations and documenting, and register works is found unsatisfactory whereas the resources with the panchayats were limited and the grievances of the people were many. Gram Panchayats (GPs) aim to improve the condition of the villagers and to make them self-sufficient. Gram Sabha (GS) prepare Gram Panchayat Development Plan (GPDP) and provides platform to fulfil the local developmental needs. GS meetings vary in number in different states but the attendees are low in most of them. Approval of works & activities are funded through GP based on GPDP. Details of GPDP formulation Process is not well understood by the people and even by Elected Representatives & thus the estimation approved is insufficient. Low education qualification and low knowledge of Sarpanch or Elected Representatives (ERs) to utilize FFC affects the participation in Gram Sabha (GS) [3].

GPs also receive grants for programmes such as Swachh Bharat Mission (SBM), National Rural Livelihood Mission (NRLM), Mahatma Gandhi National Rural Employment Guarantee Scheme (MGNREGS) and National Social Assistance Programme (NSAP) etc. One of the objectives of the government has been to encourage convergence of two or more programmes having similar objectives. However, only 31 (26%) of the visited GPs had some convergence. Convergence was reported in some GPs in the form of manpower, financial, or technical. Most of the convergence activities are related to roads, drainage and sanitation, water facilities, street (solar) lights and education developments [3].

2.2 Housing and Sustainability

The report by UN-Habitat under the project "Mainstreaming Sustainable Social Housing in India (MaS-SHIP), constituted of an analysis of 'the problem' of social housing provision in India, the response with which India has met this challenge in the past, how it translated into state level responses, and what could be learned from national and international case studies to positively influence the Government of India's prerogative to provide housing for all [4].

Housing goes beyond individual units of buildings to include the social infrastructure and services available to a housing neighbourhood. These infrastructure help create a sense of community among residents [5]. Housing is also an effective way to promote good governance [6] which involves policies and practices that ensure fair and equitable access to housing opportunities for all members of society. Effective governance ensures the development of infrastructure and services that support housing, such as water supply, sanitation, transportation, and healthcare. Good governance encompasses a broader set of principles, including transparency, accountability, rule of law, and effective public administration. Housing policies and practices need to be aligned with these principles to contribute to the overall goal of good governance and societal well-being.

Developing countries with limited resources tend to view sustainable development concept differently than developed countries where efforts to optimize the use of limited resources to maximize the impact has been the main agenda. Sustainable Housing Development (SHD) is influenced by factors that account for physical, social, environmental, and economic dimensions like location factors, cultural factors, waste management factors and energy consumption factors, respectively. These factors are indicators used to reflect the level of housing sustainability [12].

3 Methodology

3.1 Location

Sitapur is a district in Uttar Pradesh. It was founded by the legendary king Vikramaditya. At present it covers an area of 5,734 sq.km. It had a population of 44,74,446 in 2011. It is an agricultural district. Sitapur district is a part of Lucknow division [7].

The research study was carried out in Ismailpur village in outskirts of the Sitapur District which is marked as Tractor Terminal and Large-Scale Industry in Master Plan 2021, which falls under Khairabad block [8]. As per Census of India (2011), the population of Ismailpur village is 128 consisting of 24 households. In Ismailpur Village, over the last 10 years the transformation came rapidly with certain issues (Fig. 1).

Fig. 1 Existing situation scenarios of the study area, (**a**) Kuchha roads, (**b**) Poor drainage, (**c**) Poor electrification, (**d**) Improper disposal of garbage. (*Source*: Author)

3.2 Indicators Used in the Research Study for Survey.

The main indicators chosen for research purpose are as follows:

Kuccha roads and poor drainage	Leaching in the ground water	IAQ and Daylighting in houses
Total Dissolved Solids (TDS)	Irregular supply of electricity	

3.2.1 Kuccha Roads and Poor Drainage

The roads are predominantly earthen(kuccha) in the study area, partially made of 'Khadanja' (Herringbone brickwork) at some places. The soil type is fine loamy type [9]. Loamy soil is a soil type that consists of a balanced mixture of sand, silt, and clay particles. The water drainage properties of loamy soil are generally good, but in rainy season the water-logging is a common site in the area. Transportation disrupts even more after water-logging, as the heavy vehicles (Tractors Trolleys) get stuck in. In addition, the drains from the residences open to the roads in some areas in the village where the roads completely earthen (without khadanja), which increases the depth of the issues. Four homemakers during the survey stated that the mismanagement of the drains lead to unpeaceful environment. They complained:

> The drains are the cause of quarrel daily in the neighbourhood as there is undecided flow of kuccha drain. Also, the electrical wiring passing from the front of the house, it is very dangerous.

The drains in the area nearest to the main road, the first to be developed, were completely blocked. Some residents were living for almost a decade while others for two decades now.

3.2.2 Total Dissolved Solids

The source of water supply is privately-owned in the study area which includes borewells and handpumps as currently there is no provision of water services by Government. Even, the source of irrigation water is dependent 94% on ground water. [9]

In the survey, to check the quality of water obtained from these sources were tested for Total Dissolved Solids (TDS) by TDS Meter Digital water tester. The residents are directly consuming untreated drinking water in some households. TDS readings have been observed to be varying in study area, considerably.

75 years old Pyarelal Pushapdar, mentioned:

> The white deposits in water are visible quite often, they purchased RO as well but still the problem persists.

Ashok Kumar added to this:

Water turns yellow in colour after containing it for 1-2 days.

As per the report given by Department of water resources (Ministry of Jal Shakti), in Khairabad block measured 90 mg/lit. Nitrate presence in the ground water sample, which is often measured separately in water quality analysis because of its potential significance as a water pollutant due to its role in eutrophication and its potential adverse health effects at high concentrations. Permissible Limit of Nitrate as per BIS 10500–2012 is 45 mg/lit [9]. The presence of nitrates in groundwater is a concern because high levels of nitrates can have adverse health effects when consumed, lead to a condition known as methemoglobinemia, or "blue baby syndrome," which affects the blood's ability to carry oxygen. Agriculture, industry, sewage, septic tanks & landfills are the main sources. [17]

3.2.3 Leaching in the Ground Water

The lack of sewage systems can indeed contribute to the widespread use of septic tanks in areas where centralized wastewater treatment infrastructure is absent. While septic tanks provide a decentralized solution to wastewater treatment, it is important to note that their effectiveness depends on proper design, construction, and regular maintenance. In areas where septic tanks are prevalent, addressing issues related to their maintenance, periodic desludging, and environmental impact is crucial for ensuring sustainable sanitation practices.

In Ismailpur village, the infrastructure for cleaning of septic tanks is inadequate, which poses significant challenges to sanitation and public health. Desludging of septic tanks is crucial to prevent issues such as overflows, contamination of groundwater due to leaching, and the spread of waterborne diseases and it is recommended to be done half yearly or yearly. Small domestic tanks, for economic reasons, may be cleaned at least once in 2 years as per the Indian Standard Codes 2470 (Part 1). The lining of the water-proofed cement plaster is crucial in the septic tanks, even the concrete septic tanks erode with time and leak waste water resulting in contamination of groundwater. Since the concrete is generally weak in waterproofing, a waterproofing agent is used to prevent water penetration and leakage [18]. While the concept of leaching was explored during the survey, a noticeable lack of awareness and knowledge was observed in the community. Similarly, the notion of rainwater harvesting appeared to be unfamiliar to the residents as well. One of the survey objectives was to foster awareness within the community regarding water conservation. The improper disposal of solid waste, such as throwing it in open areas or incinerating it without proper controls, can have significant environmental, health, and social consequences. Blockage of drainage system due to huge volume of garbage and polythene bags leads to flooding [10] and degrades the recharging capacity and quality of water.

3.2.4 Irregular Supply of Electricity

The extension of electrification by the residents, in the context of urbanisation, poses issues like illegal hooking, negligence of standards height and wiring methods. These guidelines are provided by *UP Vidyut Niyamak Aayog* (Electric Public Service Commission) informed by retired government official. In 2007, Delhi government handled similar issue of the expansion of electricity distribution for unplanned settlements to 3 private companies, resort to a range of anti-theft technical devices, high-voltage distribution system. In dense areas where encroachment is high, the low-tension aerial cables were used in a way to reduce hooking. Large-scale metering with tamper resistant equipment to reduce human intervention to avoid high corruption-prone practice. Customer Services were made for reliable and easier to apply for [11].

In Ismailpur village, a recent transformer back charge caused electrical damage, posing risks like fires, equipment damage, and safety hazards. Tragically, a person and a pet dog lost their lives due to mishandling of 11 kV high voltage lines.

3.2.5 IAQ and Daylighting in Houses

Indoor Air Quality is a fundamental aspect of maintaining a healthy indoor environment. It plays a crucial role in mitigating indoor air pollution, promoting respiratory health, preventing the "Sick Building Syndrome," and contributing to overall well-being. According to the World Health Organization (WHO), household air pollution was responsible for an estimated 3.2 million deaths per year in 2020, including over 237,000 deaths of children under the age of 5. To promote the well-being of residents, it is essential to recognize that the lack of building regulations can significantly impact the overall safety and comfort of indoor environments. The setbacks and open areas which are mandated by building codes which promote natural ventilation and daylighting as well. Unfortunately, some individuals choose to forego these setbacks in favor of maximizing built-up spaces, a practice that can have detrimental effects on their health (Fig. 2).

The Pradhan Mantri Awaas Yojana-Urban Scheme was executed for housing in Ismailpur Village. However, the planning of the housing was subpar, executed as per the residents. The structures featured dead walls on three sides, severely compromising the ventilation within the residences. The illustrated figure serves as an example of this inadequately planned approach. Additionally, the buildings were erected without any setbacks, and shared foundation walls left no room for proper ventilation, windows on the front wall only. The residents residing in the PMAY-U housing raised concerns about drainage problems, and improper road connections in the Ismailpur. This area located in Ismailpur west of the railway tracks. The government official after construction of PMAY houses verified the toilet facilities, kitchen, and a room, as told by the residents. It was informed and confirmed by a government official that funds are transferred in piecemeal as the construction progresses.

Fig. 2 Planning of House under PMAY(U) by the resident. (*Source*: Author)

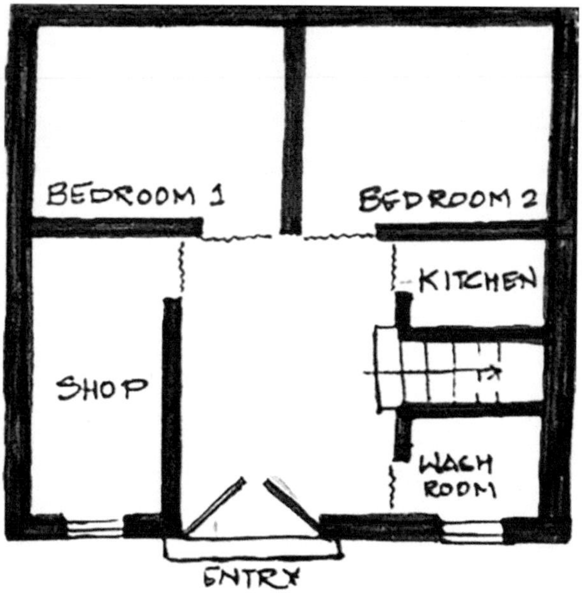

3.3 Survey Analysis

The survey, conducted during rainy season in three days in the study area through a combination of questionnaires and face-to-face interviews with random sampling of residents, spanned diverse regions exhibiting varying rural–urban characteristics and cultural practices. The distribution of households surveyed over three days based on their proximity to the road, with day 1 covering the farthest and day 3 the nearest households adjoining the main road, which developed before 2003. The timing of the survey, immediately following heavy rainfall on the day prior to the commencement (referred to as 1 and marked in Google Image), facilitated discussions with residents about the prevalent issues related to roads and drains. The rainy season, being the primary period for such concerns to emerge, provided a context for addressing these issues with the community.

Out of 55 households, the surveyed 40 households utilize the designated roads, which include earthen roads (marked in orange) and a recently built road with interlocking bricks (highlighted in yellow) which was supposed to be built connecting the main road to the inwards, funded by the State Finance Commission. Rest 15 households were either unoccupied or under-construction. Those living in closer proximity to the railway tracks prefer using the adjacent earthen passage, predominantly on foot. The analysis of the survey is done on SPSS Software Version 20 to study Pearson correlation between the variables (Fig. 3).

Fig. 3 Google image of the study area

4 Results and Discussion

1. The correlation between the income and the building construction materials are statistically significant which means the lower income group residents are living in semi-pucca houses made of the corrugated galvanised iron (CGI) sheets for roofing, due to lack of finances which is a common practice in the study area. Unfortunately, this housing pattern among the lower-income group is indicative of compromised living conditions, significantly impacting their overall well-being as the climate is composite (extreme hot and extreme cold both).
2. The source of water supply is consistently found as privately owned borewell or handpump or both which implies dependency on groundwater is very high as there is no source of government supply. The statistically significant negative correlation between the location of the residence from the main road and the TDS readings indicates that the area surveyed on day 3 has the highest readings as it was the oldest in terms of development. It also hints as the time passes; the development will increase; the TDS readings will also go up in the farthest places as well which are a couple of years older, at present.
3. The distribution of sanitation facilities in the surveyed area reveals a varied landscape, with 70% relying on individual septic tanks, 23% utilizing multi-household septic tanks, 5% depending on pit toilets, and 2% still practicing open

defecation. The survey questionnaire delved into government scheme assistance, revealing that although the Swachh Bharat Mission (SBM) is mandated in the study area, it must be effectively implemented and has not reached the study area yet.

4. The consent among residents regarding the lack of awareness about the necessity of desludging and the absence of segregation between greywater and blackwater in drains raises significant concerns. This implies that the drainage systems in the area are indiscriminately carrying both blackwater (contaminated with faecal matter) and greywater (from household activities like bathing and washing).
5. The recharging of ground water is dependent on seasonal rain water and absorption from the drain water percolating in the ground without any further treatment as there is no rainwater harvesting measures implemented neither by residents nor by government yet. It seems that the untreated drain water absorption and poor septage management since there is no proper lining in septic tank is polluting the ground water because of leaching which is the sole source of water supply in the study area. Thus, justifies the rapid degrading water quality, where the report [9] states the ground water resource as safe in quantity. The report also mentions that Ground water abstraction should be controlled and recharge methods should be implemented.
6. A mixed scenario was observed in terms of garbage disposal and waste segregation, with 60% of households disposing of waste in vacant plots and the remaining incinerating it. Additionally, 67.5% of households are engaged in waste segregation. Both the practices are adding to the environmental degradation. The Material Recovery Facility (MRF) centre was under progression as per the report [13] whereas the news report [14] states that the MRF centres are non-functional right now in Sitapur and garbage is getting piled up in vacant plots.
7. 85% Agreement on the impact of roads and drains: The high agreement percentage indicates a strong awareness among most residents regarding the significant impact of proper roads and drains on their daily lives. 15% Resigned acceptance of existing situations could be influenced by factors such as long-standing issues, lack of confidence in improvements (see Fig. 4).

1. 42.5% Strongly Agreement on Electric Pole Danger, suggests a significant portion of the community perceives a real and immediate danger posed by the improper wirings and electric poles. While slightly less than the strong agreement group, the 30% agreement still represents a substantial portion acknowledging the danger. 27.5% Relatively Unconcerned (Nearest to Main Road): The households receiving regular supply of electricity which is officially under Nagar Palika, mostly those closer to the main road, show a lower level of concern. The governance of the area farther away from main road has shifted from GP to Nagar Palika Parishad (see Fig. 5).

According to the official website of Sitapur District Nagara Palika, both the Atal Mission for Rejuvenation and Urban Transformation (AMRUT) and Swachh Bharat Mission (SBM) Schemes have been launched and are operational. The report [4]

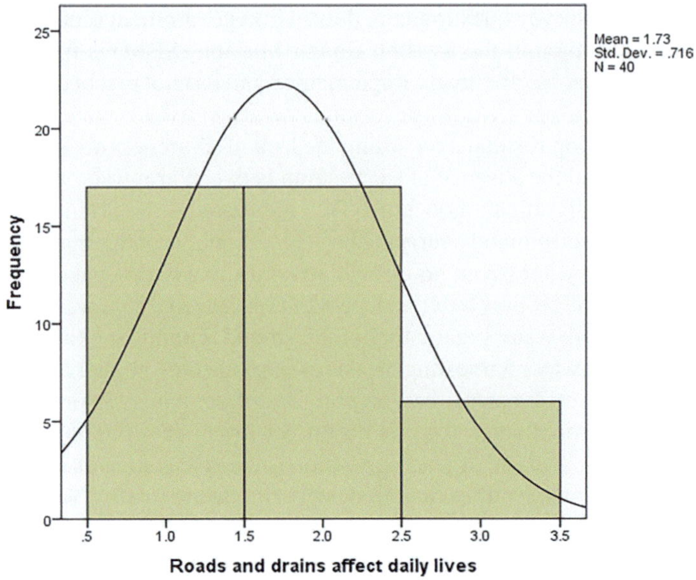

Fig. 4 Graph showing the importance of roads and drains as per the residents' survey. (*Source*: Survey Data SPSS Analysis)

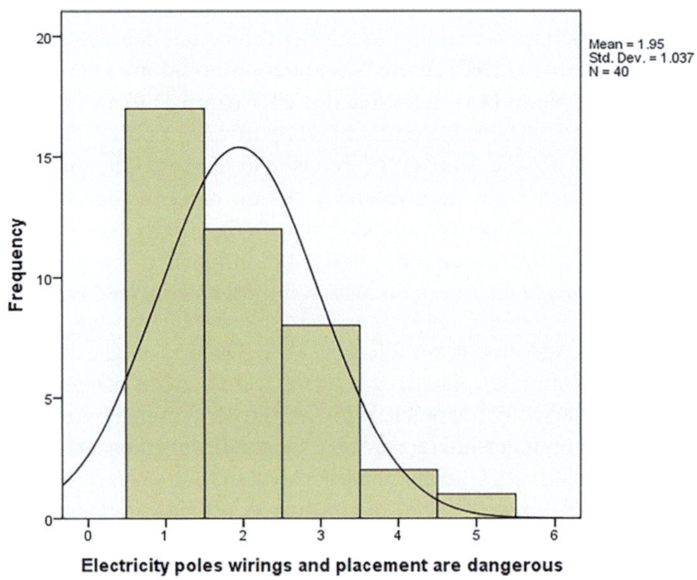

Fig. 5 Graph showing the danger level of the electrical wirings as per the survey. (*Source*: Survey Data SPSS Analysis)

suggested that the AMRUT scheme, focusing on basic services and infrastructure, should take precedence before starting construction under the PMAY scheme. The ongoing convergence of schemes, such as MNREGS with SBM and PMAY, is a positive step and will bridge the gaps for funding the activities.

5 Conclusion

A consistent gap was noted in the implementation status of certain policies and standards within the study area. The PMAY scheme has been swiftly executed over the past 2–3 years, yet there remains room for improvement, particularly concerning ventilation and daylighting. Conducting a post-occupancy evaluation of these residences would enhance their quality. The SBM has yet to be implemented in the study area, which is crucial considering the poor cleanliness and maintenance of the community. Encouraging Community Awareness and Participation could aid in managing these issues, addressing the lack of community belongingness. Additionally, the AMRUT scheme has not been initiated in the area. Recent projects such as drain construction and interlocking pavement have been financed by the 15th Finance Commission, which are basically designed for Light Moto Vehicles and pedestrians. Roads are primarily designed as per the live load of heavy vehicles, including tractors, indicating a necessity for cement concrete roads in the study area. The poorly managed electricity connections involve wiring cables from transformers without adhering to the necessary standards and guidelines, resulting in electricity losses, irregular supply and hazardous wiring conditions that pose a significant risk to individuals' safety.

6 Policy Recommendations

Development Authorities have been following different set of building regulations adopted from Regulations of Building Operations (R.B.O.) Act 1958, National building and the bye-laws framed at local-level by individual authorities [15]. The UP government has asked the housing and urban development departments to evolve standardised bylaws for the 756 local bodies in a bid to regulate haphazard development activity. At present, only 221 local bodies adhere to the rules framed by the housing department [16].

The survey emphasizes the need for coordinated efforts, considering the convergence of schemes like AMRUT, SBM, PMAY, and MNREGS to address housing, sanitation, and infrastructure comprehensively. The connectivity of roads, water supply, its drainage, Sewage Treatment Plant (STP), and Faecal Sludge Treatment Plant (FSTP) the provision of basic services makes a raw land into serviced land along with electrification guided by Deen Dayal Upadhayay Gram Jyoti (DDUGY) program and thereafter the PMAY program provides for construction of houses for those who are in need, keeping the track of Indoor Air Quality (IAQ) in the guidelines, with the help of MNREGS for construction labours. The SBM program encourages cleanliness and maintains the community Open Defecation Free. The Septage Management policies and treatment facilities should be prioritised in India.

The study underscores the holistic development and maintenance of services and infrastructure development with housing construction for sustainable development with the help of Government Policies and Programmes for financial support, through convergence.

Acknowledgements This research is conducted as a part of Masters' Thesis. I owe my gratitude to the Government Officials in the study area, at different level who shared all the needful information. I am deeply appreciative of the residents of the study area who cooperated and participated in surveys and interviews. Special thanks are due to my supervisor, whose guidance and support were instrumental throughout this research.

References

1. Sharma, Raveesh. Issues with Local Governance in India—Explained, pointwise. Forum IAS. https://forumias.com/blog/issues-with-local-governance-in-india-explained-pointwise/
2. Sustainable Development Goals and Gram Panchayats. UNDP. https://www.undp.org/sites/g/files/zskgke326/files/migration/in/Gram-Panchayat-Brochure.pdf
3. Joe W, Chakravarty S, Prakash A, Thakur AK, Kumari A, Sharma M, Shrivastava P, Sharma R, Kumar R, Surwade S, Jain V, Singh V (2020) Utilization and impact evaluation of 14th finance commission grants to gram panchayats in selected Indian states. s.l.: Ministry of Panchayati Raj
4. Herda G, Rani S, Caleb PR, Gupta R, Behal M, Gregg M, Hazra S (2017) Sustainable social housing in India: definition, challenges, and opportunities—Technical Report, Oxford Brookes University, Development Alternatives, The Energy and Resources Institute and UN-Habitat. Oxford. ISBN: 978–0–9929299–8
5. Okpoechi CU, Ebiringa CN, Ibemere LC., Emma-Ochu CA (2020) Public Infrastructure and Sustainable Urban Mass Housing in NigeriaAfr J Hous Sustain Dev 1, 1. http://ujh.unilag.edu.ng/index.php/ajhsd
6. Kissick D, Leibson D, Kogul M, Bachmann J, Anderson J, Eckert J (2006) Housing for All: essential to economic, social, and civic development the international housing coalition
7. https://www.mapsofindia.com/maps/uttarpradesh/districts/sitapur.htm
8. Master Plan/Regulated Area Map https://uptownplanning.gov.in/site/writereaddata/siteContent/201711081138100153Sitapur-Master-Plan.pdf
9. Vidya Bhooshan, Madhukar Singh (2022–23) Aquifer mapping and management of ground water resources. Sitapur District, Uttar Pradesh: Central Ground Water Board (CWGB), Ministry of Jal Shakti, Government of India
10. Sharma P (2015) Integrated drainage management plan for urban flooding: a case study of Lucknow City, Uttar Pradesh, India. https://doi.org/10.13140/RG.2.2.23611.57128
11. Criqui L (2015) Infrastructure urbanism: Roadmaps for servicing unplanned urbanisation in emerging cities. Habitat Int 47:93–102. https://doi.org/10.1016/j.habitatint.2015.01.015
12. Ming Yip N, Mohamad J, Hong Ching G (2017) Indicators of Sustainable Housing Development (SHD): a review and conceptual framework. Int J Sci & Eng Res, 8(9). http://www.ijser.org
13. Development works carried out in the previous year 2019–20 and the year 2020–21 under review https://sitapur.nic.in/developments-works/

14. Amrit Vichar, https://www.amritvichar.com/article/391841/sitapur--material-recovery-facility-center-is-of-no-use--garbage-piled-up-in-vacant-plots-of-the-city#gsc.tab=0
15. Major Achievements in housing sector by Awaas Bandhu https://www.awasbandhu.in/?page_id=87
16. TOI, https://timesofindia.indiatimes.com/city/lucknow/local-bodies-to-have-standard-building-bylaws/articleshow/95593025.cms
17. Abascal E, Gómez-Coma L, Ortiz I, Ortiz AX (2017) Global diagnosis of nitrate pollution in groundwater and review of removal technologies. In Sci Total Environ 810. Elsevier B.V. https://doi.org/10.1016/j.scitotenv.2021.152233
18. Google Patents. https://patents.google.com/patent/KR101246420B1/en

Policy Innovations for Livable Urban Settlements: Crafting Sustainable Futures Through Innovative Governance in Andhra Pradesh

Muggalla Chandrika

Abstract Urban settlement faces unprecedented challenges in the pursuit of sustainable development. This study navigates the complex interplay between policies and governance frameworks to uncover innovative methods to transform settlements into thriving, equitable, and resilient spaces for promoting sustainable futures. This research tries to explore the impact of governance and policy dynamics, employing a multidisciplinary lens considering environmental sustainability, social inclusivity, and economic vitality components of urban policy innovation in shaping the livability of settlements, with a specific focus on Andhra Pradesh. This research identifies key initiatives, experiments, and transformative policies and the role of local governance and technological interventions in crafting a sustainable urban fabric that have contributed to enhancing the livability of settlements in Andhra Pradesh. Furthermore, the research engages in comparative assessments, merits and demerits of Andhra Pradesh's policy and governance practices. And concludes by offering actionable insights to shape sustainable futures for Andhra Pradesh's settlements by highlighting successful policy innovations and governance strategies.

Keywords Sustainable development · Policy · Governance · Andhra Pradesh · Livable settlements

1 Introduction

Urban settlements are engines of economic growth, as hubs for innovation and growth due to the dynamic interplay of diverse elements in their environments [1], but also pose environmental and social challenges. As urban population grow, it poses both positive and negative effects. The established social, technological, political, economic, and natural infrastructures in cities support and drive human activities, addressing current challenges and seeking new values whereas the negative effects of

M. Chandrika (✉)
Vijayawada, India
e-mail: cmuggalla@gmail.com; cmuggalla@cspgap.gov.in

A. K. Sharma and A. Joshi (eds.), *Proceedings of the 3rd International Conference on Opportunities and Challenges for a Resilient Future*, Advances in 21st Century Human Settlements, https://doi.org/10.1007/978-981-96-0931-4_17

urbanization issues include inefficiencies in public service delivery, informal settlements, rising pollution levels, infrastructural deficits, financial shortfalls, and a lack of integrated planning frameworks [2]. The world's urban landscape is in flux, with rapid growth underway. While Asia, despite of its lower urbanization level, hosts 53% of the world's urban dwellers. Asia and Africa are urbanizing swiftly, projected to absorb nearly 90% of the anticipated 2.5 billion increase in urban population by 2050. Notably, India, China, and Nigeria are anticipated to contribute 37% of global urban population growth between 2014 and 2050 [3]. As the world continues to urbanize, challenges within cities will intensify, particularly in India where the pace of urbanization is fast as it is expected to house half of its population in urban areas by the year 2040. The increasing pace of urbanization in India has not been matched by adequate planning, sector policies, governance, and infrastructure development [4].

As populations increasingly gravitate towards urbanization, the consequences of events in towns and cities extend far beyond the urban scale, with broader-reaching impacts on social, political, and economic realms [5]. Developing sustainable urban strategies is urgent, it involves not only reducing environmental impacts but also creating livable, healthy, and inclusive urban settlements that improve residents' well-being [6]. The world faces pressing challenges, demanding innovative governance and policy responses. Amidst rapid urbanization, this research explores how local governance initiatives shape the livability of Andhra Pradesh settlements. Understanding these localized efforts is critical for building sustainable urban futures in a region experiencing explosive growth. Therefore, the integration of policy frameworks is crucial in shaping and transforming urban and regional development agendas.

2 Literature Review

A review of urban governance literature reveals a gap in understanding the nuanced interactions between governance structures, policies, and the effects of urbanization on livable settlements.

2.1 A Theory of Governance

Governance is one of the key pillars of the sustainability agendas promoted by agencies such as the World Bank, national governments, and nongovernmental organizations [5]. Governance is a term that is widely, but not always precisely used but there is no doubt that the appeal of the term 'governance' in contemporary discussion of governing [7]. Governance is the comprehensive framework of governing, spanning government entities, markets, and networks. The diverse literature on governance showcases its multifaceted nature, drawing from various disciplines. So, the next time you hear "governance," remember—it's not just about politicians and policies, it extends to families, tribes, corporations, and territories. It's the invisible structure guiding our everyday lives [8].

Whereas Raco, 2020 stated that governance is inherently political and often reflected in organizing and managing politics. He made efforts to define "good governance" which emphasizes qualities like participation, transparency, accountability, effectiveness, and equity. It ensures that political, social, and economic decisions are based on broad societal consensus and that the voices of the marginalized are heard in resource allocation. He stated that urban governance focuses on how government functions in cities and towns and the interactions between state entities and civil society, encompassing citizens, communities, private sector actors, and voluntary organizations [5].

Livability refers to the quality of life in any human living environment. Livable environments encompass both physical and social well-being aspects to ensure a thriving and purposeful human existence. The term "Livability" as a noun means "the property of being livable" and it can be defined as "suitability for human living," as an adjective. Livable" means "fit or suitable to live in or with," it can be explained as well as "can be lived" in simple words. However, the degree of livability depends on the community-specific values and context such as locally dominant economic, social, and cultural backgrounds [9]. Kim et al., 2021 argued that improving the livability of cities is a win–win situation for everyone involved. Although liability is broad in concept, safety, quality of built environments, walkability, the convenience of public facilities, access to transport, and natural environments are keys to livability [1]. The interplay between governance and livability underscores the importance of innovative policy approaches in shaping sustainable urban futures as highlighted by Raco, 2020 [5]. Good governance plays a pivotal role in ensuring that urban policies effectively address the diverse needs of communities, as well as prioritize the voices of marginalized groups. Therefore, in pursuit of crafting sustainable urban settlements, policy innovations that integrate governance principles with a focus on enhancing livability are essential. In essence, the convergence of governance and livability underscores the need for innovative policy approaches that prioritize inclusivity, sustainability, and the enhancement of human well-being in any environments.

3 Methodology

The methodology adopted for this study entails a qualitative case study approach, concentrating on the urban governance framework of Andhra Pradesh. Data collection is conducted through a combination of interviews and document analysis. The analytical framework employed encompasses multiple dimensions, including environmental, social, and economic indicators. The objective of this research is to investigate the governance policies implemented by Andhra Pradesh aimed at fostering more livable settlements. Through a systematic evaluation of the governance system, this study assesses the progress made in aligning governance practices with the Sustainable Development Goals (SDGs).

4 Innovative Governance Policies in Andhra Pradesh

Our investigation reveals a nuanced tapestry of governance innovations in Andhra Pradesh. Case studies illuminate successful initiatives and showcase the impact of innovative governance on livability dynamics. A key strategic intervention by the state involves enhancing living standards for citizens by providing permanent housing, social upliftment, economic upliftment, and other essential services.

Andhra Pradesh, the 8th largest state in India, comprises 26 districts and spans an area of 1,62,970 sq km with the country's 2nd longest coastline at 974 km, it holds 4.09% (10th position) of the nation's total population with a population density of 304 persons per sq km in 2011. The sex ratio improved from 983 in 2001 to 997 in 2011, surpassing the national figure of 943. AP exhibited a literacy rate of 67.35% in 2011, marking a notable rise from 62.07% in 2001. However, this figure remains below the national average of 72.98%. In the fiscal year 2022–23, AP recorded nominal GSDP at Rs. 13,17,728 Cr., reflecting a growth of 16.22% from the previous year, and the Per Capita Income for 2022–23 is estimated at Rs. 2,19,518, showing an increase of Rs. 26,931 (13.98%) compared to 2021–22. Urbanization plays a crucial role in fostering economic growth. In 2011, the urban population in the state constituted 29.47% of the total, a notable increase from 24.13% in 2001. Visakhapatnam led among districts with the highest urban population at 47.45%, followed by Krishna district at 40.81% [10].

Parallel to the Andhra Pradesh state frameworks, there are six central government schemes to develop urban areas. The objectives under each scheme directly contribute toward the targets under SDG 11. Those six central government schemes are Atal Mission for Rejuvenation and Urban Transformation (AMRUT) 500 selected cities, Pradhan Mantri Awas Yojana (PMAY), Smart Cities Mission, Swachh Bharat Mission or Clean India Mission, Heritage City Development and Augmentation Yojana (HRIDAY), Deendayal Antyodaya Yojana [2]. While there exists a significant alignment between these initiatives, concerns arise regarding the proliferation of multiple planning documents with divergent visions resulting from these programs. Central government has adopted a piecemeal and reactionary approach, introducing schemes and programs in response to specific events without achieving coherence among various policies or projects [4].

4.1 Andhra Pradesh: A Hub of Groundbreaking and Innovative Poverty Alleviation Programs to Improve Livability

Andhra Pradesh is recognized for its legacy in pioneering and implementing innovative poverty alleviation programs. In addition to general economic policies, targeted initiatives, including women's empowerment, housing, infrastructure development, pensions, land distribution, and health insurance improve the livability standards of the populace irrespective of factors such as caste, creed, religion, gender, or political affiliation. At the heart of this governance strategy lies the concept of "NAVARATHNALU," which is implemented through Grama/Ward Sachivalayas.

The Good Governance Initiatives and priority programs implemented under Navaratnalu

NAVARATHNALU, a key governance concept, is implemented through Grama/Ward Sachivalayas. To ensure efficient service delivery, the government deployed Grama/Ward Volunteers in both rural and urban areas, with 2.65 lakh volunteers in total with one Grama volunteer per 50 households in rural areas and one ward volunteer per 70–100 households in urban areas. This initiative, aims to provide timely and transparent services to eligible households, eliminating corruption at all administrative levels. Additionally, 15,004 Village/Ward Secretariats were established from October 2, 2019, employing 1,25,160 functionaries to deliver welfare schemes and services. Ward volunteers address public service requests. In 2020, they play a vital role in delivering key government welfare schemes to every household, including Social Security pensions, YSR Cheyutha, Jagananna Ammavodi, Jagananna Thodu, Jagananna Vidya Deevena, Jagananna Vasathi Deevena, YSR Chedodu, YSR Arogyasree, and YSR Aasara. The AP Government has put in place several new-age programs that merit replication in other States. Some of the key initiatives are:

Food Security

a. Public Distribution System: In Andhra Pradesh, combating food insecurity takes an innovative turn with doorstep delivery of rice and essential commodities to its 1.46 crore Below Poverty Line families. 9260 Mobile Dispensing Units, operated by empowered youth of Scheduled Castes, Tribes, Backward Classes, and Minorities, bring subsidized rice and goods directly to beneficiaries. This initiative not only ensures food security but also fosters sustainable livelihoods for these young entrepreneurs. Backed by a robust Aadhaar-linked, e-PDS system, the state guarantees transparency and efficiency in distribution.
b. Jagananna Gorumudda (PM-POSHAN): Jagananna Gorumudda revamped mid-day meal program, stands out as a comprehensive system focused on ensuring quality, nutritious, and delicious meals for every child in Andhra Pradesh. Since 2020, enhancements include a revised menu with increased cooking costs, five eggs per week, and peanut-jaggery chikki. To ensure quality, real-time

monitoring is implemented through the Integrated MDM Monitoring System (IMMS). Further boosting quality, Andhra Pradesh is pioneering an AI-based solution developed by TCS to identify good and bad quality food across all schools. Additionally, a dedicated toll-free number (14,417) with a call center handles complaints, showcasing the government's commitment to a well-rounded program.

c. YSR Sampoorna Poshana: To address the challenges of anemia and malnutrition among pregnant and lactating mothers, as well as children aged 6 months to 72 months, the state of Andhra Pradesh has implemented the YSR Sampoorna Poshana Plus(+) scheme in 77 Tribal and sub-plan mandals, and the YSR Sampoorna Poshana scheme in the remaining mandals across the state. The program focuses on providing supplementary nutrition through Anganwadi centers, under this scheme beneficiaries were provided one full meal at Anganwadi Centers and a monthly Take Home Ration Kit. Children aged 36 months to 72 months are offered a hot cooked mid-day meal, including an egg and 100 ml of milk at Anganwadi Centers, while children aged 6 months to 36 months receive Take Home Nutrition.

Education

Andhra Pradesh is deeply committed to achieving inclusive and quality education at all levels, aligning with the SDG for education. The state is dedicated to addressing enrollment and dropout rates, improving school infrastructure, and enhancing learning outcomes. Recognizing education as a fundamental component of human development, the government is working to establish a system that ensures accessible, inclusive, and responsive education, especially for students from disadvantaged backgrounds. The state introduced innovative schemes like:

a. Jagananna Vidya Deevena: Under this scheme, the government provides reimbursement of full fee (i.e. Tuition Fee, Special Fees, Other Fees & Exam Fees) to all eligible students studying ITI, Polytechnic, Degree and above level courses. The amount will be credited into the bank account of the mother of the respective student in four installments.
b. Jagananna Vasathi Deevena**:** Under this scheme, government provides scholarships to all eligible poverty line category of students; Rs. 20,000/- per student studying degree and above level courses; Rs. 10,000/- per student per year for ITI students and Rs. 15,000/- per student per year for polytechnic eligible student for food and hostel expenses.
c. Ammavodi: Andhra Pradesh's Jagananna Amma Vodi scheme paints a promising picture for the future of education. It provides an annual grant of ₹13,000 to economically disadvantaged mothers or recognized guardians for sending their children to school from classes 1st to 12th. The primary objective of the scheme is to improve access to education, promote equity, ensure quality learning, boost attendance and retention rates, and enhance overall child development. An additional amount of Rs. 2,000 to improve school infrastructure and hygiene. Overall, Jagananna Ammavodi is a commendable initiative that is helping to improve the

educational opportunities and living conditions of underprivileged children in Andhra Pradesh

d. Jagananna Videshi Vidya Deevena: To provide financial assistance to eligible students belonging to SC/ST/BC/Minority/EBC including Kapu (OC) to pursue higher studies in top 200 universities as per latest QS world university rankings.
e. Jagananna Vidya Kanuka Kit (JVK Kit): Under this scheme, supplying student kits to the students from 1st to 10th classes consisting of uniform cloth - 3 pairs, text books, set of note books, 1 pair shoe, 2 pairs of socks, 1 school bag, 1 belt & 3 masks.
f. YSR Vidyonnathi: Providing financial assistance to the candidates for UPSC Civil Service Examination Prelims-cum-Mains and interview. Students will be paid a stipend Rs. 12,000/- per month for 9 months.
g. Manabadi Naadu-Nedu: The Up-gradation of Infrastructure facilities in 10 sectors i.e. Toilets, Drinking Water, Major & Minor Works, Electrification, Painting, Furniture, Green Boards, English Labs, Kitchen Sheds, and Additional Class Rooms.

Health

The state is committed to ensuring universal health coverage, the cornerstone of health services is the "Dr. YSR Aarogyasri Scheme," a flagship initiative plays a pivotal role by offering end-to-end cashless healthcare services to families identified as below the poverty line, covering 3,255 procedures across 2,249 network hospitals with a financial limit of Rs. 5 lakhs per family per annum. The "YSR AarogyaAasara" scheme compensates for post-operative recovery period loss of wages, providing Rs. 225 per day up to a maximum of Rs. 5,000 per month. A significant stride in mental health services is the introduction of "Tele MANAS Cell" providing 24×7 tele-counseling accessible to individuals in remote areas. The "Family Physician Concept" is another innovative step, aiming to provide healthcare services to rural populations through YSR Health Clinics. Additionally, 104 Mobile Medical Units under the "Family Physician Concept" provide integrated primary healthcare services.

To provide better medical services to the urban people, the State government has decided to set up 528 YSR Urban Health Centres with 344 new centres and 184 renovations. The aim of setting up the health centres is to provide medical services to all at a commute of two kilometers. Furthermore, the establishment of 560 YSR Urban Clinics is underway to strengthen health infrastructure at the ward level.

Other major schemes Enhancing Social Inclusivity

a. YSR Pension Kanuka: A comprehensive initiative aimed at providing financial assistance to uplift the poor and vulnerable sections of society. The scheme encompasses diverse categories such as old age persons, widows, weavers, artists, transgender individuals, and patients with various medical conditions. The government has raised the pension amount, the enhanced amounts, effective from January 1, 2023, are Rs.2,750 per month for certain categories and Rs.3,000 per month for disabled persons. Furthermore, the government provides financial assistance to individuals facing critical medical conditions such as Thalassemia to organ transplant patients, with monthly pension ranging from Rs.3,000 to Rs.10,000. The state government has implemented an innovative 'Pensions at Door Delivery System,' involving 2.66 lakh Village and Ward Volunteers in the disbursement of pensions to eligible households.

b. YSR Vahanamitra: A significant initiative by the government, playing a crucial role in supporting the livelihoods of drivers by providing financial assistance of Rs. 10,000 to self-owned auto, taxi, and maxi-cab drivers, to cover various expenses, including insurance, fitness certificate, vehicle repairs, and other related costs. This scheme not only addresses the financial needs of individual drivers but also contributes to the overall welfare of the auto, taxi, and maxi-cab driver community, fostering economic stability and sustainability in their professions.

c. YSR Cheyutha: A flagship initiative in the state, aims to provide substantial financial assistance of Rs. 18,750 over a period of four years, amounting to a total of Rs. 75,000 to women in the age group of 45 to 60 years belonging to SC, ST, BC, and Minority communities aiming to empower women, enhance livelihood opportunities, and elevate standards of living at the household level, contributing to sustainable economic development. The scheme not only offers financial support but also focuses on empowering women through the establishment of shops in collaboration with prominent companies such as HUL, ITC, P&G, and Reliance.

d. YSR Aasara: A crucial component of the "Navaratnalu" initiative, is aimed at reimbursing the entire outstanding bank loan of Self-Help Groups (SHGs) as of April 11, 2019, in four installments. The program, designed to create a sustainable impact on the economic well-being, enhanced livelihood opportunities and wealth creation at the household level, focuses on uplifting the standards of living for SHG women in both rural and urban areas. As part of this scheme, an amount of Rs. 15,000 Crore is allocated for waiving the DWCRA loan, benefiting 89 lakh DWCRA women in the state.

e. Jagananna Thodu: In the pursuit of alleviating financial hardships for marginalized sections, the Jagananna Thodu scheme has been instrumental in providing crucial support to hawkers and caste-based workers. The scheme, designed to extend financial assistance of Rs. 10,000 to hawkers, plays a vital role in mitigating economic challenges. This strategic initiative demonstrates the government's commitment to fostering economic resilience among vulnerable sections of society.

f. Jagananna Chedodu: Government is committed to providing financial assistance of Rs. 10,000 per annum to the individuals engaged in professions such as Barber (Nayee Brahmins), Washermen (Rajakas), and Tailoring, who are living below the poverty line. This initiative stands as a testament to the government's commitment to uplift individuals in these professions and improve their overall economic well-being helping them to invest in their businesses, purchase equipment, and improve their living standards.

Housing

a. Pedalandariki Illu: A flagship housing program under "Navaratnalu" focused on providing house sites and housing. The government has undertaken the construction of 30.20 lakh houses in three phases, with 15.60 lakh houses in Phase-I and 3.03 lakh houses in Phase-II ensuring that everyone has access to a home.
b. YSR Jagananna Saswatha Bhu Hakku Mariyu Bhu Raksha Pathakam: Andhra Pradesh is the first state in the country to conduct a comprehensive re-survey using modern and advanced survey technologies. The objectives include ensuring landowners' title security through "Bhu Hakku Pathram" and securing boundaries with "Bhu Raksha Survey Stones," provided free of cost, it includes 40 lakh government and private properties in 123 urban areas.

Other major governance initiatives apart from Navaratnalu

Gadapa Gadapaku Mana Prabhutvam: From 11th May 2022 onwards in which the concerned MLAs made visits to all households (HH) Village/Ward Secretariats to create awareness among the public about the schemes and programs. The government sanctioned Rs.20.00 lakhs to every secretariat for the high-priority works identified by MLAs during GGMP visits.

Jaganannaku Chebudham: A universal grievance redressal helpline that allows citizens to directly raise their issue to the Chief Minister's Office. Citizens can use this helpline by calling 1902 to register individual or household-level grievances, follow up on the status of a registered grievance, enquire about government services and schemes [11].

Jagannana Suraksha: This program was organized for a month to solve welfare schemes and public problems at a satisfactory level. Through these special camps, 11 types of certificates like birth, death, caste, CCRC, ration card division, household division, income, etc., were provided free of charge [12].

Jagananna Arogya Suraksha: A healthcare initiative launched by the Government of Andhra Pradesh to ensure the best healthcare to every family in the state in 45 days, 3922 health camps were organized in urban areas which provided free medical screening, free doctor consultation, and free medicines to all citizens right at their doorsteps.

Major Urban Centric Initiatives by Government of Andhra Pradesh

Clean Andhra Pradesh Programme (CLAP): Program, launched on October 2, 2021, aims to create Bin Free, Litter Free, and Garbage Free cities. It focuses on 100% waste segregation, door-to-door garbage collection, scientific waste treatment, and visual cleanliness. In the first phase, CLAP started in 45 Urban Local Bodies (ULBs), including Corporations and Municipalities, while the second phase covers the remaining 78 ULBs. Essential infrastructure like garbage auto tippers, transfer stations, and three-colored bins has been provided, with 1.21 crores bins distributed to households.

Jagananna Haritha Nagaralu: Aims to enhance the aesthetic appeal and cleanliness of cities in the state by introducing greenery along municipal roads, medians, and roadsides. The program consists of two key components, 1. Greening & Beautification of Roads and Avenues. Activities include plantation and development of central medians, avenue plantation, traffic island development, and mural arts in 45 ULBs (Grade-I and above Municipalities and all Corporations). 2. Smart Street Component: Involves identifying one suitable road in each ULB, initially, with a minimum length of 2 to 3 Kms, to be developed as a Model Road in 46 ULBs. Aiming to transform the identified streets into Mobility Corridors, making them motorized, non-motorized transportation (NMT), and pedestrian-friendly rather than solely focused on vehicular movement.

5 Governance Transformation Localization of SDGs: Adoption of SDGs in Andhra Pradesh

Governance preferences are shaped by a unique blend of institutional frameworks, political norms, cultural traditions, geographical factors, and the resulting economic, social, and environmental conditions. Implementing any significant transformation necessitates facilitating swift progress toward sustainability objectives. Andhra Pradesh was one of the front runners in SDG localization, mapping of government schemes to SDGs [2].

Andhra Pradesh, in tandem with the nation, has embraced the Global Sustainable Development Goals (SDGs) 2030, with a commitment to guiding and propelling social and economic development policies. The state has integrated all Central/State government schemes, including the prestigious "Navaratnalu" program, with the 17 SDGs. Moreover, the Annual State Budget is strategically aligned with these SDGs.

The implementation and monitoring of SDGs in Andhra Pradesh demonstrate maximum granularity, leveraging the innovative feature of Village and Ward Secretariats. Notably, AP stands out as one of the few states where SDG targets have been meticulously drilled down to the Village and Ward level, showcasing a commitment to localizing sustainable development efforts (Table 1).

6 Conclusion

In summary, this study underscores the innovative governance and policy initiatives implemented in Andhra Pradesh, emphasizing their potential to improve livability and make substantial contributions to SDG achievement. While the governance transformation for SDG localization in Andhra Pradesh shows promise, it remains in its developmental stages. Notably, the state government's policies have predominantly aimed at providing social and monetary benefits across both urban and rural areas, rather than focusing on holistic urban-centric development strategies. While these initiatives have their advantages, they also present challenges to the state's

Table 1 Examining Andhra Pradesh's governance policies through the lens of sustainable development goals

Scheme name	SDG	environmental sustainability	social inclusivity	economic vitality	Inference/ Remarks/ Challenges/ Opportunities/Lacunas (based on stakeholder interviews)
Public Distribution System (Mee Intiki Mee Ration)	1, 2				Overall, the doorstep delivery PDS presents a mixed bag of pros and cons. At the same time, it offers potential benefits for convenience, food security, and empowerment, concerns about the dependency on the government for essential commodities may lead to complacency and hinder efforts towards self-sufficiency or alternative livelihoods. Additionally, it was observed that the distribution often takes place at junctions/major hotspots within the locality, rather than direct distribution at the doorstep of the beneficiaries.
Pedalandariki illu (Housing for all poor)	1, 5, 8, 10				Phase-I and Phase-II have already delivered houses to a significant number of beneficiaries, with Phase-III currently underway.

(continued)

Table 1 (continued)

YSR Pension Kanuka	1, 3				Challenge could be ensuring the accurate identification for inclusion and deletion of eligible beneficiaries. Moreover, maintaining the sustainability of the scheme amidst budgetary constraints and evolving socio-economic dynamics poses another challenge.
YSR Cheyutha	1, 5, 10				During a personal interview with a beneficiary, an unexpected response emerged, indicating that the individual had utilized the financial assistance under the Cheyutha scheme for shopping (apparel) rather than investing it towards achieving economic prosperity. Therefore, gradually decreasing the grant amount while encouraging self-sufficiency through training programs, business mentorship, and market research assistance could foster long-term sustainability.
YSR Asara	1, 5, 10				Waiving outstanding loans significantly reduces the financial burden on SHGs, offer-

(continued)

Table 1 (continued)

					ing a fresh start and boosting their economic activity. Whereas, large-scale loan waiver raises concerns about long-term financial sustainability and potential fiscal burden on the state. Beyond financial aid, equipping SHG women with relevant skills and market access support could bolster their entrepreneurial success.
Jagananna Ammavodi	1, 4				Despite its positive intentions, the scheme may encounter certain challenges such as long-term financial viability of the program which needs careful consideration to ensure consistent support. To enhance the scheme's effectiveness, there could be a need for robust monitoring and evaluation mechanisms to track its impact on attendance, retention, and learning outcomes. Linking the grant installments to attendance or academic progress could encourage engagement and better out-

(continued)

Table 1 (continued)

					comes.
YSR Arogyasri	1, 3				Overall, Andhra Pradesh's healthcare initiatives demonstrate a strong commitment to universal access and equitable care. By strengthening quality of care across a vast network of hospitals can solidify its healthcare system
Jagananna Gorumudda	1, 2, 3				Overall, Jagananna Gorumudda shows tremendous potential in improving child nutrition and well-being. While nutrition is crucial, considering complementary hygiene, sanitation, and health education programs could optimize the scheme's impact.
YSR Sampoorna Poshana,	1, 2, 3				However, despite positive strides, the scheme faces significant challenges in implementation. Additionally, robust monitoring and evaluation mechanisms are needed to track the effectiveness of the scheme and address quality control concerns. Expanding the coverage of the YSR Sampoorna Poshana scheme to include

(continued)

Table 1 (continued)

					non-Anganwadi children would indeed be a significant step to enhance its effectiveness in combating malnutrition.
Jagananna Vidya Deevena, (Full fee Reimbursement)	4				Despite its positive intentions, concerns arise regarding sustainability and potential misuse. The scheme's cost could strain state finances, and lack of stringent eligibility criteria or monitoring mechanisms might lead to misuse of funds. To improve, the scheme could consider stricter eligibility criteria based on income and academic merit. Introducing performance-based disbursement or linking it to scholarships could incentivize academic excellence.
Mana Badi - Nadu-Nedu	4, 6				Manabadi Naadu-Nedu's potential to transform Andhra Pradesh's educational landscape is undeniable. However, it also faces certain challenges notably the timely execution of infrastructure upgrades. Moreover, fostering community participa-

(continued)

Table 1 (continued)

					tion and stakeholder engagement can help ensure the infrastructure upgradation can catalyze the program.
CLAP	6, 11				While CLAP paints a promising picture of a cleaner and healthier future for the state, the objectives set, such as source segregation at the household level, have unfortunately not materialized as expected in on-ground implementation.
Jagannana Suraksha	1, 10				The primary survey revealed that the program primarily focused on distributing certificates, particularly income and caste certificates, as well as birth and death certificates.
Jaganannaku Chebudham	1, 10				Overall, Jaganannaku Chebudham stands as a unique and valuable initiative for empowering citizens and facilitating direct communication with their government.

economy, exemplified by the notable increase in total debt from Rs. 3,78,087 Cr. in 2021–22 to Rs. 4,26,234 Cr. in 2022–23. Therefore, there's an urgent need for a shift towards urban-centric policies that prioritize comprehensive development approaches to foster truly livable settlements.

References

1. Kim HM, Sabri S, Kent A (2021) Smart cities as a platform for technological and social innovation in productivity, sustainability, and livability: A conceptual framework. Smart Cities Technol Soc Innov, 9–28. https://doi.org/10.1016/B978-0-12-818886-6.00002-2

2. Kandpal R, Okitasari M (2023) Governance transformation towards localisation of sustainable development goal 11 in India. World Dev Sustain 2:100069. https://doi.org/10.1016/j.wds.2023.100069
3. Porter Z (2015) Moving Day: Urbanization in India and China. Glob Major E-J, 6(1), 16–27. https://www.american.edu/cas/economics/ejournal/upload/porter_accessible.pdf
4. Praharaj S, Han JH, Hawken S (2018) Urban innovation through policy integration: Critical perspectives from 100 smart cities mission in India. City Cult Soc 12:35–43. https://doi.org/10.1016/j.ccs.2017.06.004
5. Raco M (2020) Governance, Urban. Int Encycl Hum Geogr, Second Ed, 253-–258. https://doi.org/10.1016/B978-0-08-102295-5.10650-X
6. Shen L, Ochoa JJ, Bao H (2023) Strategies for sustainable urban development—exploring innovative approaches for a livable future. Buildings 13(3):764. https://doi.org/10.3390/buildings13030764
7. Colebatch HK (2017) Making sense of governance. Policy and Society 33(4):307–316. https://doi.org/10.1016/j.polsoc.2014.10.001
8. Bevir M (2013) A Theory of Governance. University of California Press
9. Ahmed NO, El-Halafawy AM, Amin AM (2019) A critical review of urban livability. Eur J Sustain Dev, 8(1), 165–182. https://doi.org/10.14207/ejsd.2019.v8n1p165
10. Government of Andhra Pradesh (2023) Socio Economic Survey 2022–23. https://apfinance.gov.in/downloads/FINALBOOKSES-2023FINAL.pdf
11. Jaganannaku Chebudham. (n.d.). Jaganannaku Chebudham. Retrieved January 30, 2024, from https://jaganannakuchebudham.ap.gov.in/
12. Jagananna Suraksha for the people of AP. Free from 1st July, good opportunity! (2023). Samayam Telugu. Retrieved January 30, 2024, from https://telugu.samayam.com/andhra-pradesh/news/ap-cm-ys-jagan-mohan-reddy-launches-jagananna-suraksha/articleshow/101212869.cms

Formulating Framework for the Resilience of Indigenous Communities

Sarthak Goel and Amanjeet Kaur

Abstract Indigenous architecture serves as a crucial repository of culture and heritage, encapsulating the distinctive essence of regions and communities. Over time, indigenous communities confront ongoing challenges such as cultural loss, displacement from their lands, economic inequalities, environmental pressures, health disparities, legal injustices, discrimination, social disruption, and political marginalization, collectively impacting their well-being and resilience. To date, there are numerous laws and policies aimed at indigenous groups, but there is no unified framework outlining a systematic process for their upliftment. This research paper emphasizes the pressing requirement to empower indigenous communities through the exploration of diverse avenues. The methodology employed in this study primarily comprises an extensive literature review using qualitative content analysis, which encompasses a comprehensive examination of the diverse laws and policies specifically designed to safeguard and promote the well-being of indigenous communities. Following this, an in-depth analysis and comparative assessment of these legal frameworks are conducted to extract insights into their efficacy and potential areas for improvement. The primary objective of this research is to pinpoint gaps in existing indigenous laws and policies. Through this exploration, the study aims to provide recommendations for a unified framework that can be adopted by a range of stakeholders, including policymakers, community leaders, and organizations.

Keywords Indigenous communities · Legal frameworks · Empowerment · Qualitative content analysis · Comparative assessment

S. Goel (✉) · A. Kaur
Department of Architecture, National Institute of Technology, Hamirpur, Himachal Pradesh, India
e-mail: 196045@nith.ac.in; ar.sarthakgoel@gmail.com

A. K. Sharma and A. Joshi (eds.), *Proceedings of the 3rd International Conference on Opportunities and Challenges for a Resilient Future*, Advances in 21st Century Human Settlements, https://doi.org/10.1007/978-981-96-0931-4_18

1 Introduction

Indigenous Peoples stand as the bearers and practitioners of rich cultures and unique ways of connecting with both people and the environment. Their social, cultural, economic, and political characteristics set them apart from the dominant societies surrounding them. Despite the diversity in their cultures, Indigenous Peoples globally face common challenges in safeguarding their rights as distinct communities. For years, they have tirelessly sought recognition for their identities, ways of life, and their fundamental rights to traditional lands, territories, and natural resources. Unfortunately, history reveals a consistent violation of these rights [1]. In recent times, there has been a notable shift in the discourse surrounding the challenges confronting indigenous communities. This shift, known as juridification, has gained momentum since the 1980s. It entails addressing indigenous issues through legal frameworks, particularly in courts. Beyond merely creating new laws, this signifies a change in how nations perceive themselves and their citizens. Indigenous groups are increasingly resorting to legal avenues to secure recognition and protection of their rights. This transition is transformative, not only reshaping the organizational dynamics of these groups but also playing a pivotal role in defining their self-identity. However, the process of juridification is not uniform, and its outcomes often remain unclear [2]. By critically examining existing laws and policies affecting indigenous communities, the goal is to pinpoint potential shortcomings and provide recommendations for improvement. As rightly said by Dr. Manmohan Singh that "There has been a systemic failure in giving tribals a stake in the modern economic processes that inexorably intrude into their living spaces. The systematic exploitation and social and economic abuse of our tribal communities can no longer be tolerated."

2 Methodology

The methodology for this research paper employed a comprehensive approach to delve into the historical evolution of rights for indigenous communities. The process commenced with an in-depth literature review to chart the developmental trajectory of these rights. In pursuit of the primary research objective, current laws and policies were scrutinized, drawn from relevant documents, books, journal papers, websites, and manuals. Concurrently, diverse sources were explored to identify various rights of Indigenous Communities. The amassed raw data underwent analysis using a structured Qualitative Content Analysis (QCA) process, facilitating the extraction of rights covered under specific laws and categorization of the list of identified rights into various categories or themes through valid inference and interpretation. The compilation of data involved the utilization of various sources, ranging from general internet searches to scholarly databases like Google Scholar, Web of Sciences, and The Lens, to gather historical insights and scholarly perspectives. The organized data was presented in tabular form, facilitating a comparative analysis of categories and subcategories of identified rights with those covered under existing laws. The examination of similarities and differences within the tabular data served as a crucial step in identifying gaps in current laws, forming the foundation for conclusive statements and recommendations intended for stakeholders.

3 Historical Evolution of Indigenous Rights

The early twentieth century witnessed compelling attempts by Indigenous leaders, such as Haudenosaunee Chief Deskaheh and Maori religious leader T.W. Ratana, to advocate for their people's right to self-determination and traditional lands before international forums like the League of Nations [3]. The conceptualization of indigenous identity took a significant step forward in 1981 with Jose R. Martinez Cobo's study, providing a working definition that emphasized historical continuity, distinctiveness from prevailing societies, and the determination to preserve cultural patterns, territories, and ethnic identity. This definition set the stage for international efforts to recognize and protect the rights of Indigenous Peoples [4]. The establishment of the Working Group on Indigenous Populations (WGIP) in 1982 created a platform for Indigenous Peoples to share experiences and concerns. The International Labour Organization (ILO) also played a pivotal role, culminating in the Indigenous and Tribal Peoples Convention, 1989 (No. 169), the only treaty exclusively addressing the rights of these communities [1]. The United Nations (UN) has been actively engaged in promoting Indigenous Peoples' rights. The International Year of the World's Indigenous People in 1993 and the subsequent International Decade of the World's Indigenous Peoples (1995–2004) marked a commitment to fostering a new relationship between states and Indigenous Peoples based on mutual respect [3]. The UN Permanent Forum on Indigenous Issues and the appointment of a Special Rapporteur on the rights of Indigenous Peoples further underscored this commitment [5, 6]. In 2007, the UN General Assembly adopted the Declaration on the Rights of Indigenous Peoples, providing a robust protection of Indigenous Peoples' rights on a global scale [7].

The historical evolution of Indigenous rights is not confined to international forums but extends to individual nations. In the Indian context, debates on indigeneity gained momentum in the 1980s, challenging the official stance that denies the existence of Indigenous Peoples. From 1985 Indian delegates started participating in UNWGIP meetings, advocating for the recognition of Scheduled Tribes as indigenous [8]. These debates in India, with a complex history of positive discrimination policies and territorial concessions, contribute to the broader narrative of the global evolution of Indigenous Peoples' rights.

4 State of the Art: Review of Indigenous Communities Rights Laws and Policies

The legal system, a construct moulded by social and historical contexts, holds significant influence in shaping societal values and norms [9]. The Government of India has enacted various laws and initiatives to recognize and safeguard the rights of indigenous communities; nevertheless, the paper concentrates on laws rather than schemes because schemes are geared to solve a specific issue rather than the entire problem. In India, a diverse array of laws has evolved over the years, reflecting the nation's dedication to the welfare of its indigenous communities. From the enactment of the Scheduled Castes and Scheduled Tribes (Prevention of Atrocities) Act in 1989 to subsequent legislations such as the Panchayat (Extension to Scheduled Areas) Act, 1996, and the Biological Diversity Act in 2002, the legal landscape underscores a commitment to addressing the multifaceted challenges faced by indigenous populations. The introduction of the Forest Rights Act and the Draft of the National Tribal Policy in 2006 further emphasizes the dynamic nature of the legal framework. Amid this complex web of laws and policies, key focus areas emerge, spanning land and resource rights, cultural preservation, representation in decision-making processes, healthcare and education disparities, economic empowerment, and protection against discrimination and displacement. However, as a responsible developing nation, there is a pressing need to broaden perspectives and address concerns impacting indigenous communities that may not be adequately addressed by existing laws. The legal landscape of indigenous rights encompasses broader categories such as self-determination, land and resources, culture and language, equality, fundamental freedoms, protection from harm, health and well-being, development and environment, spiritual and cultural life, and legal and procedural rights. To identify gaps and propose recommendations for the holistic well-being of indigenous communities, a detailed scrutiny of existing Indian laws and policies is paramount.

4.1 The Scheduled Castes and Scheduled Tribes (Prevention of Atrocities) Act in 1989

The Scheduled Castes and Scheduled Tribes (Prevention of Atrocities) Act, enacted in 1989 (POA), serves as a critical legislative instrument aimed at safeguarding the rights of the most marginalized sections of Indian society—the 'Dalits' or legally recognized as 'Scheduled Castes' and the 'Tribals' or legally recognized as 'Scheduled Tribes.' Recognizing the historical and pervasive atrocities committed against these communities, the Act provides essential protection against discrimination and violence. The legislative intent, as outlined during its introduction in Lok Sabha, emphasizes the Indian state's commitment to delivering justice to the Dalit class and eradicating the deeply rooted practice of untouchability. The Act is structured into three categories: the first addresses criminal law, establishing criminal liability for specifically defined crimes and expanding on penalizations outlined in the Indian Penal Code (IPC). The second category focuses on relief and compensation for victims of atrocities, aiming to redress the harm inflicted. The third category establishes special authorities tasked with enforcement and monitoring to ensure the effective implementation of the Act. Through its comprehensive framework, the SC/ST (Prevention of Atrocities) Act plays a crucial role in combating discrimination and promoting justice for these historically marginalized communities [10–12].

4.2 *The Panchayat (Extension to Scheduled Areas) Act, 1996*

The Panchayats (Extension to the Scheduled Areas) Act, 1996 (PESA) addresses the imperative of integrating tribal communities into mainstream development while preserving their cultural identity. Enacted in response to the recommendations of the Bhu ria Committee, PESA extends the provisions of Part IX of the Constitution to Scheduled Areas. Key features highlighted by the Bhuria Committee include the establishment of Gram Sabhas at the village level to manage essential resources, reorganization of boundaries based on ethnic considerations, granting sub- state status to tribal regions, addressing deficiencies in the Land Acquisition Act, ensuring participative democracy, and giving tribal communities control over economic resources. PESA empowers autonomous district councils in central Indian tribal tracts, securing local consent in land acquisition and rehabilitation, and regulating land and resource usage for industries. This legislation signifies a significant move towards tribal self-rule, recognizing the distinct governance needs of Scheduled Areas in nine states, including Andhra Pradesh, Chhattisgarh, Gujarat, Himachal Pradesh, Jharkhand, Maharashtra, Madhya Pradesh, Orissa, and Rajasthan, all heavily populated by tribal communities [13, 14].

4.3 *Biological Diversity Act, 2002*

The Biological Diversity Act of 2002 stands as a crucial framework from the perspective of indigenous communities, recognizing the intrinsic link between biodiversity and traditional knowledge. India, with its rich cultural diversity, is home to various indigenous communities that have inherently contributed to the conservation and sustainable use of biological resources through their traditional practices. Aligned with the Convention on Biological Diversity, the Act acknowledges the vital role of local and indigenous communities in maintaining biodiversity. It establishes the National Biodiversity Authority (NBA), State Biodiversity Boards (SBB), and Biodiversity Management Committees (BMCs) to ensure the active participation of local communities in decision-making processes related to bioresources within their jurisdictions. By prioritizing the protection of traditional knowledge, the Act aims to empower indigenous communities as custodians of biodiversity. The Act also emphasizes equitable benefit-sharing, recognizing the valuable role of local communities in conserving biological resources. Through the establishment of National, State, and Local Biodiversity Funds, the Act provides a mechanism for channelling resources into conservation efforts that involve and benefit indigenous communities. Moreover, by requiring approval from the NBA for intellectual property rights related to bioresources, the Act ensures that indigenous knowledge remains protected and acknowledged. Overall, the Biological Diversity Act underscores the importance of recognizing and respecting the roles of indigenous communities in preserving India's unique biodiversity and traditional knowledge [15–17].

4.4 Forest Rights Act, 2006

The Forest Rights Act of 2006 represents a significant departure from previous forest legislation, signalling a paradigm shift in recognizing and safeguarding the rights of forest-dwelling communities. Initially drafted as the Scheduled Tribes (Recognition of Forest Right) Bill 2005 by the Ministry of Tribal Affairs, it faced opposition from civil society and NGOs. Subsequently, a revised draft, titled the Scheduled Tribes and Other Traditional Forest Dwellers (Recognition of Forest Right) Bill 2006, was introduced, addressing concerns and widening its scope. Enacted on December 18, 2006, the Act aims to acknowledge and secure the rights of forest-dwelling communities, fostering their active involvement in forest and wildlife conservation. Key objectives include granting land rights to these communities, providing the right to use, collect, and dispose of minor forest produce, and recognizing rights on land used for communal purposes. Notably, the Act introduces the right to access biodiversity and community rights to intellectual property and traditional knowledge, representing a pivotal dimension of the legislation [18].

4.5 Draft National Policy on Tribals, 2006

The Nehruvian Panchasheel, established in 1952, outlines five principles guiding tribal administration in India, emphasizing unique development, land rights, minimal external influence, preservation of institutions, and quality of life measurement. Acknowledging the need for a more detailed approach, the Indian government established the Ministry of Tribal Affairs in October 1999 to hasten tribal development. In response to the generalities of the Nehruvian Panchasheel, the Ministry started crafting a draft National Policy on Tribals, aiming for a specific framework for the socio-economic development and empowerment of Scheduled Tribes. Informed by feedback from tribal leaders, states, organizations, and NGOs, the Ministry aims to finalize the policy. The National Policy recognizes that a majority of Scheduled Tribes face challenges like poverty, low literacy rates, malnutrition, and vulnerability to displacement. It acknowledges tribes as custodians of indigenous knowledge. The policy seeks to concretely address these issues and outlines measures to preserve and promote tribal cultural heritage [19].

The National Tribal Policy in India, drafted in 2006, is yet to be finalized, posing challenges to holistic development for tribal communities. Despite recent expressions of commitment, the absence of a concrete timeline and effective implementation hinders the realization of the policy's potential impact. The Ministry of Tribal Affairs drafted the policy in 2006, but revisions were needed due to legislative and policy changes. A High-Level Committee (HLC) formed in 2013 submitted recommendations in 2014, but their implementation remains pending, with no specified timeframe [20].

5 Discussion and Results

The following two sections discuss the criteria and indicators used to evaluate indigenous populations' rights within legal frameworks, followed by a detailed comparative analysis of existing legal frameworks.

5.1 Criteria and Indicators List

The United Nations Declaration on the Rights of Indigenous Peoples (UNDRIP), adopted by the UN General Assembly in 2007 after more than two decades of negotiations between Indigenous Peoples and States, stands as the most recent and comprehensive expression of indigenous aspirations. Widely supported by over 140 countries, even the four countries initially opposing it have since reversed their positions. UNDRIP delineates both individual and collective rights of indigenous peoples, encompassing aspects like culture, identity, language, employment, health, and education. While not legally binding internationally, it constitutes a crucial framework for establishing norms against human rights violations and addressing discrimination [7, 21, 22]. The 46 articles of the declaration have been analyzed through qualitative content analysis as part of the scope of this research paper, placing rights under various categories with recommendations from experts. These categories (Table 2) serve as a valuable resource for formulating recommendations and drawing conclusions in the field.

5.2 Comparative Analysis of Existing Legal Frameworks

A comparative analysis is being conducted in Table 3 to identify the gaps in the existing laws and policies affecting the rights of indigenous communities. This comparison is made between the list of identified rights by UNDRIP, categorized under various categories (Table 2), and the rights addressed by the existing laws (Table 1).

The analysis reveals variations in the focus areas among different laws, with some rights receiving attention in each law, while others remain unaddressed. Different laws and policies emphasize distinct sets of rights. Key rights, including Land, Territory, and Resources, Cultural Identity and Expression, Equality and Non-discrimination, Right to Life, Physical and Mental Integrity, Right to Nationality, Protection from Harm, and Spiritual Relationship with Lands, are universally recognized and addressed by all laws. However, specific rights, such as Right to Self-Determination, Participation in Decision-Making, Self-Governance, Right to Freedom of Expression, Right to All Social and Health Services, Protection and Conservation of Traditional Plants, and Right to Maintain Traditional Health Practices, exhibit variations across the analyzed policies. Notably, some rights, like Free, Prior and Informed Consent, Education in Their Own Language, Media and Communication Rights, and Right to Access Financial and Technical Assistance, are addressed by only one out of the five laws. And few rights including Conservation of Cultural Heritage, Development with Identity, and Maintaining and Developing Contacts and Relations are not addressed by any. The observed pattern indicates that there is a level of consistency in addressing certain traditional and common rights across the analyzed legal frameworks. These fundamental rights, such as Land, Territory, and Resources, Cultural Identity, and Protection from Harm, are recognized uniformly. However, the analysis also reveals a crucial gap – none of the existing policies undergoes an in-depth examination of all the challenges comprehensively faced by indigenous communities. Highlighting a specific instance, the right to education in their native language is addressed by only 1 out of 5 laws, which is currently in the draft stage (as part of the National Tribal Policy). Despite being included in laws, the right to education in one's native language is crucial for the upliftment of indigenous communities. Education in one's language enhances comfort and performance. Moreover, The Right To Education Act of 2009, a consequential legislation under Article 21-A, guarantees every child the right to satisfactory and equitable full-time elementary education in a formal school, adhering to essential norms and standards [23]. For instance, a highly intellectual tribal individual may face unequal situations due to language barriers, impacting their ability to produce expected outcomes. In other words, while there is some convergence on core issues, the laws fall short of providing a thorough and detailed exploration of the diverse challenges encountered by indigenous communities as a whole. This underscores the need for a more comprehensive and nuanced legal framework that can effectively address the multifaceted issues affecting these communities.

Table 1 Indigenous community rights covered under diverse existing Laws. (*Source*: By Author)

Laws/Policies	Focused rights
POA ACT, 1989	Right to Dignity and Non-Discrimination
	Projection of Property and Livelihood
	Right to Personal Security
	Right to Freedom from Forced Labor
	Right to Cultural and Religious Practices
	Voting Rights
	Protection from abuse of legal processes
PESA, 1996	Devolution of Power and Authority to Gram Sabha and Panchayats
	Conformity with Customary Law and Practices
	Participatory Democracy
	Protection of Natural Resources
	Right to Social and Economic development
	Protection of tribal traditions, beliefs and culture
	Local Self-Governance
	Permission from Gram Sabhas for land acquistion
	Gram Sabha rights over minor forest produce, land restoration, and money lending control
Biological diversity act, 2002	Conservation and Sustainable use of Biological Diversity
	Recognition and Protection of Traditional Knowledge and Customary practices
	Fair and equitable benefit sharing mechanism
	Consultation with Loacal people in Decision making
	Prior informed consent
	Capacity Building and Empowerment
	Access to Biological Resources, Healthcare, Education and other infrastructure
	Intellectual Property Rights
	Enhanced income generating opprtunities
Forest rights act, 2006	Right to hold and live in the forest land
	Right to collect and use minor Forest Produce (NISTAR)
	Traditional seasonal resource access of nomadic or pastoralist communities
	Rights to community tenures of habitat and habitation
	Rights in or over disputes lands
	Rights for conversion of Pattas/leases/grants on forest lands to titles
	Rights of settlement and conversion of all forest villages into revenue villages

(continued)

Table 1 (continued)

Laws/Policies	Focused rights
	Rights to protect, conserve or manage any community forest resource
	Rights recognised by State/District/Region as Tribal Laws
	Right of access to biodiversity and community right to intellectual property anf traditional knowledge
	Traditional right enjoyed by the forest dweller (except hunting/trapping)
	Right to in situ rehabilitation
Draft national tribal policy, 2006	Right to Formal Eduction
	Preserve, Protect and Promote Traditional Wisdom
	Right to protect Healthcare Facilities
	Proper Mechanism for Displacement and Resettlement
	Forest Rights
	Guidelines for Shifting Cultivation
	Tackling tribal land alienation
	Intellectual Property Rights
	Preserving and documenting Tribal Languages
	Focuses on Primitive Tribal Groups problems
	Regulations for better Administration
	Research and monitoring
	Participatory Approach
	Assimilate tribals into country's mainstream

6 Conclusions and Recommendations

Delving into the legal frameworks surrounding the rights of Indigenous communities reveals the potential to inspire more inclusive and culturally relevant approaches to governance. The laws and policies should act as crucial guidelines, offering protection, empowerment, and a pathway to global justice for tribal people. However, the analysis uncovers significant shortcomings in the existing legal landscape, primarily characterized by limited scope and focus. While fundamental rights are consistently addressed, there is a noticeable lack of in-depth analysis and coverage of the comprehensive challenges faced by indigenous communities. The current legal framework tends to concentrate on a set of traditional common rights but falls short of delving into the intricacies of broader issues affecting these communities.

The recommendations emerging from this analysis call for a holistic and inclusive approach in formulating legal frameworks and policies for indigenous communities. Broadening the scope of existing laws to encompass a more comprehensive set of rights and addressing the diverse challenges faced by these groups is

Table 2 Identified rights of indigenous communities as per UNDRIP. (*Source*: By Author)

Categories	Identified rights	Explanation
Self-Determination and Governance	Right to self-determination	The right of indigenous communities to determine their own political status, pursue economic, social, and cultural development, and manage their own affairs
	Free, Prior and Informed Consent (FPIC)	ensures the voluntary, prior, and well-informed agreement of indigenous communities in architectural projects or any other activity affecting their lands or rights
	Participation In decision-making	The active participation of indigenous communities in architectural decisions, policies, and programs that affect their lives
	Autonomy and self-government	Authority to self-govern in architectural constructions or any internal affairs, and allows to establish internal structures and policies
Land, territory, and resources	Ownership and control of lands, territories and resources	Authority over their architectural territory and resources, allowing them to determine land use and development
	Protection from dispossession and exploitation	Safeguarding architectural assets, preventing unauthorized seizure or displacement from indigenous lands
	Redress for land rights violations	Mechanisms to seek remedies for architectural injustices related to land rights, supporting fair resolution processes
Culture and language	Cultural identity and expression	Ensures the practice, revitalization, and architectural reflection of their cultural traditions, customs, languages and ceremonies
	Maintainance and Protection of cultural heritage	Integrates architectural measures to safeguard and showcase the tangible and intangible cultural heritage of indigenous communities
	Education in their own languages	Advocates for architectural spaces that facilitate education in indigenous languages, promoting linguistic diversity and cultural inclusivity
	Media and communication rights	Promotes architectural spaces that facilitate indigenous media, fostering cultural exchange without discrimination
Equality and Non-Discrimination	Freedom from any kind of discrimination	Giudes architects to design spaces free fram any kind of discrimination based on their identity, culture, or traditions

(continued)

Table 2 (continued)

Categories	Identified rights	Explanation
	Access of Justice	Advocates for fair and effective remedies, addressing infrigements on individual and collective rights within the built environment or any other domain
Fundamental freedoms	Right to life, physical and mental integrity	Creating an environment that values and protects people's lives, well-being, and personal liberties
	Right to nationality/ belong to an Indigenous Community	Contributes to the sense of identity and belongingness
	Right to freedom of expression	Encourages architectural spaces that facilitate open expression, diverse opinions, and peaceful gatherings within the community
Protection from harm	Freedom from any act of violence	Protection from genocide and cultural destruction by establishing safe and secure physical spaces
	Right not to be Forcefully assimilated	Crafting environments that recognize and preserve the particular cultural identity and customs, protecting them from assimilation pressures
Health and Well-being	Right to all social and health services	Creating infrastructure that promotes well-being and provides equitable access to healthcare facilities
	Protection and Conservation to Traditional Medicinal Plants	Measures to safeguard and conserve traditional medicinal plants may include the construction of spaces such as traditional plant gardens
	Right to maintain Traditional Health Practices	Integrating culturally relevant places into healthcare facilities, allowing indigenous groups to maintan their traditional health practices
Development and environment	Development with Identity	Emphasizes maintaining a symbiotic relationship between development activities and the preservation of cultural identity through thoughful and culturally sensitive mechanisms or design
	Environment protection	Adopting Practices that minimize environmental impact, promote biodiversity, and contribute to the overall well-being of the ecosystem
Spiritual life	Spiritual relationship with their lands	Maintain and strengthen their distinctive spiritual relationship with their lands, territories, water, and resources

(continued)

Table 2 (continued)

Categories	Identified rights	Explanation
Networking and financing	Maintain and develop contacts, relations and cooperation	Facilitate cross-cultural interactions and collaborations
	Right to access to financial and technical assistance	Supports the development of the built environment or any other domain through financial and technical support

imperative. Enhanced collaboration between legal institutions, indigenous representatives, and relevant stakeholders is essential to ensure the thorough consideration and integration of the perspectives and concerns of these communities into legislative and policy-making processes. Emphasizing a participatory and consultative model, which includes Free, Prior and Informed Consent (FPIC), will foster a more equitable and respectful relationship between indigenous communities and governing bodies [22]. Furthermore, it is crucial to create awareness and sensitivity among legal professionals, policymakers, and the general public regarding indigenous issues, fostering a culture of respect, understanding, and support for the rights and well-being of indigenous communities. The ultimate recommendation involves formulating a unified national policy addressing all the rights of indigenous communities, as discussed in the list of indicators, instead of creating various laws. This policy can be context-specific, with region-specific modifications, offering a more comprehensive and tailored approach to safeguarding the rights of indigenous communities.

Ultimately, this research paper provides a curated list of essential rights for integration into formulation of laws, contributing to a comprehensive and inclusive approach to indigenous community welfare. Additionally, it fosters understanding among stakeholders, emphasizing the imperative for inclusive policy development by pinpointing critical gaps.

Table 3 Comparative analysis of existing legal frameworks (*Source*: By Author)

Categories	Identified rights	POA act, 1989	PESA Act, 1996	Biological Diversity Act, 2002	Forest Rights Act, 2006	Draft national tribals policy, 2006
Self-Determination and governance	Right to Self-determination	●	✔	✔	●	●
	Free, Prior and Informed Consent (FPIC)	●	●	✔	●	●
	Participation in decision-making	●	✔	✔	●	✔
	Autonomy and self-government	●	✔	●	●	●
Land, territory, and resources	Ownership and control and lands, territories, and resources	✔	✔	✔	✔	✔
	Protection from dispossession and exploitation	✔	✔	✔	✔	✔
	Redress for land rights violations	✔	✔	✔	✔	✔
Culture and language	Cultural identity and expression	✔	✔	✔	✔	✔
	Maintainance, Protection and Conservation of cultural heritage	●	●	●	●	●
	Education in their own languages	●	●	●	●	✔
	Media and communication rights	●	●	●	●	✔
Equality and Non-Discrimination	Freedom from any kind of discrimination	✔	✔	✔	✔	✔
	Access of Justice	✔	✔	✔	✔	✔
Fundamental freedoms	Right to life, physical and mental integrity	✔	✔	✔	✔	✔
	Right to nationality/belong to an Indigenous Community	✔	✔	✔	✔	✔

(continued)

Table 3 (continued)

Categories	Identified rights	POA act, 1989	PESA Act, 1996	Biological Diversity Act, 2002	Forest Rights Act, 2006	Draft national tribals policy, 2006
	Right to freedom of expression	●	✔	✔	●	✔
Protection from harm	Freedom from any act of violence	✔	✔	✔	✔	✔
	Right not to Forcefully assimilated	✔	✔	✔	✔	✔
Health and Well-being	Right to all social and health services	●	●	✔	●	✔
	Protection and Conservation of Traditional Medicinal Plants	●	●	✔	✔	✔
	Right to maintain Traditional Health Practices	●	●	✔	●	✔
Development and environment	Development with identity	●	●	●	●	●
	Environmental protection	●	✔	✔	●	●
Spiritual life	Spiritual relationship with their lands	✔	✔	✔	✔	✔
Networking and financing	Maintain and develop contacts, relations and cooperation	●	●	●	●	●
	Right to access to financial and technical assistance	●	●	●	●	✔

References

1. Indigenous and tribal peoples (Indigenous and tribal peoples). (n.d.). Retrieved January 26, 2024, from https://www.ilo.org/global/topics/indigenous-tribal/lang-en/index.htm
2. Kingsbury, B. (n.d.). Indigenous Peoples
3. Indigenous peoples at the united nations. United nations for indigenous peoples. (n.d.). Retrieved January 25, 2024, from https://www.un.org/development/desa/indigenouspeoples/about-us.html

4. Martínez Cobo Study. United Nations For Indigenous Peoples. (n.d.). Retrieved January 26, 2024, from https://www.un.org/development/desa/indigenouspeoples/publications/2014/09/martinez-cobo-study/
5. Special Rapporteur on the rights of Indigenous Peoples. OHCHR. (n.d.). Retrieved January 26, 2024, from https://www.ohchr.org/en/special-procedures/sr-indigenous-peoples
6. United Nations Permanent Forum on Indigenous Issues (UNPFII) | Division for Inclusive Social Development (DISD). (n.d.). Retrieved January 26, 2024, from https://social.desa.un.org/issues/indigenous-peoples/unpfii
7. Candelaria SM (2012) International Labour Organisation. Manila Office., Project to Promote ILO Policy on Indigenous and Tribal Peoples., & United Nations. General Assembly. Comparative analysis on the ILO Indigenous and Tribal Peoples Convention no. 169, UN Declaration on the Rights of Indigenous Peoples (UNDRIP), and Indigenous Peoples' Rights Act (IPRA) of the Philippines. International Labour Organization, ILO Country Office of the Philippines
8. Shah A (2007) The dark side of Indigeneity?: Indigenous people, rights and development in India. Hist Compass 5(6):1806–1832. https://doi.org/10.1111/j.1478-0542.2007.00471.x
9. Shrinkhal R (2019) Evolution of Indigenous rights under international law: analysis from TWAIL perspective. Orient Anthropol 19(1):7–24. https://doi.org/10.1177/0972558X19835387
10. Acharya P, Acharya P (2020) An analysis of the scheduled castes and scheduled tribes (prevention of atrocities) Act, 1989. SSRN Electron J. https://doi.org/10.2139/SSRN.3732709
11. Ministry of Tribal Affairs—Government of India. (n.d.). Retrieved January 27, 2024, from https://tribal.nic.in/KnowledgeHub.aspx
12. The Scheduled Castes and the Scheduled Tribes (Prevention of Atrocities) Act, 1989. (n.d.)
13. Deep A (2020) Tribals and panchayats with reference to the pesa act, 1996: a critical analysis. In JETIR2006359 J Emerg Technol Innov Res, 7. JETIR. www.jetir.org
14. Panchayat (Extension to Scheduled Areas). (1996). http://socialissuesindia.wordpress.com/
15. International journal of multidisciplinary research and development. (n.d.). www.wbbb.gov.in
16. Ministry of law and justice (Legislative Department). (n.d.)
17. Venkataraman K (2008) RIS Research and Information System for developing countries access and benefit sharing and the biological diversity Act of India: A Progress Report. Asian Biotechnol Dev Rev, 10(3), 69–80. www.ris.org.in/abdr.html
18. Act, rules and guidelines. (n.d.)
19. National policy on tribals. (n.d.). Retrieved January 27, 2024, from https://www.mcrhrdi.gov.in/87fc/policies/NATIONAL%20POLICY%20ON%20TRIBALS.pdf
20. Press Information Bureau. (n.d.). Retrieved January 27, 2024, from https://pib.gov.in/PressReleasePage.aspx?PRID=1578780
21. Engle K (2011) On fragile architecture: The un declaration on the rights of indigenous peoples in the context of human rights. Eur J Int Law 22(1):141–163. https://doi.org/10.1093/ejil/chr019
22. United Nations Declaration on the Rights of Indigenous Peoples United Nations. (n.d.)
23. School education. Government of India, Ministry of Education. (n.d.). Retrieved January 31, 2024, from https://www.education.gov.in/rte

Community Empowerment and Participation: A Path to Sustainable Housing Development for Industrial Workers in Noida, Uttar Pradesh, India

Manisha and Bhanu M. Marwaha

Abstract Noida is the satellite city of Delhi and a hub for manufacturing industries, with a substantial workforce primarily composed of industrial workers, making up 29.47 percent of the total workforce. However, the current housing options available for industrial workers within the city are mostly unplanned settlements like urban villages, squatter settlements, unauthorized colonies, etc. In order to achieve sustainability in industrial workers' housing, this study examines the transforming impact of community empowerment and participation. It goes beyond traditional housing provision, focusing on broader sustainability dimensions such as social, economic, and cultural well-being. The research employs a mixed-method approach, combining qualitative and quantitative techniques. It analyses existing housing projects that actively involve communities in decision-making, planning, and management. Additionally, surveys and interviews gather insights from workers and stakeholders in Noida for identifying the issues related to parameters of community empowerment and participation in the sustainability of industrial workers housing. The study reveals that community-based housing delivery (CBHD) significantly enhances housing sustainability. Involving industrial workers in the process fosters social cohesion, responsible environmental practices, and economic resilience. This participation leads to more sustainable living environments, highlighting the value of community-driven initiatives. This research offers a fresh perspective by emphasizing the transformative potential of community engagement in industrial worker housing. It expands the understanding of sustainability by including the broader aspects of community involvement, enriching the discourse on housing and community development. The study provides practical insights for policymakers, urban planners, and housing stakeholders. It underscores the importance of involving industrial workers in shaping their housing environments. These insights are invaluable for promoting holistic sustainability in housing solutions and fostering inclusive living environments for industrial workers in other developing countries.

Manisha (✉) · B. M. Marwaha
Department of Architecture, National Institute of Technology, Hamirpur, Himanchal Pradesh 177005, India
e-mail: manisha_2019@nith.ac.in

A. K. Sharma and A. Joshi (eds.), *Proceedings of the 3rd International Conference on Opportunities and Challenges for a Resilient Future*, Advances in 21st Century Human Settlements, https://doi.org/10.1007/978-981-96-0931-4_19

Keywords Community empowerment · Quality of housing · Sustainable housing · Industrial workers · Community-based housing

1 Introduction

Community empowerment is the process of enabling communities to increase control over their lives. As such, facilitating empowerment at the community level necessitates collective and coordinated efforts from a variety of groups and social institution to enhance economic resources (i.e., housing, access to clean water and sanitation) availability and access to health and social services are important for reducing vulnerability [1]. It is important to remember that social sustainability of housing is not just about producing better quality houses and environments, but also about fostering community capacity, building bonds and trust between responsible leadership and citizenship, and engaging people in the process of city building and realizing their right to be involved and make decisions [2]. Community empowerment emphasizes that sustainable development and positive social change are most impactful when driven by the community itself. Empowered communities actively engage, take ownership, and influence decisions affecting their well-being, often through leadership development and increased social capital. Access to education and resources is key in this process, enabling communities to address challenges and create sustainable solutions. Participatory planning, involving locals in housing projects, is crucial for sustainable community development, ensuring that social, financial, and environmental needs are met.

In the context of Noida's industrial development, the pursuit of sustainability extends beyond economic progress to encompass the well-being of the workforce. The intricate relationship between industrial growth and housing in this region presents diverse challenges. Recognizing the vital connection between quality housing and community empowerment, this research explores the transformative potential of involving communities in shaping sustainable housing for industrial workers. Noida, a prominent industrial hub in the National Capital Region, has experienced rapid industrialization, creating economic opportunities but also escalating the demand for housing, especially for the labour force. Challenges in the housing scenario include inadequate infrastructure and substandard living conditions, necessitating a thorough examination and innovative solutions. This research underscores the importance of understanding industrial workers' housing in Noida through the lenses of community empowerment and participation, aiming to provide insights for policymakers, urban planners, and stakeholders involved in the intersection of industrial growth and human settlements.

1.1 Research Objectives

The overarching goal of this research is to unravel the potential of community empowerment and participation in fostering sustainable industrial workers' housing in Noida. The specific objectives include:

- To understand the role of community Empowerment and Participation for sustainable housing development.
- To explore existing models of community empowerment and participation in the context of housing development.
- To identifying the issues related to parameters of community empowerment and participation in sustainable housing development for industrial workers in Noida.
- To provide recommendations for policy improvements and community-based strategies to enhance of industrial workers' housing in Noida.

1.2 Scope and Limitations

Study will mainly focus Economically weaker migrant industrial workers working in manufacturing industries in Noida.

2 Literature Review

2.1 Community Empowerment and Participation in Housing Development

Community empowerment in housing development is influenced by several key factors. One important factor is the provision of flexible and affordable loans for low-income citizens, which enables them to own housing units and reduces their financial burden [3]. Another factor is the creative process of the inhabitants in low-income settlements, who use public spaces as places of communication and build unique housing types with used materials [4]. Access to resources and acquisition of skills are also crucial for empowerment, as seen in the case of Nthutukoville in South Africa [5]. Community participation plays a crucial role in achieving desired goals and creating conditions for economic and social progress. It involves people's involvement in decision-making, program implementation, and evaluation, as well as sharing in the benefits of development programs [6]. The institutionalization and assertion of discursive rights are important in empowering marginalized groups. This requires interactions between diverse actors and structures of governance to change opportunities for resistance and build substantive rights [7]. Additionally, adequate housing delivery has the potential to empower individuals by giving them more control and self-determination over their lives, breaking the cycle of poverty [6].

Overall, community empowerment in housing development is influenced by factors such as community participation, the institutionalization of rights, and the provision of adequate housing. The community empowerment approach, through self-help schemes and partnerships, strengthens the capacity of communities to reduce disaster risk and mobilize resources for rehabilitation and reconstruction efforts. By focusing on community empowerment, housing development can be more inclusive, sustainable, and responsive to the needs and aspirations of the people it serves.

2.2 *Community Empowerment and Participation Theories in Community Development*

Community empowerment in the context of housing development involves strategies and models that aim to involve and empower community members in decision-making processes, improve their living conditions, and foster a sense of ownership and pride in their neighborhoods. Several existing models and approaches have been implemented around the world. Here are some examples like Community Land Trusts (CLTs) are nonprofit organizations that acquire and hold land for the benefit of a community while removing it from the speculative real estate market. They often work in collaboration with residents to develop and manage affordable housing. CLTs empower communities by giving them control over the land, ensuring long-term affordability, and involving residents in decision-making about land use and development [8]. Community Development Corporations (CDCs) are nonprofit organizations dedicated to community development, including affordable housing. They often work closely with residents to identify needs and implement solutions. CDCs empower communities by involving residents in the planning and development process, ensuring that projects align with the community's vision and needs [9]. Tenant Management Organizations (TMOs) are organizations formed by tenants to collectively manage and maintain their housing developments [10]. TMOs can implement sustainability initiatives within the housing community, such as energy-efficient retrofits, waste reduction programs, and community gardens. In Community–Based Housing Cooperatives model, community members come together to form cooperatives that collectively own and manage housing units, ensuring affordability and resident control. Community-based housing cooperatives empower residents by providing an alternative to traditional housing models, promoting community engagement, and fostering a sense of shared responsibility [11].

Public participation plays an important role in the success and failure of these projects [12]. The Ministry of Construction and Housing in Iraq collaborated with the Habitat Organization to upgrade deteriorated residential neighborhoods through community participation. In Iraq, community participation policies have been found to aid in the rehabilitation of damaged residential areas and the achievement of long-term development [13]. Participation models in sustainable housing aim to involve participants in decision-making and social learning, leading towards social sustainability. Additionally, aligning the objectives of housing associations and funding agencies can reduce barriers and take advantage of opportunities for participation [14] Community participation is seen as an empowering approach to achieving the Sustainable Development Goals in housing development, with positive outcomes reported in participatory projects [7]. Participatory modeling is crucial in addressing interconnected economic, social, and environmental challenges in sustainability science, particularly in sustainable housing projects. Challenges in implementing participatory processes include the need for more inclusive, deliberative, and decentralized techniques. Successful participatory models emphasize community engagement, ensuring housing projects align with residents' values and contribute to sustainable, socially inclusive developments.

2.3 Case Studies of Successful Empowerment Programs

Here are a few examples of successful community empowerment initiatives and participatory casestudies in sustainable housing projects: Baugruppen, or building groups, are collective housing initiatives in Germany where groups of individuals come together to plan, design, and build their own sustainable housing projects. Residents are actively involved in decision-making from the early planning stages, selecting architects and making choices about sustainable design features. This approach fosters a strong sense of community and ensures that the housing project aligns with the residents' values [15]. The False Creek South neighbourhood redevelopment in Vancouver involved extensive community engagement in the planning and design of sustainable, mixed-income housing. The participatory model included workshops, community meetings, and collaborative planning sessions. Residents were actively involved in decision-making processes, ensuring that the development met both sustainability and community needs [16]. Lilac (Low Impact Living Affordable Community) is a cohousing project in Leeds where residents collectively own and manage eco-friendly homes. Residents were involved in the design process, choosing sustainable building materials and energy-efficient technologies [17]. Dongtan, a planned eco-city in China, incorporated extensive public engagement and participation in its development process. Through town hall meetings, workshops, and consultations, residents and stakeholders provided input on sustainable design, transportation systems, and community infrastructure. The participatory model aimed to create a city that reflected the values and needs of its future inhabitants [18].

These case studies highlight the success of community empowerment initiatives and participatory models in sustainable housing projects. Common themes include active resident involvement in decision-making, collaboration with architects and planners, and a focus on sustainable design principles. Successful projects often prioritize both environmental sustainability and the well-being of the community, fostering a sense of ownership and pride among residents.

3 Industrial Worker's Housing Scenario in Noida, Uttar Pradesh

Noida is located in the state of Uttar Pradesh, within the Gautama Buddha Nagar district, India. Noida's economic landscape is dominated by industrial activities, fueling rapid urbanization with various industrial and commercial enterprises. The city's systematic industrial development began in 1998, with 3587 small-scale units and 235 medium/large units. Over the years, Noida has grown substantially, from 3800 hectares in 2001 to 7789 hectares in 2011, and aims for further expansion to 15,279.9 hectares by 2031. The city's growth trajectory is directed southward, influenced by existing and planned industrial zones. In 1991, industrial workers constituted 29.47% of the Noida Urban Area workforce, impacting housing demand with 35.66% earmarked for residential areas. Various housing projects, including "Shramik Kunj," address the needs of the working class, constructing 31,035 residential units, with private builders contributing 52%, village communities and Noida authority each providing 25%. Housing options for industrial workers include urban villages squatter settlements and unauthorized colonies [19] (Fig. 1).

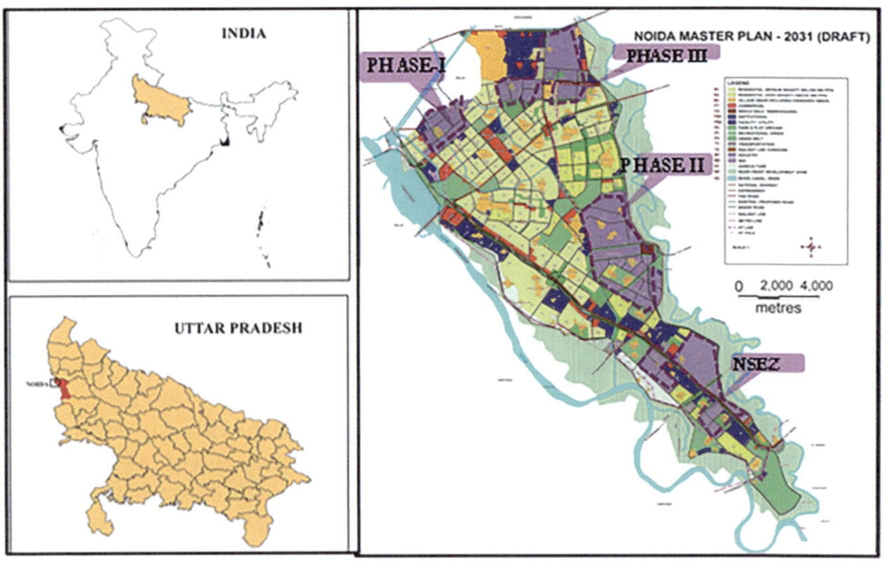

Fig. 1 Location map of Noida, *Source* Google maps and Noida master plan

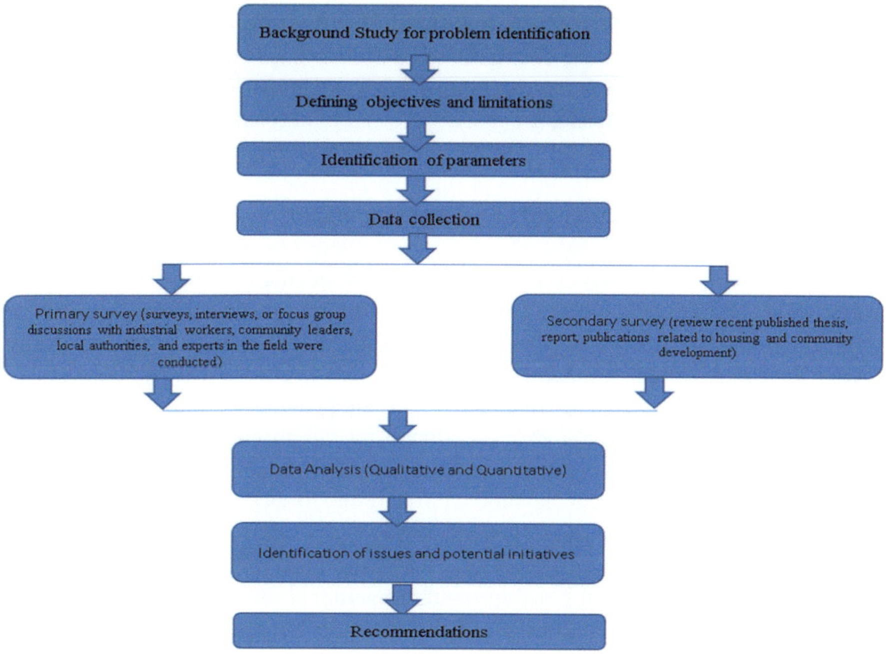

Fig. 2 Research methodology

4 Methodology

Evaluating community empowerment and participation in industrial workers' housing necessitates a comprehensive approach that considers various indicators. These indicators are crucial for assessing the community's engagement, empowerment, and overall well-being in industrial workers' housing. To pinpoint the most relevant issues, surveys, interviews, and focus group discussions were conducted with industrial workers, community leaders, local authorities, and experts in the field. Additionally, a review of recent academic publications and reports on housing and community development was undertaken to align the research with the latest developments. The research employed snowball sampling, a non-probability sampling technique, for interviews with 250 workers, and data collected from the primary survey were analysed using MS Excel 2013 (Fig. 2).

5 Analysis and Discussion

- **Availability of formal Affordable rental housing options for industrial workers**: In Noida, there was no formal rental housing provided by the employers of industries and government, [19]. Most of the workers were currently living in villages and unauthorized colonies (kind of private plotted housing) on rent. Security of tenure is lacking due to the rental housing arrangement Workers typically share overcrowded accommodations, with multiple people sharing one room where the condition of houses are just livable and maintenance was a big problem. In current housing options, they did not have proper sewage and drainage facilities, as well as there, were no facilities of open spaces. Issues with sewage, drainage, parking, and roads are common in these areas, impacting the quality of life.

- **Job Security**: Noida provides diverse employment opportunities, yet industrial workers face job insecurity and limited career advancement due to prevalent casual employment practices. Industrialists often offer temporary positions, resulting in irregular income for workers and hindering their ability to invest in housing, contributing to financial insecurity.
- **Financial Assistance**: The financial challenges faced by industrial workers employed on a casual basis are exacerbated by the absence of crucial financial assistance, such as housing rent allowances. Without such assistance, industrial workers on casual contracts may find it challenging to secure stable and affordable housing, leading to increased financial strain.
- **Awareness**: Industrial workers often lack awareness of housing schemes and programs as well as rental housing regulations that could potentially assist them in securing better living conditions only 4% workers have the awareness regarding housing schemes 96% do not have it.
- **Training and Capacity Building**: In Noida 73% workers did not get training for skill development. Without proper training, industrial workers may struggle to advance, leading to professional stagnation and limited career opportunities. Casual employment often comes with limited skill development, increasing job insecurity amid evolving market demands. Untrained workers may be less efficient, impacting overall productivity. Lack of training hinders social mobility, making it difficult for workers to escape low-income brackets and perpetuating social inequalities.
- **Transparent institutional mechanism**: 96% workers said that there is lack of transparent institutional mechanism due to which they were not able to take advantage of different housing scheme of government.
- **Regular feedback Mechanism**: 95% workers said that there is no feedback mechanism. Robust and continuous feedback mechanism is vital for ensuring the effectiveness and adaptability of industrial workers' housing initiatives. A transparent and responsive feedback system allows for ongoing improvements and adjustments based on evolving needs and preferences (Fig. 3).

Key issues identified include a lack of formal affordable housing, inadequate basic services, low income, job insecurity, and a need for community-driven solutions etc. These challenges underscore the necessity for involving industrial workers in shaping their housing solutions.

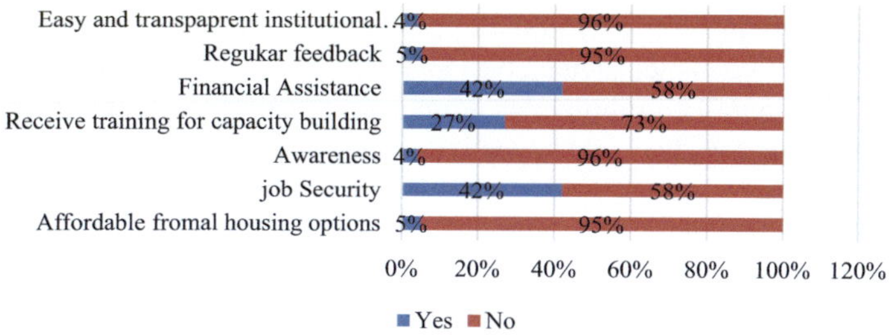

Fig. 3 Primary survey

6 Recommendations for Identified Issues

To achieve sustainable industrial workers' housing in Noida, Uttar Pradesh, through community empowerment and participation, consider the following recommendations for identified issues:

- Encourage industries for creating formal jobs that are inclusive and offer equal opportunities to the workers by providing fiscal incentives.
- Noida Authority should submit a research proposal to the central government under the Affordable Rental Housing sub-scheme of the Pradhan Mantri Awas Yojna, launched in May 2020. Collaboration with industrial entities can lead to targeted housing solutions that align with the specific needs and preferences of migrant workers in the industrial sector as well as facilitate employee participation in housing planning.
- Implementing skill development initiatives for migrant industrial workers can enhance their employability, promoting long-term economic stability. Additionally, developing capacity-building programs for community leaders and members equips them with the skills and knowledge needed to actively contribute to the sustainable development of housing projects.
- Regular awareness campaigns and collaborative efforts with NGOs and government agencies are essential to reach a wider audience and secure resources for sustainable housing initiatives, rental housing regulations and housing programs and schemes. Additionally, organizing workshops and educational programs fosters open communication and community involvement, enhancing active participation in the development process.
- Industrialist should provide reliable and accessible transportation facilities. This enables workers to make informed choices about housing locations, reducing the need for housing to be situated very close to workplaces.
- Formulates policies to encourage enterprises to offer rental subsidies to migrant workers. Provides financial literacy training to industrial workers, empowering them to understand and engage in financial aspects related to housing, mortgages, and community funding models.

- Involve workers in industrial housing planning through surveys, focus groups, or committees to enhance satisfaction and well-being in living spaces. Establish community planning committees that include industrial workers, local residents, and experts to ensure diverse perspectives and active participation in the decision-making process.
- Encourage the formation of community-led maintenance groups to ensure the long-term upkeep and sustainability of housing projects. This fosters a sense of ownership and responsibility.
- Ensure transparency in decision-making processes related to housing projects. Regularly update the community on project developments and involve them in key decisions.
- Establish mechanisms for regular feedback from the community. This ensures that adjustments can be made based on evolving needs and preferences

7 Conclusion

In conclusion fostering community empowerment and participation is crucial for sustainable industrial workers' housing in Noida. Engaging workers, local communities, and stakeholders in planning creates tailored, environmentally conscious solutions. Recommendations emphasize collaboration through workshops and transparent decision-making. Worker involvement cultivates pride and resilient living environments. Partnerships with employers, government, NGOs, and financial institutions are essential for success. Overall, community empowerment and participation aligns with broader goals of sustainability, equity, and economic development, building a shared vision for a better future in Noida.

References

1. Abu Talib IF, Takim R, Mohammad MF, Hassan PF (2018) Community empowerment through rehabilitation and reconstruction in social sector of Kuala Krai, Kelantan, Malaysia. Procedia Eng 212:294–301. https://doi.org/10.1016/J.PROENG.2018.01.038
2. Golubchikov O, Badyina A (2012) Sustainable housing for sustainable cities, no October
3. Al-Jawari DHAKSM (2023) Empowerment and its impact on affordable housing sustainable planning: a case study of Al-Sultan housing complex in Al-Najaf. https://doi.org/10.1063/5.0135982
4. P B, Sutanto H (2017) Community empowerment strategy by sustainable built environment planning in Urban Kampong. Iopscience.Iop.Org, 8(February):68–74, https://doi.org/10.1088/1755-1315
5. Ndinda C (2022) Housing delivery in Nthutukoville, South Africa: successes and problems for women. J Int Women's Stud 5(1)
6. Howell KL (2016) Planning for empowerment: upending the traditional approach to planning for affordable housing in the face of gentrification. Plan Theory Pract 17(2):210–226. https://doi.org/10.1080/14649357.2016.1156729

7. Aule TT, Bin Moh'd Jusan M, Ayoosu MI (2019) Outcomes of community participation in housing development: an update review. Int J Sci Res Sci Eng Technol 6(6):208–218. https://doi.org/10.32628/ijsrset196642
8. Greenstein R, Sungu-eryilmaz Y (2007) Community land trusts
9. Walker C Community development corporations and their changing support systems
10. Co-operative Councils Innovation Network (2018) Community-led housing: a key role for local authorities the. [Online]. https://www.councils.coop/wp-content/uploads/2018/01/Community_Led_housing_Report_2017_web_version.pdf
11. Carroli L (2021) Participatory research and planning in practice. Urban Res & Pract 14(1):107–109
12. Jiwane A (2020) Public participation in housing projects of developing countries: capacities and confines. 2020 2nd Int Sustain Resil Conf Technol Innov Build Des. https://doi.org/10.1109/IEEECONF51154.2020.9319944
13. Golić K, Kosorić V, Lau SK (2020) A framework for early stages of socially sustainable renovation of multifamily buildings with occupants' participation. Sustain 12(21):1–23. https://doi.org/10.3390/su12218823
14. Hale L (2012) Participation in sustainable housing a case study of participation in a Swedish housing cooperative's sustainability efforts. [Online]. http://lup.lub.lu.se/luur/download?func=downloadFile&recordOId=3129225&fileOId=3129226
15. Trevisan R, Figueira A (2019) Baugruppen: the German model of cohousing and its constitutive variables. VIRUS J (December). [Online]. https://www.academia.edu/41299765/Baugruppen_the_German_model_of_cohousing_and_its_constitutive_variables
16. Levitt K (2021) Future of false creek south: advancing a conceptual development plan and addressing lease expiries
17. Council LC (2017) LILAC (Low impact low living affordable community)
18. Wang X (2018) Tomorrow's eco-city in China improving eco-city development though a culture of collaborative communication thesis submitted in accordance with the requirements of the university of Liverpool for the degree of doctor of philosophy by
19. Noida development authority (2020) Noida master plan-2031. 1

Present Scenario of Industrial Sector in Himachal Pradesh: Problems and Prospects

Ar. Bhupender Panwar and Ar. Amanjeet Kaur

Abstract Since its inception as a state in 1971, Himachal Pradesh has relied heavily on horticulture and tourism to generate revenue. The government of India announced a special industrial package of incentives for setting up new industries in the state during the period 2003–14, attracting about 1074 large and medium enterprises as reported by the Industries Department, Govt. of H.P. Industrial growth took a constant decline 2013–14 onwards & continued despite the fact that the special package was renounced in 2014 with some modification (Mohd S (2017) Impact of special industrial package on the growth of Himachal Pradesh). The paper assesses the problems and prospects of industrial cities in Himachal Pradesh using secondary data sources, including online and print sources, from 1990–91 to 2020–21. During the evaluation period, medium and large-scale enterprises contributed significantly to employment and the number of industries. The challenges related to planning perspective faced by industrial towns in the state of Himachal Pradesh highlighted in the study. The problems were identified through literature review and ranked in order of significance by taking expert opinion using delphi method. Further solutions were explored through literature and suggested in the related context.

Keywords Industrial sector · Small and medium scale industries · Government policies

1 Introduction

Himachal Pradesh was regarded as a tourism and horticulture state in the past, as there were limited registered industries at the time of its establishment as a state. Those include two industries located in Nahan, one brewery in Kasauli, Tea factory in Palampur, Salt mine in Darang and four small gun factories in Mandi [1]. The government of Himachal Pradesh took steps for promoting industrial development

Ar. B. Panwar (✉) · Ar. A. Kaur
Department of Architecture, National Institute of Technology, Hamirpur, HP 177005, India
e-mail: 22rar003@nith.ac.in

A. K. Sharma and A. Joshi (eds.), *Proceedings of the 3rd International Conference on Opportunities and Challenges for a Resilient Future*, Advances in 21st Century Human Settlements, https://doi.org/10.1007/978-981-96-0931-4_20

including the state industrial policy particularly from 1990 onwards with the help of industrial policies and schemes. According to a report by Himachal Pradesh Human Development, the growth of the industrial sector showed significant improvement in Gross State Domestic Product (from 12.18% in 1995–96 to 14.38 in 1999–00) [2].

1.1 Prominent Industrial Areas of Himachal Pradesh

Himachal Pradesh is host to various manufacturing industries that contribute to the state's economy. As per the [2], the major contribution is of the pharmaceuticals industries located in Baddi and Paonta Sahib, engineering industries in Amb, Parwanoo and Sansarpur Terrace, electronics in Chambaghat and Tahilwal, Gagret. These prominent industrial areas can be located on a map of Himachal Pradesh in Fig. 1. It can be observed that the industrial areas have been emerged in the districts located on the borders of the neighboring states i.e., Punjab, Haryana, Chandigarh and Uttar Pradesh.

1.2 Schemes and Incentives that Helped in the Growth of Industrial Sector

The effect of special package incentives can be observed in the statistics showcasing its impact. The number of medium-scale and large-scale units registered in 2002–03 were only 3, that went to 35 in 2004–05 and 64 in 2005–06 and this growth continued to 2010 as seen in Fig. 3. The growth of industrial sector has significantly declined after 2014–15 when the special package incentives were called off by the government [1]. It was the main factor in Himachal's industrial boom in 2003, when the central government announced special packages for the state of Himachal Pradesh in the form of tax and excise duty concessions.

Some of the incentives include: 100% excise duty exemption to industries for 10 years when they start production, 100% income tax exemption for the first 5 years, after that 30% for companies, 25% for additional 5 years when they start production in the whole state [3]. The Deen Dayal Hathkarga Protsahan Yojna changed the funding ratio between the Indian government and the states from 50:50 to 90:10. Under the Prime Minister Rojgar Yojna (PMRY) government made provision for 15% subsidies of the total cost of the project with a maximum limit of up to 15,000 per entrepreneur [4] (Table 1).

From the above information it can be said that the center special incentive scheme was responsible for the huge growth in the industrial sector from 2003–04 onwards.

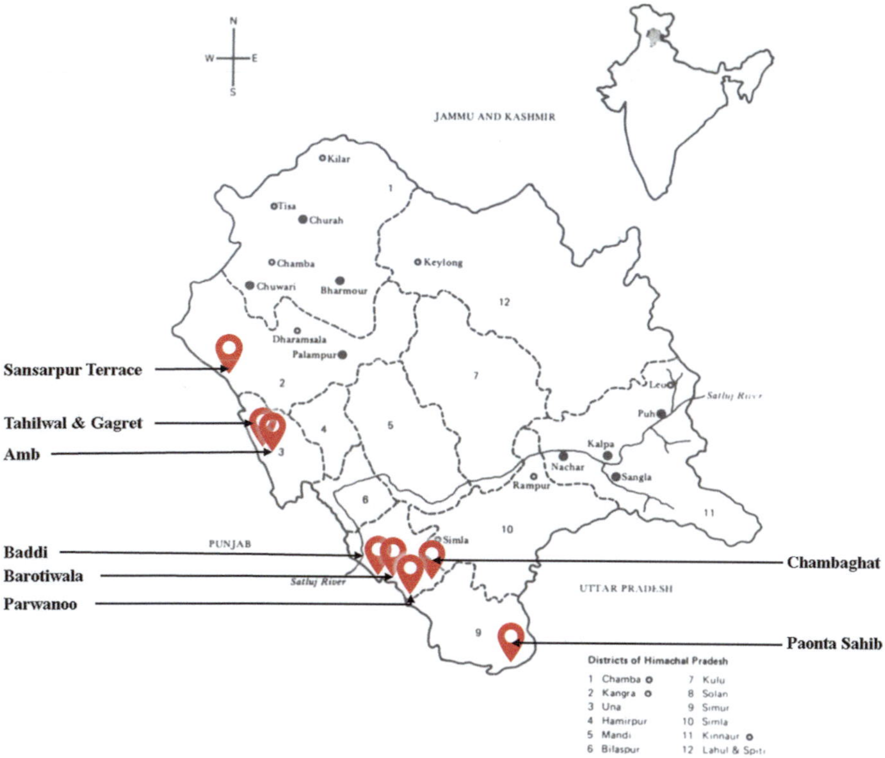

Fig. 1 Map of Himachal Pradesh showing the prominent industrial areas of Himachal Pradesh generated by author in the basis of locations given in state industrial profile, Govt. of Himachal Pradesh, various issues

Table 1 Number and, employment of small, medium & large-scale industries in Himachal during 1990–91 to 2020–21

Year	Small scale industries		Large & medium scale industries	
	No. of units	Employment (Nos.)	No. of units	Employment (Nos.)
Upto 31.02.1991	20,545	86,227	110	15,125
1991–00	7500	33,391	72	13,805
2000–10	8356	69,959	262	23,153
2010–21	4345	36,870	73	11,939

1.3 Inter-District Analysis of Small, Medium and Large-Scale Industries in Himachal Pradesh

Understanding the variations among districts regarding the concentration of small, medium, and large-scale businesses during the period under review will be made

Table 2 District wise distribution of industries in Himachal Pradesh upto 2020–21

District	Factories (No.)			Employment (Nos.)		
	2019	2020	2021	2019	2020	2021
Bilaspur	103	104	104	4942	4992	4992
Chamba	88	91	91	2688	2778	2778
Hamirpur	200	200	200	1001	1001	1001
Kangra	496	498	498	10,915	11,065	11,065
Kinnaur	14	14	14	2302	2302	2302
Kullu	205	208	208	5163	5283	5283
Lahual & Spiti	0	0	0	0	0	0
Mandi	182	185	185	4175	4345	4345
Shimla	183	186	188	7099	7299	7649
Sirmour	703	722	727	55,122	56,812	57,182
Solan	2696	2737	2777	235,858	239,648	245,428
Una	303	308	313	23,755	24,025	24,555

easier with the aid of inter-district analysis. The contribution of small-, medium-, and large-scale industries in each district of Himachal during the 2019–21 fiscal year is shown in Table 2. It shows a high rate of inter-district disparity about the various components of growth of Small, Medium and Large Industries in the state of Himachal Pradesh during the year 2019–21 [5]. Table 2 indicates that the main industrial centers in Himachal are Solan, Sirmaur, Kangra and Una contributing to 82% of the total industries as per the Statistical Report, Department of Economics and Statistics [6].

2 Objectives of the Study

In light of the preceding context, the current study investigates the growth trend and challenges confronting Himachal Pradesh's industrial sector during the evaluation period. More specifically, the study's major objectives are:

To analyze industrial sector trends in Himachal Pradesh between 1991 and 2021.

To identify the main obstacles to the industrial sector's expansion in Himachal Pradesh and the challenges it faces.

To assess the requirement for infrastructure improvement as a component of spatial planning that will assist in reducing the growth disparity among industrial sectors.

3 Material and Methods

In order to achieve the study's goals, secondary data sources including books, journals, government reports (from the Himachal Pradesh government's Industrial Department and the Central Statistical Organization (CSO) and data released by the Economic and Political Weekly Research Foundation (EPWRF) were consulted for information on different facets of the state's industrial development. As a result, the gathered data were presented using basic statistical tools and tabular data.

Firstly, the trends from 1991–2021 are studied and examined to analyze the present scenario of industrial sector. Secondly, the factors that are acting as barrier in the growth of the industrial sector in Himachal Pradesh were identified on the basis of the literature review. The factors that were listed out of papers were sent to fifteen experts including urban planners, architects, industrial association presidents and researchers in the same field for ranking based on its significance in growth of industrial sector, using the Delphi method. The Delphi method, known for its iterative and anonymous expert consensus process, allows for a more refined judgment in assessing the significance of factors influencing industrial sector growth.

Following steps were involved in the research.

3.1 Factor Identification

Initially, factors hindering industrial sector growth in Himachal Pradesh were identified through a thorough literature review.

3.2 Expert Panel Selection

A panel of fifteen experts, consisting of urban planners, architects, industrial association presidents, and researchers in the field, was assembled.

3.3 Questionnaire Design

A structured questionnaire was formulated, listing the identified factors. Experts were asked to rank these factors based on their perceived significance in influencing industrial sector growth.

3.4 Round One

The first round involved distributing the questionnaire to the expert panel. Responses were collected anonymously.

3.5 Analysis and Feedback

After analyzing the responses, a summary report highlighting the aggregated rankings and comments was shared with the experts.

3.6 Round Two

Experts were provided with the summary report and had the opportunity to revise their rankings based on the collective feedback.

3.7 Consensus Building

Iterative rounds continued until a consensus was reached among the experts regarding the relative importance of the identified factors.

3.8 Final Report

The final results, incorporating the refined rankings and any areas of consensus or dissent among the experts, were included in the study report.

4 Results

According to the analysis based on the trends shown in Table 1, Himachal Pradesh's small-, medium-, and large-scale industries all saw significant growth starting in 2003. However, a decline was noted starting in 2014. During the same time frame, the Centre Special Package was in effect. It is evident that the expansion of the industrial sector was significantly aided by the Centre Special Package. As seen in Figs. 2 and 3, the industrial sector is currently in a declining stage. Expert assistance

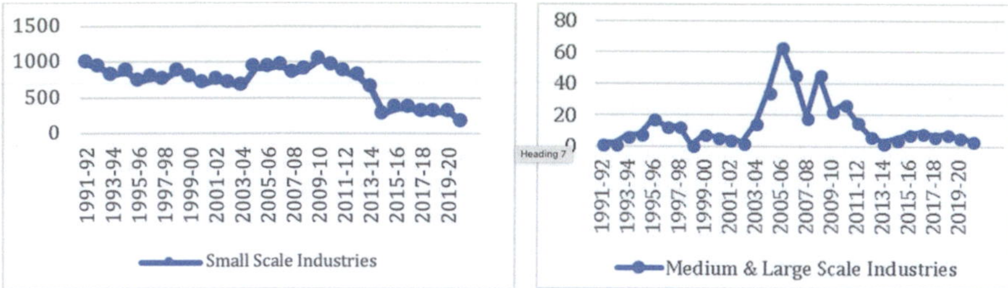

Fig. 2 Trend of small-scale industries in Himachal Pradesh from 1991 to 2020

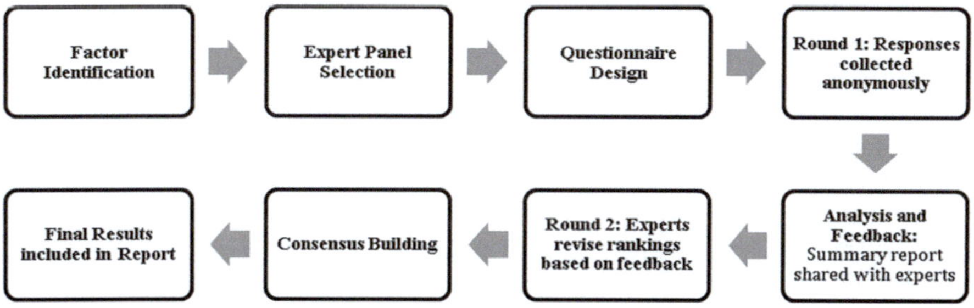

Fig. 3 Stepwise process followed for factor ranking

was sought to identify the issues pertaining to this matter that are impeding the state's industrial sector's growth. We've talked about these issues below.

4.1 *Problems Faced by the Industrial Sector of Himachal Pradesh*

The industrial sector is experiencing issues that have been brought to light from a planner's perspective based on published articles and reports, as the decline in the sector's growth has been overserved in recent years:

Inter-district disparity of Industrial Growth. As per the data given in Table 2, in present the state is facing a huge disparity of industries between the District. For e.g. four districts i.e., Solan, Sirmaur, Kangra and Una alone occupies about 82% of the total industries. According to [1, 4] the main reasons are the lack of connectivity, basic necessities and availability of raw materials. The government of Himachal Pradesh have pointed out the potential industrial hub in the Chap. 16 of Development report [2]. However, this will be possible only the issues will be resolved.

Transportation Problems. Industries in Himachal Pradesh often encounter transportation challenges due to the lack of a reliable railway network. The entire transportation system relies heavily on roads, leading to inefficiencies and increased costs

Table 3 Ranking of problems affecting the growth of industrial sector

S. No	Description	Rank
1	Lack of skilled and qualified personnel	01
2	Issue of unreliable transportation network	02
3	Lack of basic infrastructure for industries	03
4	Lack of participation in skill development centers by private organizations	04
5	Frequent Increment of electricity rates	05

[7]. In an interview with [8], Vice-chairman of the Himachal Pradesh Chamber of Commerce and Industries, Arun Rawat, stated that the absence of railway tracks is the main issue facing the industry since all transportation is dependent on roads, which drives up transportation costs and increases the price of goods.

Infrastructural Facilities. The productivity and profitability of industries are greatly impacted when basic utilities like power and water are scarce. According to [9], there are significant infrastructure deficiencies and reliability issues in many places, posing a significant threat to innovate. Secondly the power rates were one of the key attractions for the industries but due to the withdrawal of centre special package, the tariffs came closer to the neighboring states (Table 3).

The scarcity of qualified personnel. Innovation requires a skilled workforce, and India has a large reservoir of human resources, but there is a scarcity of skilled workers in Himachal Pradesh. According to [10], in order to fulfill the requirements, skilled or semi-skilled labour are imported from other parts of the country. As per the article published by [11], lack of skilled and regular labour is top barrier in the growth of industrial sector in Himachal Pradesh. The migration of skilled personnel to other states is one of the main reasons behind this problem. Secondly, the incompetence of vocational and training institutes to generate skilled workforce as per the requirement of local industries is another reason [5].

On the basis of the ranking given by the experts, the table showcases the problems in the rank order where first rank defines the problem that is affecting the most in growth of industrial sector. According to the rankings given by experts, lack of skilled and regular personnel is affecting the most in growth followed by unreliable transportation network.

4.2 Potential Solutions

The industrial sector in Himachal Pradesh, like many regions, faces a myriad of challenges that hinder its economic growth. Recognizing the challenges as opportunities for improvement, there exist potential solutions that can pave the way for a thriving industrial landscape in the state. The potential solutions for the discussed problems have been discussed on the basis of the references given in Table 4. By addressing these issues strategically, Himachal Pradesh can unlock its economic potential and foster a conducive environment for industrial growth.

Promoting new industrial hubs in the state. As examined above the development of industrial sector is taking place in the periphery districts of Himachal Pradesh. There is need to promote new industrial hubs in the other parts on the state to lower the disparity between the districts. The new industries must be promoted on the basis of traditional methods, knowledge and availability of raw material. According to the [2], the state government has recommended the potential industrial area in the state refer to Table 5. The main requirement to develop will be upgrade to the reliable and all-season connectivity between these clusters, existing industrial hub and neighboring states. Secondly, the infrastructural facilities shall be provided to attract the investors to the state.

Boosting Connectivity between the industrial towns. To address connectivity issues, the government should invest in improving road and trains networks. With the help of Public–Private Partnerships (PPPs), the potential industries areas in the state should be connected with multiple and reliable transportation systems. This should

Table 4 Potential solutions

S. No	Solutions	Sources
1	Improving Transportation Connectivity	[10, 12–17]
2	Skill Development Institutes	[9, 11, 14, 18–20]
3	Proposing New Industrial Clusters	[18, 21–25]

Table 5 Recommended Industrial Clusters with activities in Himachal Pradesh

S. No	Description	Areas	Name of activity
1	Bilaspur	Lethwin Bhadrog	Potters weavers
2	Hamirpur & Nadaun		Shawl making, furniture, hosiery, leather products, card board boxes & herbal processing
3	Kangra	Kandror	Engineering/agriculture implements
4	Kullu	Nagri	Wood carving
5	Kinnaur	Tapri	Metal fabrication, Shawl weaving bamboo craft, pattu making etc
6	Sirmaur	Rajgarh	Fruit processing
7	Lahaul & Spiti	Udiapur	Weaving of woolen patti
8	Mandi	Ner chowk	Automobile
9	Shimla	Ramnagar	Wooden and steel furniture
10	Sirmaur	Rajgarh	Fruit processing
11	Bilaspur	Sangarh Shillai	Limestone weaving

include the expansion and maintenance of roads multi model connectivity between the industrial zones, ensuring smooth transportation of goods and raw materials.

Skill Development Programs. To ensure a skilled workforce, the government should collaborate with educational institutions and private organizations such as industries, to offer specialized training programs that are related to the types of industries present in the state. For this the social infrastructure should be given equal importance while preparation of development plans. For e.g., as per a report by [26], Maruti Suzuki India Limited (MSIL) signed a memorandum of agreement (MoA) with the Government of Haryana and Gujrat in which they have set up four Japan-India Institute for Manufacturing (JIM), in two in each state, at Uchana and Kansala, in Haryana and at Gandhi Nagar and Mehsana in Gujrat which helped in upskilling the local workforce and making them industry ready.

5 Discussion and Conclusions

A significant factor in the economic growth of developing nations like India is industrialization. In order to examine the current situation of the industrial sector in Himachal Pradesh, the trends of the evaluation period i.e. from 1990–2021 were analysed and found that the sector flourished with the help of incentives given in centre's special package. But the small, medium and large scale is under declining stage since the withdrawal of incentives from 2014. Himachal itself has huge potential to grow as an industrial state. It can accommodate those industries that might not be possible in other states due to its topographical and climatic conditions, which makes it favourable for specific types of industries.

An analysis of print and online literature was used to determine the problems the state's industrial sector was facing. As discussed in the paper, unavailability of qualified and skilled personnel, lack of connectivity and infrastructure facilities, the industrial sector of the state is facing a decline in the growth. According to the opinion given by the experts under delphi method, the lack of skilled and qualified personnel is the most significant problem that is restricting the growth of industrial sector followed by the scarcity of transport and rail connectivity. By investing in improving connectivity and infrastructure facilities with the help of private partnerships, Himachal Pradesh can attract more industries and revive its existing industrial sector.

It's a dire need to unlock the potential of the state by resolving the issues as discussed in the paper. The development plans must be prepared by the government focusing on proposing new industrial hubs considering the type of industries that flourish in the state along with the upgradation of infrastructure. The development of social infrastructure, such as skill development institutes and industrial training institutions, must be oriented towards the types of industries that are thriving in the region, along with participation from the private sector, especially industries. It will enable a skilled workforce to meet the specific needs and demands of the thriving industries in the region, leading to economic growth and development in Himachal Pradesh, ultimately contributing to the overall progress and prosperity of the state.

References

1. Kumar S, Pattanaik F (2017) Growth and structure of industrial sector in Himachal Pradesh. https://www.researchgate.net/publication/323072226
2. Planning Commission of India (2021) Himachal Pradesh development report (2021). http://planningcommission.gov.in/plans/stateplan/sdr_hp/sdr_hpch21.pdf
3. Department of Industries (2004) Industrial policy
4. Mohd S, Singh R, Jai, Sharma K (2017) Impact of special industrial package on the growth of Himachal Pradesh. www.tjprc.org
5. Chaudhary R, Janjhua Y, Thakur S (2016) Problems faced by entrepreneurs of micro, Smalland medium enterprises: a case study of district Solan in Himachal Pradesh. JIMS8M: J Indian Manag & Strat 21(2):34. https://doi.org/10.5958/0973-9343.2016.00014.4
6. Department of Economics and Statistics (2020) Statistical abstract of Himachal Pradesh 2019–20. https://himachalservices.nic.in/economics/pdf/StatisticalAbstract_2019_20.pdf
7. Pettinger T (2018) Problem of declining industries. Economicshelp.Org. https://www.economicshelp.org/blog/19037/business/problems-of-declining-industries/
8. Bodh A (2013) Industrial development keeps away from Himachal Pradesh. Times of India. https://timesofindia.indiatimes.com/india/industrial-development-keeps-away-from-himachal-pradesh/articleshow/20532164.cms
9. Sharma PK, Gandhok T, Chaudhary S, Sharma D (2021) Difficulties facing Himachal Pradesh in today's sector. Turk Online J Qual Inq (TOJQI) 12(7):13733–13740
10. Kumar R, Shankhan A (2017) Spatio-temporal analysis of industrial hubs development in Himachal Pradesh: large and medium enterprises. 59 Trans Inst Indian Geographers 39(1)
11. Jamwal A, Towhidul Islam Nayim SM, Shukla RK, Agrawal R, Gupta S (2021) Assessment of barriers in lead time improvement: an exploratory study of electronics manufacturing companies in Himachal Pradesh (India). Int J Bus Syst Res 15(2):182–199. https://doi.org/10.1504/IJBSR.2021.113422
12. Arvis J, Shepherd B (2016) Measuring connectivity in a globally networked industry: the case of air transport. World Econ 39(3):369–385. https://doi.org/10.1111/twec.12363
13. Fageda X, Gonzalez-Aregall M (2017) Do all transport modes impact on industrial employment? empirical evidence from the Spanish regions. Transp Policy 55:70–78. https://doi.org/10.1016/j.tranpol.2016.12.008
14. Loreta A (2018) Industrial restructuring in small and medium sized towns from Romania–evolution background and positive perspectives with challenges. Analele Universității Din Oradea, Seria Geografie 28(2/2018):197–209. http://istgeorelint.uoradea.ro/Reviste/Anale/anale.htm
15. Mandeli K (2019) Public space and the challenge of urban transformation in cities of emerging economies: Jeddah case study. Cities 95:102409. https://doi.org/10.1016/j.cities.2019.102409
16. Maparu TS, Mazumder TN (2017) Transport infrastructure, economic development and urbanization in India (1990–2011): is there any causal relationship? Transp Res Part A: Policy Pract 100:319–336. https://doi.org/10.1016/j.tra.2017.04.033
17. Netirith N, Ji M (2022) Analysis of the efficiency of transport infrastructure connectivity and trade. Sustain 14(15):9613. https://doi.org/10.3390/su14159613
18. Dyomin M, Dmytrenko A, Chernyshev D, Ivashko O (2019) Big cities industrial territories revitalization problems and ways of their solution. In: Proceedings of the 2nd international conference on building innovations. pp 365–373. https://link.springer.com/chapter/10.1007/978-3-030-42939-3_37
19. Koneru K (2017) Women entrepreneurship in India-problems and prospects. SSRN Electron J. https://doi.org/10.2139/ssrn.3110340

20. Sheikh YA (2017) Higher education in India: challenges and opportunities. J Educ Pract 8(1):39–42. https://files.eric.ed.gov/fulltext/EJ1131773.pdf
21. Merrell I, Phillipson J, Gorton M, Cowie P (2022) Enterprise hubs as a mechanism for local economic development in rural areas. J Rural Stud 93:81–91. https://doi.org/10.1016/j.jrurstud.2022.05.016
22. Ortiz-Moya F (2020) Green growth strategies in a shrinking city: Tackling urban revitalization through environmental justice in Kitakyushu City, Japan. J Urban Aff 42(3):312–332. https://doi.org/10.1080/07352166.2018.1448225
23. Fox RG (2015) Urban culture: the industrial city. Encyclopadia Britannica. http://www.britannica.com/topic/urban-culture
24. Wang M, Cheng Z, Zhang P, Tong L, Ma Y (2014) Old industrial cities seeking new road of industrialization. World Scientific. https://doi.org/10.1142/8404
25. Xu X, Lu Y, Vogel-Heuser B, Wang L (2021) Industry 4.0 and industry 5.0—inception, conception and perception. J Manuf Syst 61:530–535. https://doi.org/10.1016/j.jmsy.2021.10.006
26. Kavipriya (2023) Maruti Suzuki to establish 4th JIM in Haryana. Telematics Wire. https://www.telematicswire.net/maruti-suzuki-to-establish-4th-jim-in-haryana/
27. Fernández Agueda B (2014) Urban restructuring in former industrial cities: urban planning strategies. Territ Mouv 23–24:3–14. https://doi.org/10.4000/tem.2527
28. Himachal Pradesh State Audit Department (2004) Annual administrative report. http://lad.hp.gov.in/sites/default/files/documents/AdminReportEng2013_A1b.pdf

Indigenous Design Solutions and Technology; Optimizing Existing Resources

AI and It's Future in Architecture

Ashiti Goyal and Jasmine Kaur

Abstract This research paper explores the potential of artificial intelligence (AI) for architects. Understanding AI for architects is important to analyse data and upgrade according to new techniques and trends. It explores for more options and opportunities available in AI in free or paid versions which will help the architects to do a lot more and better. AI gives wings to the architects to express their potential of work thoughts and new trends which is not presentable through orthodox techniques. AI grabs updated design techniques and implement it in design behaviour through tools. As architects always lack for time. To present their ideology their being always a barrier to represent them in front of others. Architects always find it difficult to express their technicality, complexity, and vision of the design in terms of art and emotion to the clients. Various tools of AI can help by saving time by completing ordinary tasks, which will be studied in this paper. The study comprises of how AI tools offers a wide range of capabilities, for fast work, customizing design, which enhance design processes, optimize construction practices, and create more intelligent and responsive built environments. This paper studies about the architect's preferences and views about AI and its future in architecture by conducting surveys and information which will aware the architects to express their potential through the new upcoming trends in profession.

Keywords Artificial Intelligence (AI) · Architects · Tools · Techniques · Design

1 Introduction

For so long, architects have structured the world by converting people's needs and dreams into functional and inspiring realities [4]. They work on complexity with an imaginative perspective that makes difficult for architects to express themselves in the form of creating and expressing ideas [8] (see Fig. 1). This always make them

A. Goyal · J. Kaur (✉)
DIT University, Dehradun, India
e-mail: jasmine071202@gmail.com

A. K. Sharma and A. Joshi (eds.), *Proceedings of the 3rd International Conference on Opportunities and Challenges for a Resilient Future*, Advances in 21st Century Human Settlements, https://doi.org/10.1007/978-981-96-0931-4_21

feel that some kind of barricade for representing themselves in a fair manner to the world. This is where Artificial Intelligence (AI) steps in making a bridge for architects in various representative aspects [3]. By easy access to express the forms of architecture helps to revolutionize the future of architecture. AI is here not only idea representation but also gives power to experiment with their thoughts and ideas which ensures retaining previous designs as well (see Fig. 2) [10]. It is very important for all the architects to analyse and upgrade themselves according to new techniques and trends which can handle their tasks easily (see Fig. 3). AI gives wings to the architects to express their potential of work thoughts and new trends which is not presentable through traditional techniques (see Fig. 4).

The main aim of the paper is to examine the uprise of Artificial Intelligence in future of architecture. It's effect on the creativity in the designing phase with respect to other fields in architecture profession [7]. This paper explores different AI tools which are going to ease the work of architects and satisfy the client by helping to create better design Outcome, faster design process.

The value of this research lies in the fact that it helps to understand the role of AI in architecture. To identify the various to AI tools available for architects and helps to examine the usage of AI tools and various building experts.

Fig. 1 Author (AI generated). An architect drafting blueprints by AI

Fig. 2 Author (AI generated). Architect practicing designs using AI laser

Fig. 3 Author (AI generated). Construction work monitoring by AI

2 Literature Review

Knowledge is very important before exploitation of architectural design. AI with it's sense of conceptual elements and knowledge unlocks the ideology of designing based on generative solutions and creativity [9]. Artificial intelligence is a field which is

Fig. 4 Author (AI generated). Experimenting different ideas with AI

newly introduced concept and is rapidly arising [4]. By infusion of AI in architectural work, they gain a range of valuable benefits. It helps in generating design options, improvements and solutions [1]. It also works great in solving conceptual problems related to designing [7]. Architects are concerned about the AI and its impact. Interviews was conducted with architects, and other people related to profession like draftsmen, owners of construction company, and architectural professors. In the interview major topics discussed are intelligent technology in buildings, Model based projects, Calculative Design, Parametricism and AI [2]. Beside this the interviews conducted concluded that AI cannot completely take over architects in any way. In future, it may enhance the working style, strategies and ideology but definitely will not going to affect majorly to the professionals [9]. Observation to a five-year period resulted the increase of 85% to the solutions related to conceptual designing using AI. The key objective to study this is to investigate ideology with relation to development in nature and form creation by prompting generation of discrete details of evolutionary shapes [7]. The AI generation of required information generally involves 3 steps which is gathering and acquiring required data, model training and result expected predictions [10].

Architecture profession have evolved a lot from early time [8]. The function was an important concept of architectural design. In 1896, Louis Sullivan stated in an article "form follows function". Functionality of a building is not based on any references, it is the combination of practical items. The functionality of the building is it's real beauty. Form really follows function on eye of deep learning, this is discovered with the use of the machine learning system [4]. Also later on in 1957, the PRONTO software was launched which welcomed the CAD and changed the face of construction industry. Architects are excited about the technology after this interview. Even at that time they concluded AI softwares does better collection of data than humans [2].

AI is spreading its establishment in many fields like clustering, better suggestion, designing solution, modelling, anomaly detection and association. By analyzing the rapid growth, it is expected that AI will play a crucial role in future projects related to architecture [3]. The findings also suggest that gen-AI finding curiously about prompts and their application in research papers and books [1]. Various books also are there now which mention the use of AI architects in robotic form for architectural designs and state about the detection by AI to use this technology. It recommends this technology for having specialized field control, flexibility to work and technology acceptance. It works on AI architectural development and it's technological ways of implementations [9]. AI prompt can resolve issues and handle problems easily as their intelligence respond to provide solution with a balance of multiple designing criteria. This makes AI an appropriate approach to solve problems with design optimization [7]. To create architectural renders and sketches, Midjourney (an AI powered program) is suggested by many the researchers. It has a potential of development of projects as it fastens the work in the beginning stages of work by an architect. Without the description of any object, AI can interpret and read it correctly. 3D buildings can be represented more beautifully using some software code based on AI [5].

AI not only in designing but also can interpret numerous data and can evaluate statistically AI with combination to BIM predicts the critical behavior of the building [6]. In construction industry by identifying leading AI tools, the study emphasizes their ability to enhance creativity with infrastructure, streamline design iterations, and facilitate team collaboration [8]. Building Information Modelling (BIM) is used exceedingly which gives a collaborative approach to architects, engineers and contractors. It presents the architectural, engineering and construction in 3D form to efficiently plan, design infrastructure. Now BIM is evolving to new form with collaboration with AI to resolve upcoming challenges by detecting building discomforts [3]. By this factor causing major discomfort are suggested by AI with their possible solutions [6].

Researchers are continuously studying about new evolving methods and AI application in architecture [3]. They are searching for discovery of the AI tools which not only automate the process but also resolve design defects [1]. For example, architects are using by advanced commercial softwares like CATIA and Grasshopper to resolve the challenges faced at initial stages of designing. One of them is Stanislas Chaillou who used AI for floor plan generative analysis and Nagi's 2017 project Discover. AI tools developed by 3XN Cophenhagen are leading to swiftly drastic shift [3]. AI is not only trying to achieve interior human comfort by suggesting the changes which supports less carbon emission and are ecofriendly [6], but also levigating itself for ease in research design and knowledge management [3].

When diffused with software it results more dynamically [8]. AI also have some generative models which increase the accuracy possibilities of results. Generative Adversarial Networks (GANs) and Latent Diffusion Models (LDMs) are generative models which has level up the learning and decoding. This leads to generate diverse solutions with creativity [10]. Other software like Vray, Lumion, Twin motion and Cinema 4D also taking support of AI with prompt extension in it with better representation of computational design [3]. Debate is ongoing about the automation of AI in architecture. Ai is potential in accelerating creative process but final touch to the design still needs human expertise [5].

3 Research Methodology

AI is a vast aspect which are catered differently in different levels. To fulfil in purpose, this study is divided into 3 levels (see Fig. 5).

Starting with a case study analysis enlightening the aspects of architecture's future with respect to artificial superior present and upcoming technology. These case studies helped to understand how these tools can be implemented. A visit to various firms was done with an unstructured interview depending on various factors like time of practicing in profession, architects according to their age group, using present technology, type of projects etc. so as to observe in detail how people use AI practically. Finally, case studies and interaction is verified in the form of structured interview by google form which was circulated among architects of different age groups and time of their practicing according to their project. This helped to understand and fulfill the aim of the research that land to a conclusion that is vital for this study.

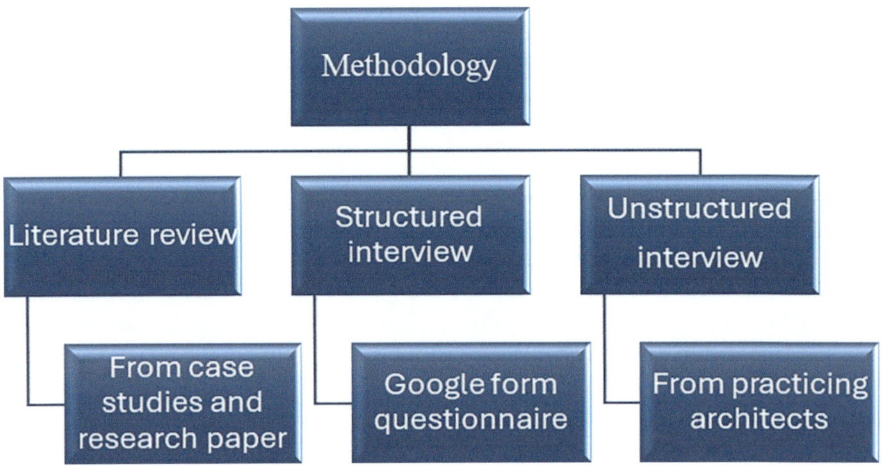

Fig. 5 Chart showing the approach of research methodology stages

4 Results and Discussions

A survey was conducted among architects and designers, to find out different AI tools and their purpose. This survey is conducted in the form of online form and live interviews of architects. In which a set of questions were prepared (see Appendix-1) and reviews of the respondents were recorded.

The online survey was conducted in which 30 people of architecture profession participated. More than 50% of respondents are between age of 25–34 and least count of respondents are between 45–55 age group. More than 50% of responders are architects and mostly they have been in this profession form last 1–5 years whereas 10–20 years is 20%. The survey recorded that very few are not familiar with the concept of AI, and more than 75% of them uses it occasionally for their work (see Fig. 6). As AI is new to this field it is surprising to know that more than 50% respondents are somewhat familiar with the concept of AI (see Fig. 7). In survey it is asked about their profession to get analysis for the respondent's response for AI (see Fig. 8). It is observed that most of them are from architecture backgrounds practicing in differently related fields.

AI is mostly being used for image creation and then for inspirations, designing of plans, 3d renderings, and also being used for 3d models,2d renderings, etc. (see Fig. 9). From survey it is recorded that 60% respondents agreed that AI will take place to future works (see Fig. 7). Responders suggested that AI open the creativity doors, and it can change the future and the way we work, whereas we can't completely depend of AI as it just provide us maximum accurate results possible according to technology. 43.8% of the responders use virtual reality, 25% use augmented reality and cloud computing. Other technologies used by the people surveyed: 3D printing—18.8%, Generative Design, algorithm, etc.—6.3%, Chat GPT—6.3%, Metaverse—6.3%. They also listed some AI they use in designing like prome AI, playground, gamma, fabrie, rerender.AI, mnml.AI, lookX.AI and Leonardo.AI etc. (Fig. 10).

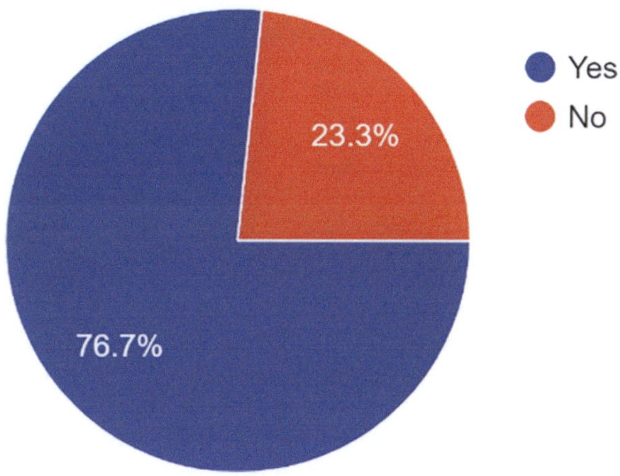

Fig. 6 Showing how many people use AI for their work

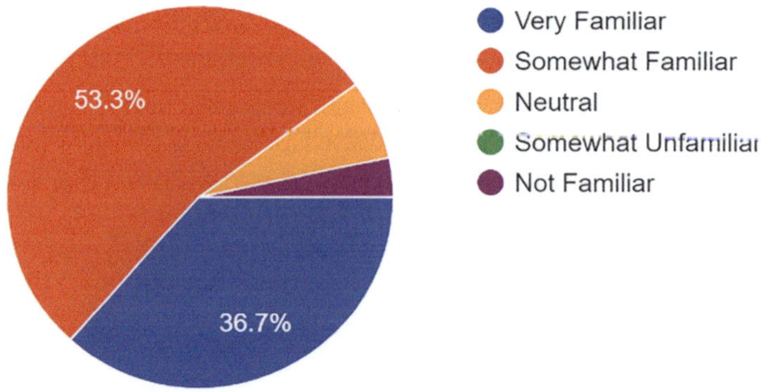

Fig. 7 Representing familiarity of users with AI

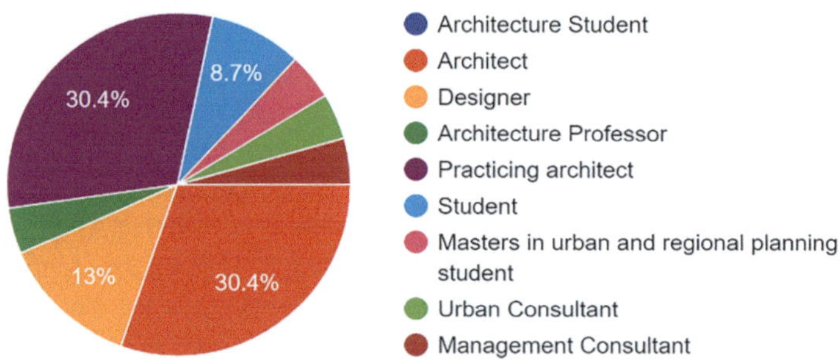

Fig. 8 The figure shown describes the profession of respondents

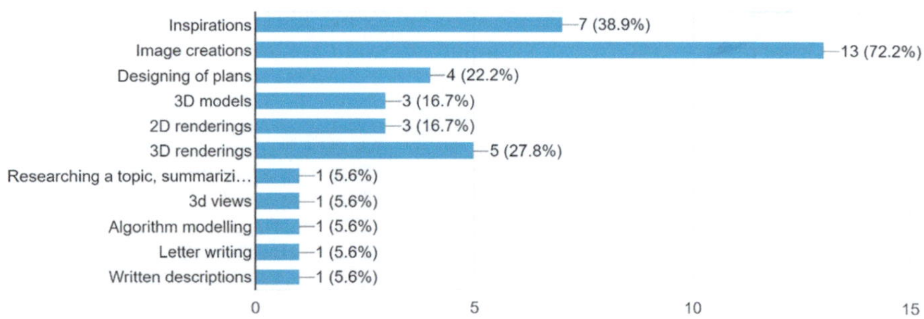

Fig. 9 Purpose for which respondents use AI

As the result of survey concluded that only 22.2% respondents agree for the complete dependency for tasks (see Fig. 11). This means there are still many factors in functioning of AI which need to be upgraded according to human needs. Many of the responders using AI suggested their most used tools in terms of free and paid some are: Dall-e, Canva, Quill bot, Claude AI, Ppt maker and Mid journey.

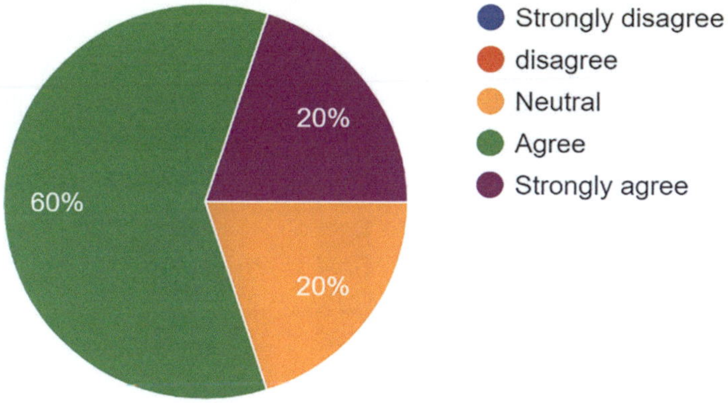

Fig. 10 Their reviews for working efficiency of AI

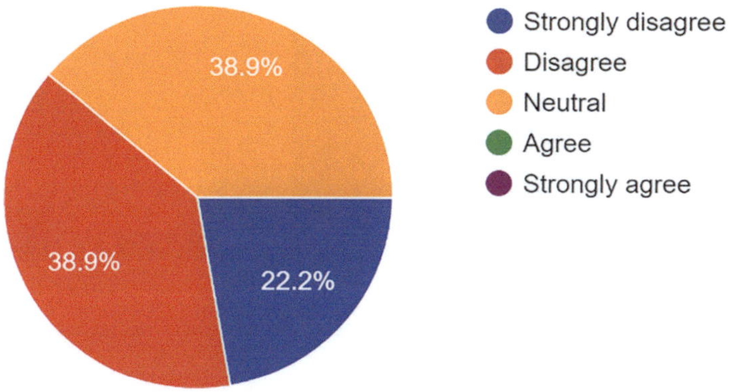

Fig. 11 Their reviews for complete dependency on AI

The survey form also made the respondents to rate the ability of AI to change the future of Architecture (see Fig. 12). It is positive to know most of them strongly agree to accept changes by AI and are ready to welcome AI as major part for professionals. They also shared their views on the AI senses to understand, cater and respond to human emotions and senses in designing sense (see Fig. 13). Most of them are neutral as the upcoming trends can only be predicted, but other than neutral, many of them are in favour that AI cannot match the human senses and will not be able to gain that ability to understand human emotions completely in designing.

Second form of survey was done in which an unstructured live interview was conducted among 10 architects and people from architecture background. They shared their views and senses for future of AI and they have a positive envision for AI impacting architectural design process. Maximum of them feels that AI will be going to be a tool which will play a crucial role in designing. It will enhance the self-creating thought process and conceptualise the design faster but before that it is

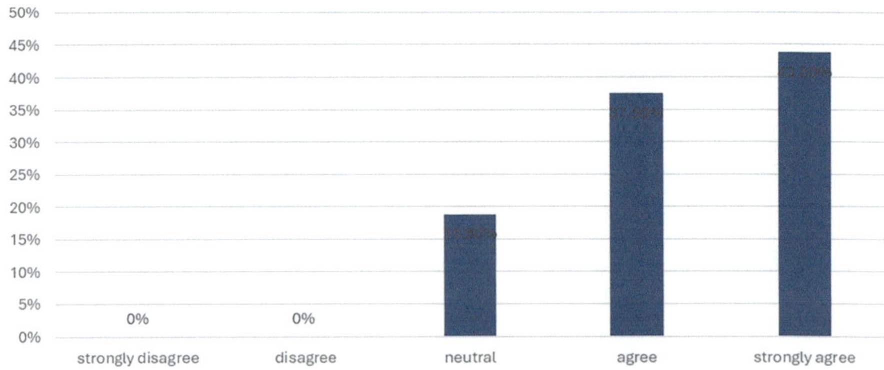

Fig. 12 Respondents rated for their believe that AI can change future of Architecture

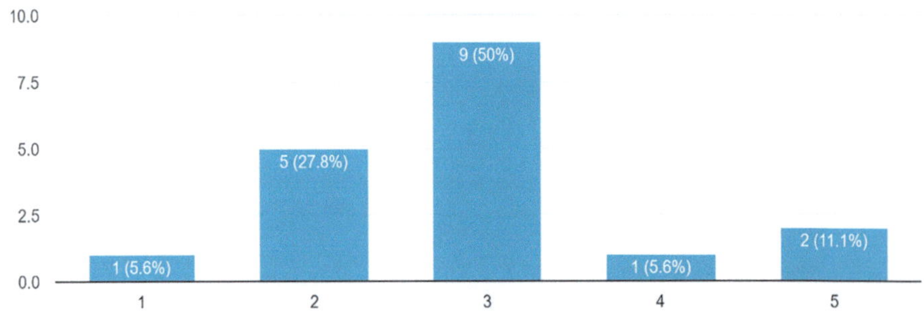

Fig. 13 Respondents rated for AI ability to understand and respond to human sense of designing

important for AI developers to know the needs of architecture profession. Biggest challenge the architects going to face in implementation of AI technologies in their practices is completely dependent on the adaptability of the architects to perceive technological development. When AI will completely steps in architecture world, it will be going to be quite challenging for them to interpret information and following them for designing.

When they are asked about their expectation about dynamical change which may happen to the profession to change working trend, they stated that the changes are unpredictable but somehow it depends on that with which level of complexity people will absorb the change. It will definitely result in dynamic change if AI will be a success in future and will definitely leads to major changes. Architects also pointed out their expectation that will AI be able to match or cross human intelligence to the complexity, technicality, human sensitivity, authenticity, emotions etc. overall they concluded that AI can match human intelligence completely in the work which is contributed by intelligence of architects. But they are very sure that AI will never be able to gain the complete human senses like emotions. Sensitivity, harmony and authenticity etc. these are the senses which architects put by their heart in their designs.

They concluded this interview by analysing the trends and understanding the upcoming shift of the strategies that AI is a power tool for architects. They will definitely take the tools in their power to change the world. AI will add to their abilities and will levitate the profession. It will be going to lead better capacities for data accumulation and ability to lead burden release activity by reducing repetitive work. Although they have strong believe that it will never match to the working efficiency and human essence in the work. It will never take over architects. In contrast, AI is a great assistant, creating opportunities for the extension of the architect's functionality. Architects are ready to accept AI the future of architecture.

5 Conclusion

This study shows the future interconnection between AI and architecture. It captures well the user requirement and create designs even with the detection of solutions to the designing problems in architect's design. However, AI still struggle to match the human touch in designing sense when is to represent sensitivity, authenticity, harmony and emotions. AI is still working in progress for the personal taste which is expected to be achievable in future. In terms of future perspective, AI is working on the design perceiving senses of humans. It not only handles large amount of data but also generates summarized information which simplifies the architects in decision-making. From handling of administrative work to managing the budget, AI is self-sufficient to perform best results. It can become a powerful tool for architects which will levitate the profession to another level.

AI will free architects to focus on complex design decisions making and client interaction. This human-AI partnership will definitely be going to bring a paradigm shift leading to a future of innovative, sustainable, and user-friendly architecture.

This study is a glimpse of the transformative potential of Artificial Intelligence (AI) in architecture, making path to future, which is full of new possibilities and AI empowerment to architects. By automating complex tasks, optimizing performance, creating solution to repetitive tasks, and generating creative design solutions, AI demonstrates the potential to become an indispensable partner in future for the architectural process.

> In a design, technicality and complexity are like a sky, art and emotions are like earth, which are connected by AI like the rain.

Acknowledgements We would like to thank Dr. Aanchal Sharma for giving us such an opportunity to conduct this research. We would also like to thank our respondents for helping us by giving the loyal reviews to complete this research paper.

Appendix 1

(1) How long have you been in your profession?
(2) Are you familiar with the concept of Artificial Intelligence (AI) in architecture?
(3) Do you use Artificial Intelligence (AI) for your work?
(4) How often you use artificial intelligence for your architectural work?
(5) What tasks do you use AI for?
(6) Do you use paid AI?
(7) Which paid AI do you use?
(8) Which free AI do you use?
(9) does the creativity doors can be opened by AI in architecture?
(10) Can we completely depend on AI?
(11) Do you use any other technology other than AI as well?
(12) Do you think AI can change future of architecture?
(13) Rate the ability of AI to perform architectural tasks?
(14) Rate the ability of current AI tools to be able to understand and respond to human emotions and designing sense?
(15) Can AI help the designers to choose or make new materials?
(16) How can AI tools analyse data to improve the sustainability and efficiency of building projects?
(17) Do you see AI taking place in the future of architecture work?
(18) Is there anything specific task you would like AI to help you with in the future of architecture?
(19) How do you envision AI impacting the architectural design process?
(20) What do you consider the biggest challenges architects will face when implementing AI technologies in their practices?
(21) Is AI going to replace architects or will be a power tool for changing the world?
(22) According to your expectations, how dynamic AI is going to be in future for architecture professionals?
(23) AI successfully fulfills technicality and complexity, but can AI take over human intelligence senses completely?
(24) Is there anything you would like to share about your thoughts on AI and it's role in architecture?

References

1. Abdulrahman Aymana YM (2024) Generative vs. non-generative AI: analyzing the effects of AI on the architectural design process, 28
2. Agha DI (2019) The impact of artificial intelligence on the future of architecture & architects. Revol Artif Intell, 18
3. Amer NA (2023) Architectural design in the light of AI concepts and applications, 18
4. As I, Pal S (2018) Artificial intelligence in architecture: generating conceptual design via deep learning. Int J Archit Comput 16(4)(1):22

5. Jaruga-Rozdolska A (2022) Artificial intelligence as part of future practices in the architect's work: MidJourney generative tool as part of a process of creating an architectural form. Architectus 3(71)(1):10
6. Krausková V (2021) Use of artificial intelligence in the field of sustainable architecture: current knowledge, 4(1):10
7. Luz Castro Pena M, Adrian C, Rodríguez-Fernández N, Santos I, Romero J (2021) Artificial intelligence applied to conceptual design. A review of its use in architecture. Autom Constr 124(1):30
8. Sharma NA (2024) Exploring the impact of AI on architectural creativity and efficiency, 18
9. Xu C-YL (2021) Extended TAM model to explore the factors that affect intention to use AI robotic architects for architectural design, 1(1):14
10. Zhang Z, Fort JM, Mateu LG (2023) Exploring the potential of artificial intelligence as a tool for architectural design: a perception study using Gaudí'sWorks, 18

Analyzing Impact of Changing Environmental Conditions on the Health Status for People of Uttarakhand

Garima Varshney, Bhavna Kumar, and Seema Praliya

Abstract Ignoring our surrounding environment; in today's scenario, we seem to be having an ordinary life, facilitated with digital technology. In contrast, the deteriorating environmental conditions are as observant scenario worldwide. Irregular weather patterns, rising temperatures, retreating global ice sheets, rising sea level, droughts, contaminated and polluted resources etc. are the result of the changing environmental conditions. This is further depleting the condition of our basic resources for survival which increases the likelihood of airborne, vector borne and waterborne diseases, and recent examples of same is COVID-19 pandemic. Environmental change also accelerates the problem of antimicrobial resistance, which is another threat to human health due to increasing cases of resistant pathogenic infections. "Uttarakhand; the land of gods" is not left untouched by the ill effects of environmental changes. In lieu of the recent developments, where trees are expurgated down, plenty of constructions works going around, and extreme rise of tourism; leading to depletion of environment of Uttarakhand, which is having ill effects on the health of people of Uttarakhand. This paper aims to analyze how the environmental changes are impacting the health conditions of the people of Uttarakhand, mainly focusing on the deteriorating conditions due to contaminated/ polluted air and water.

Keywords Airborne · Climate change · Environment · Health · Waterborne

G. Varshney · B. Kumar (✉)
Faculty of Pharmacy, DIT University, Dehradun, Uttarakhand, India
e-mail: bhavna@dituniversity.edu.in

G. Varshney
e-mail: 1000021138@dit.edu.in

S. Praliya
Faculty of Architecture, SoAPD, DIT University, Dehradun, Uttarakhand, India
e-mail: seema.praliya@dituniversity.edu.in

A. K. Sharma and A. Joshi (eds.), *Proceedings of the 3rd International Conference on Opportunities and Challenges for a Resilient Future*, Advances in 21st Century Human Settlements, https://doi.org/10.1007/978-981-96-0931-4_22

1 Introduction

Human beings exist in the sphere of environment; they are continuously surrounded by and interact with it. The biosphere, the thin membrane that encircles the planet, its mud, and the whole lot that is living, is the most cherished component of nature in relation to human beings. Our world, although it is outside of us, contains not just its image, which is both actual and imaginatively mirrored, but also its material energy, information routes and activities. Nature's existence in man's Self in the form of idealized energy and knowledge is so biological that when these extrinsic principles of nature go, man vanishes from life. We lose our lives when we lose nature's picture.

We are a part of the ecological environment, which is part of the cosmos. Human acts that breach natural principles and biosphere equilibrium threaten to bring calamity, which may be worldwide. Anthropogenic Climate change is a huge worldwide environmental and development challenge that affects all countries, regardless of size or development level [1, 2].

Climate change is characterized by the following definition: "a climatic change caused directly or indirectly by human activity that modifies the chemistry of the global environment and is in contrast to the variability in the climate observed over equivalent intervals of time". It impacts social & ecological health factors such as clean air, adequate food, safe drinking water, & safe shelter. Climatic variations may harm health of human in variety of customs, the most prevalent of which are heat waves of high frequency & severity, which leads to increase in temperature-related diseases & mortality, amplified precipitation, droughts, floods, and desertification, all of which cost people directly. High temperatures raise 'ground level ozone' and other 'climate changing pollutants', worsening cardio-respiratory, allergy, and cancer risk. Figure 1 shows the various climate changes and their impact on health systems. In addition, there has been an amplification in the transmission & spread of communicable illnesses, variations in the division of water-borne, vector-borne diseases, food-borne which impacts on catastrophe risk & hunger [3].

The Himalayas are one of the areas with complicated orographic characteristics. Some research on Uttarakhand covers trend analysis of basic climatic indicators. In Uttarakhand, rising temperatures are exacerbating forest fire outbreaks. Climatic changes due to humans directly or indirectly impacting weather & climatic boundaries at higher elevation areas & numerous current research have reported on growing prevalence of EDW (Elevation-dependent Warming). EDW has an impact on Himalayan glaciers & rivers, which are considered as the only supply of water for the mountainous region's population. Intense rainfall events, forest fires, landslides, flash floods, & water scarcity, according to community perspective, are the most serious environmental threats to the Uttarakhand mountainous region's life. Young Himalayan belt is geodynamically mobile, & Himalayan rocks are severely splintered, faulted, & joined, resulting in uncontrolled mass wasting and environmental degradation [4]. Climatic variations are affecting wellbeing in a variety of behavior, considering increased death & ill health due to severe weather actions like

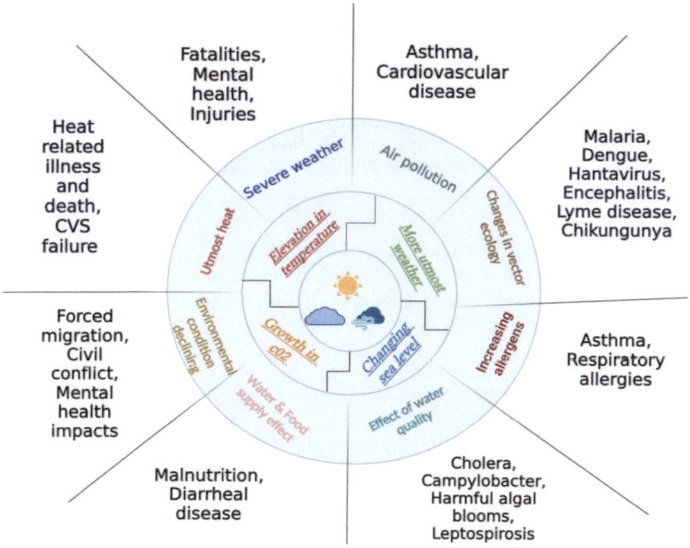

Fig. 1 Various climate changes and their impact on health system [6]

heat waves, floods, storms, disturbance of food systems, rise in zoonoses & water, food, & vector-borne diseases [5].

2 Uttarakhand Climate Profile

It is considered as one of mountainous state in the Himalaya of India. The state contains two separate climate regions: the highlands and the comparatively modest plain region. Climatic variation in plain regions is alike that in hilly regions. Hilly terrain experiences frigid winters with prolonged snowfall, abundant monsoon rains, and pleasant summers. This climate draws travelers seeking scenic beauty, adventure, or a spiritual setting. Graph 1 and 2 shows temperature and rainfall variations in the state (Figs. 2, 3).

2.1 Rainfall

The state has a comparatively high average annual rainfall of 1229 mm. According to the India Meteorological Department, rainfall often begin in the state in late April and last until September. However, rainfall intensity rises from June to September. The first week of July sees more rainfall. Rain persists through August and into the first week of September.

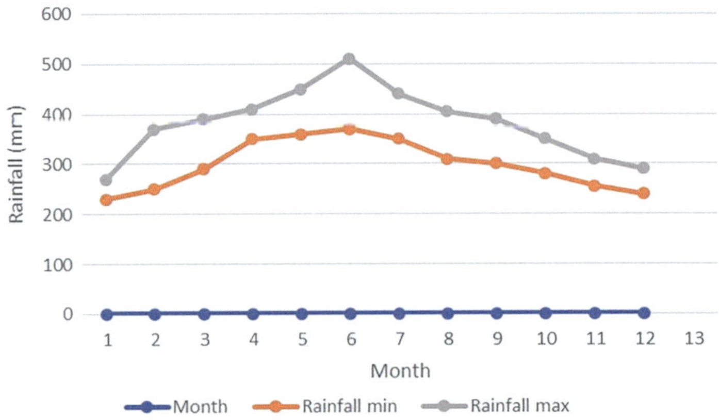

Fig. 2 Rainfall variations across year in Uttarakhand [7]

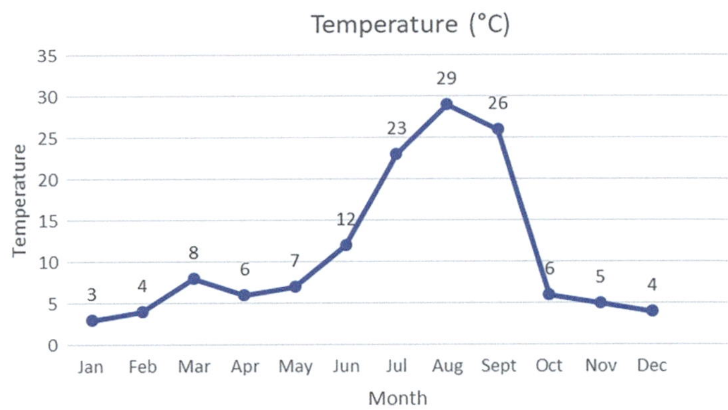

Fig. 3 Temperature variations across year in Uttarakhand [7]

2.2 *Temperature*

Summers are quite hot, with temperatures exceeding 40 degrees Celsius and high humidity. Winters may be extremely cold, ranging beneath 5 °C. The minimum recorded coldness is -5 to -7 °C, while the maximum is between 40 °C and 45 °C [8].

3 Human Health

Climatic variations have a significant impact on wellbeing physical condition. A changing climate is anticipated to have a disproportionately detrimental impact on human health. Direct implications of climatic variations on human health includes thermal extremes cause morbidity and mortality, including cardiovascular and respiratory disorders, and flooding and other natural disasters can cause mental stress, injuries, and scratch to health infrastructure due to increased frequency and severity of weather occurrences.

Tremendous weather measures caused by climatic variations, such as tsunamis, cyclones, and floods, have claimed lives of thousands and millions of affected people in numerous states of India, including Uttarakhand, Assam, West Bengal, Uttar Pradesh, Himachal Pradesh, and Gujarat. This results in a negative influence on the health of the survivors.

These calamities increase the occurrence and spread of illnesses including Japanese encephalitis, plague, malaria, chikungunya, dengue, and filariasis. These calamities also cause the emergence of water-borne illnesses such as cryptosporidiosis, amoebiasis, giardiasis, cholera, typhoid, and other infections. Drinking contaminated water and inhaling filthy air kills 900,000 Indians every year, according to WHO 2000. Approximately 0.6–0.7 millions (about 1 week 1 day) of children under 5 years of age die because of diarrheal infection every year [9].

Furthermore, indirect implications of climatic variations on human health includes changes in the range and functioning of responsible vectors & parasites, resulting in variations in geographical regions & occurrences of vector-borne illnesses such as malaria, encephalitis, and dengue.

Higher temperatures have changed the local natural science of water & food-borne infective precursors, resulting in increased occurrences of diarrhea and other infectious illnesses, including cholera. Higher temperatures promote the development of food-borne diseases, such as Salmonella [10].

4 Climate Change and Vector Borne Diseases

Vector is a biologically derived agent that aids in spreading a communicable agent from affected animal to human or another species. Arthropods including ticks, mosquitoes, fleas, flies, and lice are the widespread vectors. Vector-borne illnesses refer to diseases spread by vectors. Vectors can spread infectious illnesses actively or passively. Biological vectors, such as mosquitoes and ticks, can carry infections that can proliferate inside their body & spread to newer host, commonly via biting and mechanical vectors like flies can acquire infectious materials from exterior of their body and transfer the infectious material via physical touch [11].

Unpredictable weather conditions have resulted in untimely rainfall, excessive humidity, and variable temperatures, providing an ideal environment for vector-borne illnesses in Uttarakhand. The state is seeing an increase in diseases like dengue and malaria, particularly in the monsoon zone, which provides an ideal habitat for pathogens to grow [12]. Higher average temperatures due to greenhouse gas emission leads to severe droughts & storms. Odorless & colorless gas, carbon monoxide, and ozone are mostly created by automobile exhaust gases, including manufacture of tools, industrial work, and motors. On other hand heat waves, causing amplification in water vapors & ozone generation on earth's surface, which is connected to a number of health risks. Along with raising social awareness of mental health issues, global warming's environmental impact exposes individuals to a rise in communicable illnesses, respiratory and allergy problems [13]. To establish a relationship between vector borne diseases and climatic variations, proof of association, biological sensitivity of the vector, and changes in disease prevalence with weather-related factors such as temperature and/or precipitation should be studied [14].

Based on a report by Times of India Aug 2023, Dr. Pankaj Singh, State nodal officer of Integrated Disease Surveillance Program, Health department said Out of 423 cases of dengue, 345 patients in UK had recovered. Out of which 5% cases reported, dengue was worsening, mostly if the patient had a combined illness and only 1% of cases are leading to death. In this year approximately 12 cases of malaria have also been identified. This is all because of sudden drastic changes in the climate and the environmental conditions [12].

4.1 Major Vector-Borne Diseases

1. **Malaria:** It is caused by five Plasmodium genera: Plasmodium falciparum, P. ovale, P. vivax, P. malariae, & P. knowlesi, which are spread via female Anopheles mosquitos. Malaria transmission is widely known to be impacted thorugh environmental factors like terrain, climate, rainfall, & population socioeconomic status. As a result, tropical areas having hot temperatures, high humidity, heavy rainfall, & low elevations are ideal for longevity, mosquito breeding, & parasite sporogony [15].
2. **Dengue:** Dengue is a virus carried by mosquitoes & primary causes of arthropod-borne viral illness worldwide. It is also known as break bone fever due to the intensity of muscular spasms and joint pain, dandy fever, or seven-day fever because of the typical length of symptoms. The majority of cases are asymptomatic, although serious disease and death are possible. The virus is transmitted by Aedes mosquitoes, which are abundant in tropical and subtropical regions across the world. Dengue fever has expanded substantially in recent decades, and the sickness is now endemic in various regions of the world, including Uttarakhand, a major afflicted state [16–18].

4.2 COVID-19: A Challenge

Recent public health problems in the SARS-CoV-2 area have centered on vector-borne illnesses to improve surveillance and prevention. Although there is no evidence that COVID-19 can be transmitted through mosquitos, arthropod ectoparasites such as ticks and unnourished cat fleas may play a role in virus transmission to their human hosts due to sequence homology between ectoparasite ACE and human ACE2 proteins and their interaction with the viral Spike. Understanding SARS-CoV-2 animal reservoirs is crucial for the vector-borne COVID-19 population that has been influenced by global climate change. Recent studies have explored the impact of environmental parameters such as temperature, humidity, wind speed, and air pollution on the transmission of COVID-19 [13, 19].

4.3 Measures to Prevent Vector-Borne Diseases

Numerous methods & control measures had been used to combat emerging & re-emerging vector-borne communicable illnesses. It has been critical to intensify active surveillance, ensure excellent early diagnosis, and provide appropriate case management. Molecular methods such as genome sequencing & phylogenetic investigations may help discover new diseases with greater precision. Controlling vector-borne illnesses is one of the most difficult tasks facing the global health agenda. Current rapid & uncontrolled modernization had heightened worry about solving number of issues through organized planned motive that could be established & implemented at both the global & narrow levels [15].

The government of Uttarakhand in accordance with the National Vector Borne Disease Control Programme continued with Malaria Eradication Programme for control of malaria within the state. In this state there are 13 districts, out of which four districts Dehradun, Haridwar, Udham Singh Nagar and Nainital are in plain region & rest of nine districts are at high elevation. Accoring to the survey report Malaria is only prevalent in the plain districts of Uttarakhand state. Government's objective is to to reduce Malaria mortality by 70% along with the dengue mortality [20].

5 Climate Change and Water Borne Diseases

Water-borne infectious infections are those that occur mostly in the human digestive tract and are spread by water like fish, fruits, & vegetables. Humans face risks like missing people, injuries, and deaths from natural catastrophes. Climate change is altering the pattern of precipitation, leading to elevated sea levels, and changing seawater salinity levels, in addition to affecting multiple variables like water's surface temperature and reproduction, survival, and sustenance of viruses and bacteria in bodies of water, all of which affect human health. Furthermore, the risk of stools, in addition to cholera and typhoid, one of the most common waterborne diseases (WBDs), may potentially grow with heavy rainfall and high temperatures. [21–23].

5.1 Major Water Borne Diseases

Cholera: Cholera is caused by the bacteria Vibrio cholera and is mostly found in impoverished communities and villages with poor sanitation. High climate changes and tourism in Uttarakhand result in polluted water that spreads cholera and causes diarrhea and severe dehydration. While cholera can be fatal, only one in every ten patients has life-threatening symptoms such as nausea, vomiting, abdominal pain, and so on. It can only be managed by consuming safe and sanitary water.

Typhoid: Typhoid is a very frequent illness in impoverished nations. This extremely infectious illness is spread by contaminated food and drink. It spreads quickly due to insufficient sanitization. If left untreated, typhoid can cause major health issues and even death. Symptoms include high temperature, fatigue, muscle pains, sweating, loss of appetite, and so on.

The most effective strategy to avoid typhoid is to be vaccinated before traveling to typhoid-prone areas. To avoid this sickness, you should eat at home-cooked meals and drink purified water.

Dysentery: Dysentery is a digestive illness induced by contact with faeces and contaminated water or food. Dysentery is characterized by severe diarrhea, mucous, and blood in the stool. The preventative measures for dysentery are similar to those for the other water-borne illnesses described above [24, 25].

6 Climate Change and Air Borne Diseases

Pollution of the air is a major source of worry in big cities, where the level of pollution exceeds frequently, posing significant health related problems to wellbeing. Outdoor and interior air contaminants in the urban area environments may classified as main or secondary. Harmful gases and particulate matter are examples of primary air pollutants released directly, whereas secondary air pollutants include smog, Peroxyacyl nitrates, ozone, and so on [9]. Variations in climate are projected to had a negative impact on the air quality through a number of methods, along with raised allergen generation and allergenicity, there are also increased local levels of ozone & fine particles. Several pollutants may trigger lung illness or aggravate existing difficulties in susceptible individuals, including youngsters & the elderly. [26].

Air pollution has been linked to narrow-, moderate, and long-term effects on the wellness of people. Short-term consequences consist of irritation to eyes, nose, and throat in addition to respiratory illnesses like pneumonia and bronchitis, while long-term effects include chronic respiratory illnesses heart problems, cancer of the lungs, and even harm to other organs of body. [9, 27, 28].

6.1 Climatic Variations: Precursor to Cardiovascular Diseases

Given the pathophysiological consequences of greater heat exposure, a rise in the risk of cardiovascular illnesses with rising temperatures seems physiologically reasonable. Thermal stress triggers physiological reactions to changes in body temperature, such as amplified perspiration, increased breathing rates, vasodilation, and increased or reduced coagulation. These alterations may damage heart autonomic regulation, triggering systemic inflammation, increase local arterial pressures, and impede coagulation reactions. These homeostasis abnormalities may predispose susceptible people to atherosclerotic plaque rupturing and eventually myocardial infarction. One may conclude that practical linking of heat contact with higher respiratory illness rate of mortality is also influenced by heat's pshyiological influence on right ventricular heart failure [29, 30].

7 Temperature Change: Threat to Drug Interactions

Temperature has been proven to contain possible interactions with cardiovascular medications. It can also alter medication ADME (absorption, distribution, metabolism, and excretion) and the therapeutic response. Furthermore, heat might affect the human body's adaptive systems, increasing the chance of pharmacological adverse effects. Considering an example, temperature-induced vasodilation may considerably boost blood pressure thereby lowering benefits of these medications. It may lead to major damage, or even myocardial ischemia.

However, certain medicines can disrupt regular physiological functions in a variety of behaviors, counting alterations in sweat rate, cardiac output, peripheral vasodilation, and dehydration. It is important to highlight that the anticholinergic medications impair physiological functions, which may lower perspiration rate and produce increase in body temperature in case of severe weather conditions. Diuretics, which are widely used to treat individuals with heart failure, may aggravate the negative effects of heat by causing dehydration and an imbalance of electrolytes [31–33].

8 State Action Plan to Control Climatic Change

National Programme on Climate Change and Human Health 2022 had various objectives to control disease rate in the state of God due to fluctuations in the environment having numerous climatic faces.

Objective 1: Raising awareness on climatic variations impacting health of man among the public, vulnerable communities, healthcare practitioners, and policymakers.

Objective 2: Improve healthcare system capability to prevent illnesses and disorders caused by climatic variability.

Objective 3: Improve health preparation and response through situational analysis at every level of states.

Objective 4: Coordinating with Ministry of Health and Family Welfare for promoting collaborations, create synergy with other missions, and prioritize health in the state's climate change strategy.

Objective 5: Enhancing research in state to address evidence-based gap on climatic variations impacting wellbeing health [8].

9 Conclusion

According to the investigation, changing environmental circumstances in Uttarakhand have major and remarkable ramifications for the health of its population. Changes in precipitation patterns, temperature, climate, pollution, deforestation, air and water quality, and ecosystem alterations all contribute to health hazards, including vector-borne infections and respiratory difficulties. Developing solutions for sustainable environmental management and healthcare adaptation is critical for mitigating these effects and protecting public health in Uttarakhand. Addressing these challenges requires a comprehensive and multifaceted approach that includes public health initiatives, environmental conservation, climate adaptation strategies, community engagement, and awareness programs to protect the population's well-being and mitigate negative health impacts in the face of changing environmental conditions.

References

1. Mishra A (2014) Changing climate of Uttarakhand, India. J Geol Geosci. 3(4):163
2. Dhara VR, Schramm PJ, Luber G (2013) Climate change & infectious diseases in India: Implications for health care providers. Indian J Med Res 138(6):847
3. https://ncdc.gov.in/WriteReadData/l892s/52480150911632305398.pdf (accessed on 26–1–2024)
4. Kumar P, Patel A, Rai J, Kumar P. Environmental Challenges and Concurrent Trend of Weather Extremes over Uttarakhand Himalaya

5. https://www.who.int/news-room/fact-sheets/detail/climate-change-and-health#:~:text=Climate%20change%20is%20impacting%20health,diseases%2C%20and%20mental%20health%20issues (accessed on 26–1–2024)
6. https://www.cdc.gov/climateandhealth/effects/default.htm (Accessed on 27–1–2024)
7. Rawat D, Tiwari JK, Tiwari P, Singh H (2017) Floristic diversity of montane zone: western ramganga valley, Uttarakhand, India. J Econ Taxon Bot. 40:104–125
8. https://ncdc.mohfw.gov.in/WriteReadData/l892s/91813363891698219475.pdf (Accessed on 26–1–2024)
9. Kaur R, Pandey P (2021) Air pollution, climate change, and human health in Indian cities: a brief review. Front Sustain Cities. 13(3):705131
10. https://moef.gov.in/wp-content/uploads/2017/08/Uttarakhand-SAPCC.pdf (Accessed on 26–1–2024)
11. https://www.efsa.europa.eu/en/topics/topic/vector-borne-diseases (Accessed on 26–1–2024)
12. https://timesofindia.indiatimes.com/city/dehradun/surge-in-vector-borne-diseases-across-uttarakhand/articleshow/102885645.cms (Accessed on 26–1–2024)
13. Mojahed N, Mohammadkhani MA, Mohamadkhani A (2022) Climate crises and developing vector-borne diseases: a narrative review. Iran J Public Health 51(12):2664
14. Dhimal M, Kramer IM, Phuyal P, Budhathoki SS, Hartke J, Ahrens B, Kuch U, Groneberg DA, Nepal S, Liu QY, Huang CR (2021) Climate change and its association with the expansion of vectors and vector-borne diseases in the Hindu Kush Himalayan region: a systematic synthesis of the literature. Adv Clim Chang Res 12(3):421–429
15. Chala B, Hamde F (2021) Emerging and re-emerging vector-borne infectious diseases and the challenges for control: A review. Front Public Health 5(9):715759
16. Schaefer TJ, Panda PK, Wolford RW. Dengue fever
17. Baak-Baak CM, Cigarroa-Toledo N, Pech-May A, Cruz-Escalona GA, Cetina-Trejo RC, Tzuc-Dzul JC, Talavera-Aguilar LG, Flores-Ruiz S, Machain-Williams C, Torres-Chable OM, Blitvich BJ (2019) Entomological and virological surveillance for dengue virus in churches in Merida, Mexico. Rev Inst Med Trop Sao Paulo 14:61
18. Oliveira LN, Itria A, Lima EC (2019) Cost of illness and program of dengue: A systematic review. PLoS ONE 14(2):e0211401
19. Lam SD, Ashford P, Díaz-Sánchez S, Villar M, Gortázar C, de la Fuente J, Orengo C (2021) Arthropod ectoparasites have potential to bind SARS-CoV-2 via ACE. Viruses 13(4):708
20. https://health.uk.gov.in/pages/display/82-national-vector-borne-disease-control-programme (Accessed on 26–1–2024)
21. Jung YJ, Khant NA, Kim H, Namkoong S (2023) Impact of climate change on waterborne diseases: directions towards sustainability. Water 15(7):1298
22. Nwabor OF, Nnamonu E, Martins P, Christiana A (2016) Water and Waterborne Diseases: A Review. Int J TROPICAL DISEASE & Health. 12:1–14. https://doi.org/10.9734/IJTDH/2016/21895
23. Arya A, Bisht V, Pandey U, Danu S, Kumar S, Chauhan S, Dhami T, Khati T, Haldar P, Nayal P, Jaiswal P (2019) Knowledge regarding water borne diseases and its prevention. Asian J Nurs Educ Res. 9(4):501–503
24. https://www.metropolisindia.com/blog/preventive-healthcare/common-water-borne-diseases (Accessed on 27–1–2024)
25. https://doi.org/10.1016/B978-012373960-5.00565-7
26. Mansour SA (2023) Impact of climate change on air and water borne diseases. Air Water Borne Dis. 3(1)
27. D'Amato G, Cecchi L, D'Amato M, Annesi-Maesano I (2014) Climate change and respiratory diseases. Eur Respir Rev 23(132):161–169
28. Nardell E, Lederer P, Mishra H, Nathavitharana R, Theron G (2020) Cool but dangerous: How climate change is increasing the risk of airborne infections. Indoor Air 30(2):195
29. Peters A, Schneider A (2021) Cardiovascular risks of climate change. Nat Rev Cardiol 18(1):1–2

30. Schneider A, Rückerl R, Breitner S, Wolf K, Peters A (2017) Thermal control, weather, and aging. Current environmental health reports. 4:21–29
31. https://www.escardio.org/Journals/E-Journal-of-Cardiology-Practice/Volume-22/climate-change-and-cardiovascular-disease-the-impact-of-heat-and-heat-health-a#:~:text=Similarly%2C%20another%20study%20showed%20that,similar%20scenario%20that%20did%20not (Accessed on 27–1–2024)
32. Kuch B (2021) Der Einfluss des Klimawandels auf das Auftreten von Herz-Kreislauf-Erkrankungen. Handlungsansätze und die besondere Herausforderung durch Arzneimittelwechselwirkungen. Versorgungs-Report: Klima und Gesundheit (eds Günster C, et al.)(Medizinisch Wissenschaftliche Verlagsgesellschaft, 2021)
33. Martin-Latry K, Goumy MP, Latry P, Gabinski C, Bégaud B, Faure I, Verdoux H (2007) Psychotropic drugs use and risk of heat-related hospitalisation. Eur Psychiatry 22(6):335–338

Implementation and Challenges to the Adaptive Reuse of Heritage Houses in Lucknow

Anam Amjad and Amanjeet Kaur

Abstract Heritage buildings bridge the past, present, and future, holding cultural and historical significance. Beyond emotional attachments, they offer untapped potential for addressing contemporary needs with Adaptive Reuse as a powerful tool. This research endeavours to explore the significance of adaptive reuse within historic residences of Lucknow, with a focus on enhancing their functionality. The primary objective encompasses the compilation of successful cases of residential adaptive reuse from different perspectives, and the exploration of different approaches for the development of evaluation parameters to determine the viability of adaptive reuse within Lucknow's residential contexts. The research methodology involves an in-depth literature review followed by an analysis of alternate strategies and factors impacting adaptive reuse, along with an exploration of Lucknow's heritage. The study seeks to propose a range of tailored approaches for implementing adaptive reuse in Lucknow's historic residences, with the goal of not only preserving heritage but also enhancing the city's urban landscape.

Keywords Heritage · Preservation · Adaptive reuse · Houses

1 Introduction

In the dynamic realm of urban development, the safeguarding of heritage structures serves as an enduring testament to our cultural identity and historical legacy. These architectural marvels, seamlessly integrated into the urban fabric, bear witness to the passage of time, encapsulating narratives of bygone eras. Heritage buildings often require new functions when their original purpose becomes untenable to preserve their significance. Conserving heritage buildings and repurposing them in line with their location, size, and potential facilitates an understanding of cultural roots for subsequent generations. When the original function is no longer sustainable, adaptive

A. Amjad (✉) · A. Kaur
National Institute of Technology, Hamirpur, India
e-mail: 22mar006@nith.ac.in

A. K. Sharma and A. Joshi (eds.), *Proceedings of the 3rd International Conference on Opportunities and Challenges for a Resilient Future*, Advances in 21st Century Human Settlements, https://doi.org/10.1007/978-981-96-0931-4_23

reuse becomes inevitable, but the proposed function must align with preserving the cultural significance of the historic fabric [1].

Amidst the challenges posed by urban development and evolving needs, adaptive reuse emerges as a vital concept, breathing new life into venerable structures. Historical buildings inherently possess sustainable characteristics, making their conservation and continued usage aligned with sustainability goals [2]. The concept of this approach extends beyond preservation; it embodies a form of sustainable urban regeneration. Adaptive Reuse not only prolongs the life of buildings but also avoids demolition waste, encourages the reuse of embodied energy, and provides significant social and economic benefits to society [3].

While various studies have explored adaptive reuse approaches and methodologies for different building typologies, there exists a noticeable gap in the literature concerning heritage residential buildings, particularly those in urban areas like Lucknow, where many heritage houses remain underutilized due to issues such as private ownership, insufficient funds, and a lack of government support. This research endeavours to address two primary objectives. Firstly, it aims to provide a comprehensive review of factors influencing the adaptive reuse of heritage buildings by examining relevant studies. Secondly, it seeks to analyse this framework in the context of Lucknow's heritage homes, identifying the best approach for their adaptive reuse.

2 Methodology

Data was gathered through a literature survey, and content analysis was conducted to identify parameters influencing the adaptive reuse of heritage buildings. Successful examples of reused heritage residential buildings were also analysed to get a better understanding. Finally, the results were examined in the heritage homes of Lucknow through surveys and interviews with the stakeholders. The aim was to offer suggestions for more effective utilization considering the challenges faced by them.

2.1 Literature Review

The literature survey for this research encompasses diverse perspectives on the adaptive reuse of heritage buildings, offering multifaceted insights into this intricate domain.

Bullen [4] provided a firsthand exploration into the views and experiences of architects, developers, and building managers engaged in adaptive reuse. Yung [3] delved into sustainable development within the context of built heritage conservation, utilizing in-depth interviews to comprehend the challenges of integrating a sustainability framework into adaptive reuse projects. Dyson [5] emphasized the imperative of averting the loss of heritage buildings and identifying critical success factors for their adaptive reuse, thereby preserving tangible examples of historical periods. Günçe [1] conducted a comprehensive review of factors influencing adaptive reuse decision-making, employing a combination of literature survey and content analysis, subsequently developing a holistic model for adaptive reuse strategies for heritage buildings. Kee [6] examined the positive externalities of adaptive reuse on adjacent residential property prices, shedding light on the economic impact of this heritage management practice. Tam [7] analysed the advantages and disadvantages of adaptive reuse as a sustainable development tool in Toronto, with a particular focus on environmental, social, and economic factors. Li [8] explored research frameworks, methodologies, and assessment methods for the adaptive reuse of architectural heritage. Arfa [9] contributed a conceptual model for the adaptive reuse process of heritage buildings, delineating ten key steps from the 'initiative' stage to 'evaluation after years,' offering a theoretical foundation for subsequent studies. Vafaie [10] performed a systematic literature review to identify factors contributing to the success of heritage adaptive reuse projects, categorizing results into ten key factors such as architectural, socio-cultural, economic, and environmental aspects.

Throughout the literature review, it becomes evident that while substantial research has been conducted, there is a need for a framework specific to residential areas concerning adaptive reuse, underlining the importance of comprehending challenges for its practical implementation.

2.2 *Cases of Residential Adaptive Reuse*

The examination of cases involving the adaptive reuse of old heritage houses as shown in Table 1 provided valuable insights into the diverse parameters influencing such initiatives. This study tries to comprehensively understand the underlying parameters by delving into the imperative need for adaptive reuse, the essential modifications undertaken, the benefits derived from transformations, and the challenges encountered throughout the process.

Heritage areas bear significant importance for individuals, and therefore the dismantling of architecture with such historical significance proves to be a challenging task. Employing the technique of adaptive reuse becomes imperative in preserving these architectural gems, while concurrently facilitating the creation of functional spaces for residential, commercial, and entertainment purposes. The predominant factor identified during the examination of the case studies was revenue generation, involving the transformation of the structure into a functional space capable of generating income and the need to restore the building. In addition to this financial aspect, the buildings possessed significant heritage value and held emotional attachments to the community. Furthermore, their adaptability contributed to the promotion of tourism.

Table 1 Example of adaptive reuse in heritage residential areas

Adaptive reuse of heritage residential buildings			
Name	Original function	New function	Parameter identified
1. Kashi Art Café (Constructed in 1600s)—Kochi, Kerala	Dutch row house	Art Gallery and Café (Changed in 1997, renovated later in 2012)	Community Involvement, Revenue Generation
2. Gratitude Guest House (Constructed in 19th Century)—Puducherry, Tamil Nadu	French Colonial house	Weekend Retreat (Changed in 2004)	Heritage Value, Revenue Generation
3. Cinnamon Boutique (Constructed in 1892)—Bangalore, Karnataka	Colonial bungalow	Store (Completed in 2014)	Revenue Generation, Needed Restoration
4. Pol Houses (Constructed between 1600–1700)—Ahmedabad, Gujarat	Houses	Once occupied by many, now offers space for activities like restaurants, libraries, and information centers showcasing the city's art, culture, and heritage, for tourists	Heritage value, Place attachment, Revenue Generation, Needed Restoration
5. Haveli Dharampura (Constructed in 1887)—Delhi	Haveli	Luxury Heritage Hotel (Changed in 2016)	Heritage Value, Tourism, Needed Restoration
6. The Calcutta Bungalow (Constructed in 1926)—Kolkata, West Bengal	Residence	Heritage bed-and-breakfast (Changed after 2015)	Heritage Value, Revenue Generation
7. Limassol Castle (Constructed in the 12th Century)—Limassol, Cyprus	Castle	Museum (Changed in 1948)	Heritage Value, Community Involvement, Tourism

2.3 Framework for Adaptive Reuse

The exploration of related literature and diverse case studies in various contexts have been instrumental in comprehending different parameters associated with adaptive reuse. Several notable studies contribute nuanced perspectives to this understanding:

Günçe [1] presented a model for developing adaptive reuse strategies for heritage buildings, emphasizing a holistic approach. Their research considers factors such as actors, analysis of existing fabric, conservation actions, adaptive reuse potentials, and functional changes.

Tam [7] delved into Toronto's redevelopment, aiming to ascertain if adaptive reuse contributes to historically and culturally sustainable development. Their study introduces a comprehensive weighting and ranking system for adaptive reuse projects, assessing sustainability based on environmental, social, and economic factors.

Arfa [9] proposed a detailed model for the adaptive reuse process of heritage buildings. Their study identifies four main phases: the pre-project phase, preparation phase, implementation phase, and post-completion phase, providing a structured framework for future research in this domain.

Vafaie [10] conducted a systematic literature review classifying success factors into ten categories, encompassing architectural, structural, socio-cultural, economic, environmental, energy, authenticity, legal, management, and functional aspects.

The literature facilitated a deeper understanding of the parameters, which are subsequently classified into three sustainability pillars for this research in Table 2. This categorization aims to enhance the understanding of their impact across all dimensions of sustainable development.

2.4 Context of Residential Buildings in Lucknow

Situated on the plains of Northern India, along the banks of the Gomti River, Lucknow stands as the capital of Uttar Pradesh, the most populous state in India. Renowned for its rich Nawabi culture, traditions, and intricately carved buildings, the city reflects a unique blend of traditional and European architectural styles [11]. Different literature, comprising published research and books, provides comprehensive insights into the diverse monuments in Lucknow. This study is positioned to address the pressing need for preservation by focusing on the underused residential buildings within Lucknow.

In its urban structure, pre-colonial Lucknow demonstrated a typical Indian Islamic city. The indigenous houses in Lucknow showcased an inward-looking design, characterized by one or two square courtyards situated side by side, serving the Mardana (men's quarters) and Zenana (women's quarters). Functioning as natural ventilation systems, the courtyards facilitated air circulation throughout the house and were essential for family activities [12]. Lucknow's traditional residential buildings were categorized based on the socio-economic status of their inhabitants. The first type comprised kothis, havelis, and palaces commissioned by the Nawabs and their courtiers, while the second type belonged to middle-income families having simple layouts with courtyards surrounded by rooms. Construction materials primarily included lakhauri bricks and lime, forming thick masonry walls. Lime mortar and surkhi were used for finishing, while lime plaster adorned the elevation with intricate motifs and floral patterns [11].

Table 2 Framework for adaptive reuse of heritage residential buildings

Framework for adaptive reuse of heritage residential buildings		
Themes	Parameters	Parameter description
Social	Original Use	Architectural layout and features, Character of the Building
	New Proposed Use	Thinking of the best fit as per the surroundings, architectural layout, funding and resources, state of the building, etc
	Heritage Value	Cultural importance, People–place attachment, Place Identity
	Community Involvement	A new purpose that will benefit the community
	Role of Stakeholders	Responsibilities as well as benefits to the people involved—the owners, government, tourists, and neighbours
Economic (Funding & Resources)	Property Value	The cost of property, depending on the area it is in
	Repair Costs	Costs in the renovation, use of new material, or the same, Structural repair
	Operational and Maintenance cost	Cost of maintaining the new use in that building
	Tourism	No. of visitors, revenue generation, job opportunities
Environmental	Site and Surroundings	The nearby area, that would be affected by the new use of the building
	Energy and Resource Efficiency	Making use of traditional building techniques or adding new services to ensure the comfort of the occupants
	Materials	Reusing materials or coming up with alternative
	Indoor Quality	Comfort and safety of the occupants
	Building structural strength	The life cycle of the building, if it is safe from seismic activity, Structural decay, Interior finishes decay, or outer façade decay

2.5 Case Study

Situated on the plains of Northern India, along the banks of the Gomti River, Lucknow stands as the capital of Uttar Pradesh, the most populous state in India. Renowned for its rich Nawabi culture, traditions, and intricately carved buildings, the city reflects a unique blend of traditional and European architectural styles [11].

Iqbal Lari Manzil, Mashaqganj

Nestled near Lucknow's City Railway Station, this 1928-built residence showcases a grand, aged gate adorned in white and green hues. Featuring two gates, one leading to the upper-floor Mardana section and the other to the ground-floor Zenana, the latter has been repurposed as the primary entry point, now embellished with intricate stucco decorations. The zenana provides direct access to the family's private ground floor imambara. Within the Zenana, a staircase ascends to the upper floors, revealing an expansive terrace. A corridor links the terrace to the Mardana living quarters, where the verandah, once a traditional space has changed, transforming into a modern dining room to meet the family's evolving needs.

The property owner has embraced adaptive reuse by renting out a portion of the palace for film shoots and intends to allocate some rooms for tourists interested in Lucknow's rich history. Additionally, a school occupies the downstairs area, contributing to the property's ongoing maintenance. Notably, the garden, once a venue for gatherings, has been repurposed into a commercial parking area, reflecting a dynamic approach to adaptive reuse in response to changing circumstances (see Fig. 1).

Mahmudabad House, Qaiserbagh

Originally designated to Raja Mahmudabad after the events of the 1857 revolt, this house within the Qaiserbagh Complex has remained in the possession of his family, meticulously preserving its original appearance. Characterized by expansive courtyards that once served as bustling centers of activity, the larger front rooms played host to social gatherings during both the Nawabi and British eras. In a contemporary context, these rooms continue to fulfill their role as spaces for hospitality. Meanwhile, the upper floor has been adapted to accommodate the family's modern living needs, and the historical function of the servant quarters at the rear remains.

The owner's focus is on the preservation of the house's authentic features. The iconic yellow walls and green doors of Qaiserbagh are diligently maintained to retain their original aesthetic appeal. Special attention is given to the conservation of fixtures such as the boiler, Victorian-era basin, shower, and bathtub, with dedicated efforts aimed at their careful preservation, including the retention of vintage electrical switches and regulators. While public activities are not a regular occurrence, certain spaces are occasionally repurposed for film screenings, ensuring the ongoing vibrancy and vitality of the place.

Fig. 1 **a** Main Gate for School and Entrance of the House **b** Imambada **c** Classrooms **d** Commercial Parking Area

Despite nearly half of the house being in a state of disrepair, there is an opportunity for adaptive reuse to enhance its potential. Consideration could be given to repurposing these sections as museums or information centers dedicated to the Qaiserbagh Complex, serving both as a nod to its historical significance and contributing to its revitalization. Since the family occupies one part of the residence, the remaining portion could be utilized as homestays for individuals seeking to experience the Awadhi lifestyle (see Fig. 2).

Dilkusha Kothi

Dilkusha Kothi stands as the remnants of an eighteenth-century residence constructed in the English Baroque architectural style. Presently, only a few towers and external walls persist, serving as a tangible monument, while the extensive gardens are still there. The house endured shelling during its participation in the Siege of Lucknow in 1857.

Originally employed as a hunting retreat due to the adjacent forests, Dilkusha Kothi also functioned as a recreational venue for the Begums and female members of royal families. Unfortunately, the current state of Dilkusha Kothi is far from optimal, necessitating restoration efforts. Despite its historical significance, the site

Fig. 2 **a** Outer Facade, **b** Rooms needing Restoration, **c** Vintage electric fixtures and Boiler, **d** Still from a movie in front of the house

attracts minimal visitation, indicating a pressing need for enhanced publicity. At present, the Archaeological Society of India has undertaken initiatives to impede further deterioration, undertaking minor repairs. Although Dilkusha Kothi forms an integral part of heritage walks, its potential utilization remains under-realized, warranting consideration for alternative applications to not only augment its usage but also amplify its visibility.

One plausible proposition involves transforming Dilkusha Kothi into a museum or information center following requisite renovations while the open spaces can be used for cultural shows or as exhibition areas. Such a transformation would serve the dual purpose of educating the public about the events surrounding the 1857 revolt while ensuring the optimal utilization of the surrounding gardens for complementary purposes. This strategic repurposing aligns with the imperative to safeguard historical landmarks, fostering both increased engagement and awareness among visitors (see Fig. 3).

(a)

(b)

Fig. 3 **a** Garden in front of the Kothi, **b** Front Façade

2.6 *Inferences from Case Studies*

Through interviews and visual surveys, it became evident that property owners were making earnest efforts to utilize their properties to the best of their abilities. However, with additional support and well-thought-out policies, there is potential for enhancing the utilization of these houses. The inferences from the case study are discussed in Table 3.

Acknowledging adaptive reuse as a strategic tool holds the promise of unlocking substantial potential within these historic properties. Owners of residential properties, encouraged by a sincere commitment to both share and safeguard their ancestral legacies, demonstrate a genuine willingness to contribute to the broader community. Nonetheless, the realization of these aspirations frequently depends on external support. Consequently, it becomes imperative for the government to formulate well-informed and empathetic policies, coupled with providing necessary assistance in terms of financial resources or aid in conservation efforts.

3 Results

The analysis of the literature review, different cases of adaptive reuse, interviews with the owners and stakeholders, and site visits helped in an understanding of the challenges these houses face, as demonstrated in Table 4.

During the Nawab era, havelis flourished as opulent residences and pleasure palaces. Today, they face neglect due to various challenges. To address these issues effectively, it is crucial to understand the specific problems faced by these buildings and to implement strategies according to them. Encouraging public–private partnerships fosters collaboration, addressing the common goal of restoring these houses. Tailoring policies, such as offering subsidies for house restoration, can incentivize owners. Government agreements with owners for unused places, coupled

Table 3 Inferences from the case studies along with suggestions

Case study inferences			
Case study	Ownership	Present use	Proposed use
Iqbal Lari Manzil, Mashaqganj	Private Ownership	The ground floor of the property accommodates a school, offers space for movie shoots, and features a garden utilized for commercial car parking	The classrooms are currently in suboptimal condition and could benefit from additional support to enhance their overall quality. Furthermore, the courtyard of the premises has the potential to be restored and utilized for various student activities
Mahmudabad House, Qaiserbagh	Private Ownership	Half of the house is dedicated to the family's residence, while the remaining portion is undergoing gradual restoration and is utilized for movie shoots occasionally	The property could serve multiple purposes, such as museums dedicated to the history of Qaiserbagh. Given the owners' efforts to restore these houses to their original grandeur, they could also be utilized as homestays or restaurants that authentically showcase Awadhi culture
Dilkusha Kothi, Cantonment Area	Under Government	The garden serves as a venue for social events and is included in heritage walks. Although the building is currently undergoing restoration by the (ASI), it is not presently in a usable condition	The building has the potential to be restored and repurposed for various social events, such as exhibitions. Another possible utilization could involve transforming the building into an information center commemorating the events of the 1857 revolt

with incentives, can revitalize properties. Utilizing these structures enhances tourism and strengthens place attachment. Lucknow's cultural legacy presents opportunities for rejuvenation, including workshops showcasing traditional crafts and establishing cafes featuring local cuisine. Initiatives supporting craftsmen and investing in research and development for conservation techniques are vital for preserving architectural details.

4 Discussion and Conclusions

The framework presented in Table 2 was analysed within the heritage houses of Lucknow to discern the nuanced roles of various parameters in recommending new uses for these properties. The interpretation of the framework is outlined as follows-

Social parameters—The current example of utilisation of heritage houses, whether in the form of schools, for movie shoots, or heritage walks is not setting a benchmark or acting as encouragement for others as more can be done. The new use of the house

Table 4 Challenges faced by heritage houses and its impact

Challenges faced by heritage houses		
Challenges	Its impact	Sources
1. Multiple Ownership	Under multiple owners, a conflict arises where no individual is willing to assume complete responsibility or there is a lack of collaboration, resulting in a lack of concerted effort in providing proper care or attention	Interview with the owners, Case study examples
2. Financial Constraints	Maintaining the old houses and required renovations becomes difficult with limited funds, especially for working-class people, because of which required conservation is not done, or the house is left vacant	Interview with the owners, Literature study, Case study examples
3. Private Ownership	Most of the houses are privately owned, due to which the government doesn't help with the required funds or assistance in renovations	Interview with the owners
4. Vacant Houses because owners have relocated	Relocation due to studies or jobs has left the house unused, which leads to its deteriorated stage	Interview with the owners, Case Study Examples
5. Lack of Policies	The absence of government policies or incentives fails to motivate individuals to preserve their old houses	Interview with the owners, Literature study
6. Lack of Craftsmanship to do the required restoration	Lack of skills to preserve the ornamental details of these houses leads to problems in the conservation of these houses in their original state	Interview with the owners
7. Compatibility of modern building materials with traditional structures	In some cases, the current building materials (like concrete) do not work with the old building materials	Interview with the owners, Case Study examples

should be able to develop a sense of pride in the people involved, as it is a part of their history. Encouraging uses that benefit the community would increase the attachment of people to that place. Moreover, including activities that raise awareness about their history, such as talk shows, book readings, and the display of paintings, would foster a closer connection to the house. Addressing these factors would contribute to thinking of a better use for the building.

Economic Parameters—The restoration of houses and maintenance primarily depend on funds, which the owners of the houses often cannot afford. Policies should be crafted to provide necessary support to owners, accompanied by incentives encouraging them to undertake restoration amidst various challenges.

Environmental parameters—Challenges arise in conservation due to a lack of artistic expertise in that kind of restoration and the incompatibility of new materials with old materials. There should be a platform where people who need assistance with their restoration can find support, along with training programs for artisans.

While the research primarily focuses on Lucknow, the framework is designed to provide insights and strategies that can be applied to different regions facing comparable challenges in heritage preservation and suggest context-based solutions. The successful application of this framework can guide policymakers and heritage house owners towards informed decisions, fostering a balance between preservation and adaptation in a manner that respects both the cultural heritage and the contemporary needs of the community.

References

1. Günçe K, Misirlisoy D (2014) Adaptive reuse of military establishments as museums: conservation vs. museography. Present Def Herit 2014, Venice. https://doi.org/10.2495/DSHF140111.
2. Karimi F, Valibeig N, Memarian G, Kamari A (2022) Sustainability rating systems for historic buildings: a systematic review. Sustain 14:12448. https://doi.org/10.3390/su141912448
3. Yung EHK, Chan EHW (2012) Implementation challenges to the adaptive reuse of heritage buildings: Towards the goals of sustainable, low carbon cities. Habitat Int 36:352–361. https://doi.org/10.1016/j.habitatint.2011.11.001
4. Bullen PA, Love PED (2011) Adaptive reuse of heritage buildings. Struct Surv 29:411–421. https://doi.org/10.1108/02630801111182439
5. Dyson K, Matthews J, Love PED (2016) Critical success factors of adapting heritage buildings: an exploratory study. Built Environ Proj Asset Manag. 6:44–57. https://doi.org/10.1108/BEPAM-01-2015-0002
6. Kee T (2019) Sustainable adaptive reuse—economic impact of cultural heritage. J Cult Herit Manag Sustain Dev. 9:165–183. https://doi.org/10.1108/JCHMSD-06-2018-0044
7. Tam VWY, Hao JJL (2019) Adaptive reuse in sustainable development. Int J Constr Manag 19:509–521. https://doi.org/10.1080/15623599.2018.1459154
8. Li Y, Zhao L, Huang J, Law A (2021) Research frameworks, methodologies, and assessment methods concerning the adaptive reuse of architectural heritage: a review. Built Heritage. 5:6. https://doi.org/10.1186/s43238-021-00025-x
9. Arfa FH, Zijlstra H, Lubelli B, Quist W (2022) Adaptive reuse of heritage buildings: from a literature review to a model of practice. Hist Environ: Policy & Pract. 13:148–170. https://doi.org/10.1080/17567505.2022.2058551
10. Vafaie F, Remøy H, Gruis V (2023) Adaptive reuse of heritage buildings; a systematic literature review of success factors. Habitat Int 142:102926. https://doi.org/10.1016/j.habitatint.2023.102926

11. Aligarh Muslim University, Kamal, MA (2021) Assessment of traditional architecture of lucknow with reference to climatic responsiveness. aej. 6, 19–31. https://doi.org/10.23968/2500-0055-2021-6-1-19-31
12. Srivastava M Architecture and development as instruments for political control and marginalization in lucknow, India.

Reflections

The International Conference was a three-day event held from April 22nd to April 24th, 2024. The first day began with an inaugural ceremony, featuring the conference focus by the Honorable Chancellor and Pro Vice Chancellor. This was followed by Keynote Speaker Ar. Anne Feenstra addressing the theme of the conference and proceeded by technical sessions where presenters showcased their work. The second day continued with multiple technical sessions, addressing various issues and concerns related to the conference theme. The conference concluded on the third day with a panel discussion. The invited panelists were Dr. D.P. Uniyal, Joint Director, UCOST; Er. N.K. Yadav, Retired Chief Engineer, Irrigation Department, Uttarakhand Government; and Ar. M.S. Negi, Chairman, Indian Institute of Architects, Uttarakhand Chapter.

The 3rd International Conference aimed to engage academics, practitioners, and students at both National and International level, however, participation was limited to a national level only. The primary goal was to foster an academic and research-oriented atmosphere and to facilitate discussions related to the built environment.

The tracks of the conference were planned accordingly and focused on Ecological development, energy, optimizing existing resources, policies formulation, these key themes of discussion aimed to exploring influences that aligned with Sustainable Development Goals (SDGs).

The track on ecological and inclusive development received the maximum response, addressing a wide range of environmental issues from transformations to energy concerns, and showcasing community resilience through case studies across the nation. Notably, some papers focused on Islamic Architecture, Climatic Resilience, and Gender connotations in the Walled City of Shahjahanabad. Additionally, a few papers explored regional-scale developments such as Bellandur Lake, Indian cities, and the reinterpretation of Ghats.

Case examples from Srirangam Temple Town demonstrated resilience through inclusive development. While the track on Sustainability and Energy has been consistent for decades, it has yielded limited results. At the national level, India has initiated

A. K. Sharma and A. Joshi (eds.), *Proceedings of the 3rd International Conference on Opportunities and Challenges for a Resilient Future*, Advances in 21st Century Human Settlements, https://doi.org/10.1007/978-981-96-0931-4

numerous projects in both urban and rural areas across various geographical zones is in place. An interesting paper on warm and humid climatic regions delved into climatic factors, while the rest focused on urban contexts, exploring SDG goals, Urban Heat Islands, noise pollution, and standard thermal comfort.

The thrust of the conference, Opportunities and Challenges for a Resilient Future, is strongly underpinned by policies and governance at global, regional, and local levels. The papers presented in this track largely focused on indigenous communities, urban settlements, and rural housing. However, one presentation outlined a path to sustainable development for industrial workers in Noida, Uttar Pradesh. Given the university's location in the Himalayan region, case examples from Himachal Pradesh were appropriately and debated upon.

The intent of the conference was also to map the way forward, with presentations on architecture and the future of AI proving insightful. Additionally, the impact of the changing environment on the health status of people in Uttarakhand generated firsthand narratives, with local students from Dehradun contributing their experiences. Adaptive reuse emerged as a valuable tool to limit the carbon footprint by utilizing existing building stock and building meaningful narratives for India's rich cultural heritage.

The fruitful presentations by the keynote speaker followed by intense paper presentation and on the last day meaningful panel discussions, all led towards a successful completion of the conference. We have no hesitation in stating that the aim of the conference was achieved and we are sure that compilation of the conference research papers would help the young researchers in exploring new ideas.

Dr. Anjali Krishan Sharma

Chairperson of the International Conference.

MIX
Papier aus verantwortungsvollen Quellen
Paper from responsible sources
FSC® C105338

If you have any concerns about our products,
you can contact us on
ProductSafety@springernature.com

In case Publisher is established outside the EU,
the EU authorized representative is:
Springer Nature Customer Service Center GmbH
Europaplatz 3, 69115 Heidelberg, Germany

Printed by Libri Plureos GmbH
in Hamburg, Germany